The Forces of Nature

" An Overview of The Fundamental Equations "

Edited by Paul F. Kisak

Contents

Chapter 1

Force

For other uses, see Force (disambiguation) and Forcing (disambiguation).

In physics, a **force** is any interaction that, when unopposed, will change the motion of an object.[1] In other words, a force can cause an object with mass to change its velocity (which includes to begin moving from a state of rest), i.e., to accelerate. Force can also be described by intuitive concepts such as a push or a pull. A force has both magnitude and direction, making it a vector quantity. It is measured in the SI unit of newtons and represented by the symbol **F**.

The original form of Newton's second law states that the net force acting upon an object is equal to the rate at which its momentum changes with time. If the mass of the object is constant, this law implies that the acceleration of an object is directly proportional to the net force acting on the object, is in the direction of the net force, and is inversely proportional to the mass of the object

Related concepts to force include: thrust, which increases the velocity of an object; drag, which decreases the velocity of an object; and torque, which produces changes in rotational speed of an object. In an extended body, each part usually applies forces on the adjacent parts; the distribution of such forces through the body is the so-called mechanical stress. Pressure is a simple type of stress. Stress usually causes deformation of solid materials, or flow in fluids.

1.1 Development of the concept

Philosophers in antiquity used the concept of force in the study of stationary and moving objects and simple machines, but thinkers such as Aristotle and Archimedes retained fundamental errors in understanding force. In part this was due to an incomplete understanding of the sometimes non-obvious force of friction, and a consequently inadequate view of the nature of natural motion.[2] A fundamental error was the belief that a force is required to maintain motion, even at a constant velocity. Most of the previous misunderstandings about motion and force were eventually corrected by Galileo Galilei and Sir Isaac Newton. With his mathematical insight, Sir Isaac Newton formulated laws of motion that were not improved-on for nearly three hundred years.[3] By the early 20th century, Einstein developed a theory of relativity that correctly predicted the action of forces on objects with increasing momenta near the speed of light, and also provided insight into the forces produced by gravitation and inertia.

With modern insights into quantum mechanics and technology that can accelerate particles close to the speed of light, particle physics has devised a Standard Model to describe forces between particles smaller than atoms. The Standard Model predicts that exchanged particles called gauge bosons are the fundamental means by which forces are emitted and absorbed. Only four main interactions are known: in order of decreasing strength, they are: strong, electromagnetic, weak, and gravitational.[4]:2–10[5]:79 High-energy particle physics observations made during the 1970s and 1980s confirmed that the weak and electromagnetic forces are expressions of a more fundamental electroweak interaction.[6]

1.2 Pre-Newtonian concepts

See also: Aristotelian physics and Theory of impetus
Since antiquity the concept of force has been recognized as integral to the functioning of each of the simple machines. The mechanical advantage given by a simple machine allowed for less force to be used in exchange for that force acting over a greater distance for the same amount of work. Analysis of the characteristics of forces ultimately culminated in the work of Archimedes who was especially famous for formulating a treatment of buoyant forces inherent in fluids.[2]

Aristotle provided a philosophical discussion of the concept of a force as an integral part of Aristotelian cosmology. In Aristotle's view, the terrestrial sphere contained four elements that come to rest at different "natural places"

1

Aristotle famously described a force as anything that causes an object to undergo "unnatural motion"

therein. Aristotle believed that motionless objects on Earth, those composed mostly of the elements earth and water, to be in their natural place on the ground and that they will stay that way if left alone. He distinguished between the innate tendency of objects to find their "natural place" (e.g., for heavy bodies to fall), which led to "natural motion", and unnatural or forced motion, which required continued application of a force.[7] This theory, based on the everyday experience of how objects move, such as the constant application of a force needed to keep a cart moving, had conceptual trouble accounting for the behavior of projectiles, such as the flight of arrows. The place where the archer moves the projectile was at the start of the flight, and while the projectile sailed through the air, no discernible efficient cause acts on it. Aristotle was aware of this problem and proposed that the air displaced through the projectile's path carries the projectile to its target. This explanation demands a continuum like air for change of place in general.[8]

Aristotelian physics began facing criticism in Medieval science, first by John Philoponus in the 6th century.

The shortcomings of Aristotelian physics would not be fully corrected until the 17th century work of Galileo Galilei, who was influenced by the late Medieval idea that ob-

jects in forced motion carried an innate force of impetus. Galileo constructed an experiment in which stones and cannonballs were both rolled down an incline to disprove the Aristotelian theory of motion early in the 17th century. He showed that the bodies were accelerated by gravity to an extent that was independent of their mass and argued that objects retain their velocity unless acted on by a force, for example friction.[9]

1.3   Newtonian mechanics

Main article: Newton's laws of motion

Sir Isaac Newton sought to describe the motion of all objects using the concepts of inertia and force, and in doing so he found that they obey certain conservation laws. In 1687, Newton went on to publish his thesis *Philosophiæ Naturalis Principia Mathematica*.[3][10] In this work Newton set out three laws of motion that to this day are the way forces are described in physics.[10]

1.3.1   First law

Main article: Newton's first law

Newton's First Law of Motion states that objects continue to move in a state of constant velocity unless acted upon by an external net force or *resultant force*.[10] This law is an extension of Galileo's insight that constant velocity was associated with a lack of net force (see a more detailed description of this below). Newton proposed that every object with mass has an innate inertia that functions as the fundamental equilibrium "natural state" in place of the Aristotelian idea of the "natural state of rest". That is, the first law contradicts the intuitive Aristotelian belief that a net force is required to keep an object moving with constant velocity. By making *rest* physically indistinguishable from *non-zero constant velocity*, Newton's First Law directly connects inertia with the concept of relative velocities. Specifically, in systems where objects are moving with different velocities, it is impossible to determine which object is "in motion" and which object is "at rest". In other words, to phrase matters more technically, the laws of physics are the same in every inertial frame of reference, that is, in all frames related by a Galilean transformation.

For instance, while traveling in a moving vehicle at a constant velocity, the laws of physics do not change from being at rest. A person can throw a ball straight up in the air and catch it as it falls down without worrying about applying a force in the direction the vehicle is moving. This is true even though another person who is observing the moving vehicle pass by also observes the ball follow a curving parabolic path in the same direction as the motion of the vehicle. It is the inertia of the ball associated with its constant velocity in the direction of the vehicle's motion that ensures the ball continues to move forward even as it is thrown up and falls back down. From the perspective of the person in the car, the vehicle and everything inside of it is at rest: It is the outside world that is moving with a constant speed in the opposite direction. Since there is no experiment that can distinguish whether it is the vehicle that is at rest or the outside world that is at rest, the two situations are considered to be physically indistinguishable. Inertia therefore applies equally well to constant velocity motion as it does to rest.

The concept of inertia can be further generalized to explain the tendency of objects to continue in many different forms of constant motion, even those that are not strictly constant velocity. The rotational inertia of planet Earth is what fixes the constancy of the length of a day and the length of a year. Albert Einstein extended the principle of inertia further when he explained that reference frames subject to constant acceleration, such as those free-falling toward a gravitating object, were physically equivalent to inertial reference frames. This is why, for example, astronauts experience weightlessness when in free-fall orbit around the Earth, and why Newton's Laws of Motion are more easily discernible in such environments. If an astronaut places an object with mass in mid-air next to himself, it will remain stationary with respect to the astronaut due to its inertia. This is the same thing that would occur if the astronaut and the object were in intergalactic space with no net force of gravity acting on their shared reference frame. This principle of equivalence was one of the foundational underpinnings for the development of the general theory of relativity.[11]

1.3.2   Second law

Main article: Newton's second law

A modern statement of Newton's Second Law is a vector equation:[Note 1]

$$\vec{F} = \frac{\mathrm{d}\vec{p}}{\mathrm{d}t},$$

where $\vec{p}$ is the momentum of the system, and $\vec{F}$ is the net (vector sum) force. In equilibrium, there is zero *net* force by definition, but (balanced) forces may be present nevertheless. In contrast, the second law states an *unbalanced* force acting on an object will result in the object's momentum changing over time.[10]

Though Sir Isaac Newton's most famous equation is $\vec{F}=m\vec{a}$, he actually wrote down a different form for his second law of motion that did not use differential calculus.

By the definition of momentum,

$$\vec{F} = \frac{\mathrm{d}\vec{p}}{\mathrm{d}t} = \frac{\mathrm{d}\left(m\vec{v}\right)}{\mathrm{d}t},$$

where m is the mass and $\vec{v}$ is the velocity.[4]:9-1,9-2

Newton's second law applies only to a system of constant mass,[Note 2] and hence m may be moved outside the derivative operator. The equation then becomes

$$\vec{F} = m\frac{\mathrm{d}\vec{v}}{\mathrm{d}t}.$$

By substituting the definition of acceleration, the algebraic version of Newton's Second Law is derived:

$$\vec{F} = m\vec{a}.$$

Newton never explicitly stated the formula in the reduced form above.[12]

Newton's Second Law asserts the direct proportionality of acceleration to force and the inverse proportionality of acceleration to mass. Accelerations can be defined through kinematic measurements. However, while kinematics are well-described through reference frame analysis in advanced physics, there are still deep questions that remain as to what is the proper definition of mass. General relativity offers an equivalence between space-time and mass, but lacking a coherent theory of quantum gravity, it is unclear as to how or whether this connection is relevant on microscales. With some justification, Newton's second law can be taken as a quantitative definition of *mass* by writing the law as an equality; the relative units of force and mass then are fixed.

The use of Newton's Second Law as a *definition* of force has been disparaged in some of the more rigorous textbooks,[4]:12-1[5]:59[13] because it is essentially a mathematical truism. Notable physicists, philosophers and mathematicians who have sought a more explicit definition of the concept of force include Ernst Mach, Clifford Truesdell and Walter Noll.[14][15]

Newton's Second Law can be used to measure the strength of forces. For instance, knowledge of the masses of planets along with the accelerations of their orbits allows scientists to calculate the gravitational forces on planets.

1.3.3 Third law

Main article: Newton's third law

Newton's Third Law is a result of applying symmetry to situations where forces can be attributed to the presence of different objects. The third law means that all forces are *interactions* between different bodies,[16][Note 3] and thus that there is no such thing as a unidirectional force or a force that acts on only one body. Whenever a first body exerts a force *F* on a second body, the second body exerts a force −*F* on the first body. *F* and −*F* are equal in magnitude and opposite in direction. This law is sometimes referred to as the *action-reaction law*, with *F* called the "action" and −*F* the "reaction". The action and the reaction are simultaneous:

$$\vec{F}_{1,2} = -\vec{F}_{2,1}.$$

If object 1 and object 2 are considered to be in the same system, then the net force on the system due to the interactions between objects 1 and 2 is zero since

$$\vec{F}_{1,2} + \vec{F}_{2,1} = 0$$

$$\sum \vec{F} = 0.$$

This means that in a closed system of particles, there are no internal forces that are unbalanced. That is, the action-reaction force shared between any two objects in a closed system will not cause the center of mass of the system to accelerate. The constituent objects only accelerate with respect to each other, the system itself remains unaccelerated. Alternatively, if an external force acts on the system, then the center of mass will experience an acceleration proportional to the magnitude of the external force divided by the mass of the system.[4]:19-1[5]

Combining Newton's Second and Third Laws, it is possible to show that the linear momentum of a system is conserved. Using

$$\vec{F}_{1,2} = \frac{d\vec{p}_{1,2}}{dt} = -\vec{F}_{2,1} = -\frac{d\vec{p}_{2,1}}{dt}$$

and integrating with respect to time, the equation:

$$\Delta \vec{p}_{1,2} = -\Delta \vec{p}_{2,1}$$

is obtained. For a system that includes objects 1 and 2,

$$\sum \Delta \vec{p} = \Delta \vec{p}_{1,2} + \Delta \vec{p}_{2,1} = 0$$

which is the conservation of linear momentum.[17] Using the similar arguments, it is possible to generalize this to a system of an arbitrary number of particles. This shows that exchanging momentum between constituent objects will not affect the net momentum of a system. In general, as long as all forces are due to the interaction of objects with mass, it is possible to define a system such that net momentum is never lost nor gained.[4][5]

1.4 Special theory of relativity

In the special theory of relativity, mass and energy are equivalent (as can be seen by calculating the work required to accelerate an object). When an object's velocity increases, so does its energy and hence its mass equivalent (inertia). It thus requires more force to accelerate it the same amount than it did at a lower velocity. Newton's Second Law

$$\vec{F} = d\vec{p}/dt$$

remains valid because it is a mathematical definition.[18]:855–876 But in order to be conserved, relativistic momentum must be redefined as:

$$\vec{p} = \frac{m_0\vec{v}}{\sqrt{1 - v^2/c^2}}$$

where

> v is the velocity and
>
> c is the speed of light
>
> m_0 is the rest mass.

The relativistic expression relating force and acceleration for a particle with constant non-zero rest mass m moving in the x direction is:

$$F_x = \gamma^3 m a_x$$

$$F_y = \gamma m a_y$$

$$F_z = \gamma m a_z$$

where the Lorentz factor

$$\gamma = \frac{1}{\sqrt{1-v^2/c^2}}.\ [19]$$

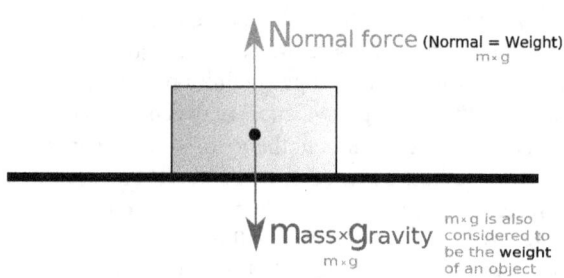

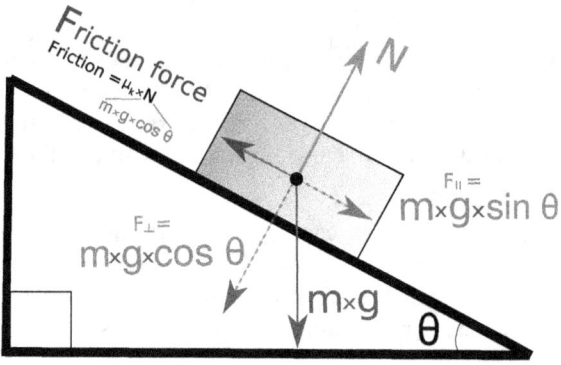

Free body diagrams of a block on a flat surface and an inclined plane. Forces are resolved and added together to determine their magnitudes and the net force.

In the early history of relativity, the expressions $\gamma^3 m$ and γm were called longitudinal and transverse mass. Relativistic force does not produce a constant acceleration, but an ever decreasing acceleration as the object approaches the speed of light. Note that γ is undefined for an object with a non-zero rest mass at the speed of light, and the theory yields no prediction at that speed.

If v is very small compared to c, then γ is very close to 1 and

$$F = ma$$

is a close approximation. Even for use in relativity, however, one can restore the form of

$$F^\mu = m A^\mu$$

through the use of four-vectors. This relation is correct in relativity when F^μ is the four-force, m is the invariant mass, and A^μ is the four-acceleration.[20]

1.5 Descriptions

Since forces are perceived as pushes or pulls, this can provide an intuitive understanding for describing forces.[3] As with other physical concepts (e.g. temperature), the intuitive understanding of forces is quantified using precise operational definitions that are consistent with direct observations and compared to a standard measurement scale. Through experimentation, it is determined that laboratory measurements of forces are fully consistent with the conceptual definition of force offered by Newtonian mechanics.

Forces act in a particular direction and have sizes dependent upon how strong the push or pull is. Because of these characteristics, forces are classified as "vector quantities". This means that forces follow a different set of mathematical rules than physical quantities that do not have direction (denoted scalar quantities). For example, when determining what happens when two forces act on the same object, it is necessary to know both the magnitude and the direction of both forces to calculate the result. If both of these pieces of information are not known for each force, the situation is ambiguous. For example, if you know that two people are pulling on the same rope with known magnitudes of force but you do not know which direction either person is pulling, it is impossible to determine what the acceleration of the rope will be. The two people could be pulling against each other as in tug of war or the two people could be pulling in the same direction. In this simple one-dimensional example, without knowing the direction of the forces it is impossible to decide whether the net force is the result of adding the two force magnitudes or subtracting one from the other. Associating forces with vectors avoids such problems.

Historically, forces were first quantitatively investigated in conditions of static equilibrium where several forces canceled each other out. Such experiments demonstrate the crucial properties that forces are additive vector quantities: they have magnitude and direction.[3] When two forces act on a point particle, the resulting force, the *resultant* (also called the *net force*), can be determined by following the parallelogram rule of vector addition: the addition of two vectors represented by sides of a parallelogram, gives an equivalent resultant vector that is equal in magnitude and direction to the transversal of the parallelogram.[4][5] The magnitude of the resultant varies from the difference of the magnitudes of the two forces to their sum, depending on the angle between their lines of action. However, if the forces are acting on an extended body, their respective lines of application must also be specified in order to account for their effects on the motion of the body.

Free-body diagrams can be used as a convenient way to keep track of forces acting on a system. Ideally, these diagrams are drawn with the angles and relative magnitudes of the force vectors preserved so that graphical vector addition can be done to determine the net force.[21]

As well as being added, forces can also be resolved into independent components at right angles to each other. A horizontal force pointing northeast can therefore be split into two forces, one pointing north, and one pointing east. Summing these component forces using vector addition yields the original force. Resolving force vectors into components of a set of basis vectors is often a more mathematically clean way to describe forces than using magnitudes and directions.[22] This is because, for orthogonal components, the components of the vector sum are uniquely determined by the scalar addition of the components of the individual vectors. Orthogonal components are independent of each other because forces acting at ninety degrees to each other have no effect on the magnitude or direction of the other. Choosing a set of orthogonal basis vectors is often done by considering what set of basis vectors will make the mathematics most convenient. Choosing a basis vector that is in the same direction as one of the forces is desirable, since that force would then have only one non-zero component.

Orthogonal force vectors can be three-dimensional with the third component being at right-angles to the other two.[4][5]

1.5.1 Equilibrium

Equilibrium occurs when the resultant force acting on a point particle is zero (that is, the vector sum of all forces is zero). When dealing with an extended body, it is also necessary that the net torque in it is 0.

There are two kinds of equilibrium: static equilibrium and dynamic equilibrium.

Static

Main articles: Statics and Static equilibrium

Static equilibrium was understood well before the invention of classical mechanics. Objects that are at rest have zero net force acting on them.[23]

The simplest case of static equilibrium occurs when two forces are equal in magnitude but opposite in direction. For example, an object on a level surface is pulled (attracted) downward toward the center of the Earth by the force of gravity. At the same time, surface forces resist the downward force with equal upward force (called the normal force). The situation is one of zero net force and no acceleration.[3]

Pushing against an object on a frictional surface can result in a situation where the object does not move because the applied force is opposed by static friction, generated between the object and the table surface. For a situation with no movement, the static friction force *exactly* balances the applied force resulting in no acceleration. The static friction increases or decreases in response to the applied force up to an upper limit determined by the characteristics of the contact between the surface and the object.[3]

A static equilibrium between two forces is the most usual way of measuring forces, using simple devices such as weighing scales and spring balances. For example, an object suspended on a vertical spring scale experiences the force of gravity acting on the object balanced by a force applied by the "spring reaction force", which equals the object's weight. Using such tools, some quantitative force laws were discovered: that the force of gravity is proportional to volume for objects of constant density (widely exploited for millennia to define standard weights); Archimedes' principle for buoyancy; Archimedes' analysis of the lever; Boyle's law for gas pressure; and Hooke's law for springs. These were all formulated and experimentally verified before Isaac Newton expounded his Three Laws of Motion.[3][4][5]

Dynamic

Main article: Dynamics (physics)

Dynamic equilibrium was first described by Galileo who

Galileo Galilei was the first to point out the inherent contradictions contained in Aristotle's description of forces.

noticed that certain assumptions of Aristotelian physics were contradicted by observations and logic. Galileo realized that simple velocity addition demands that the concept of an "absolute rest frame" did not exist. Galileo concluded that motion in a constant velocity was completely equivalent to rest. This was contrary to Aristotle's notion of a "natural state" of rest that objects with mass naturally approached. Simple experiments showed that Galileo's understanding of the equivalence of constant velocity and rest were correct. For example, if a mariner dropped a cannonball from the crow's nest of a ship moving at a constant velocity, Aristotelian physics would have the cannonball fall straight down while the ship moved beneath it. Thus, in an Aristotelian universe, the falling cannonball would land behind the foot of the mast of a moving ship. However, when this experiment is actually conducted, the cannonball always falls at the foot of the mast, as if the cannonball knows to travel with the ship despite being separated from it. Since there is no forward horizontal force being applied on the cannonball as it falls, the only conclusion left is that the cannonball continues to move with the same velocity as the boat as it falls. Thus, no force is required to keep the cannonball moving at the constant forward velocity.[9]

Moreover, any object traveling at a constant velocity must be subject to zero net force (resultant force). This is the definition of dynamic equilibrium: when all the forces on an object balance but it still moves at a constant velocity.

A simple case of dynamic equilibrium occurs in constant velocity motion across a surface with kinetic friction. In such a situation, a force is applied in the direction of motion while the kinetic friction force exactly opposes the applied force. This results in zero net force, but since the object started with a non-zero velocity, it continues to move with a non-zero velocity. Aristotle misinterpreted this motion as being caused by the applied force. However, when kinetic friction is taken into consideration it is clear that there is no net force causing constant velocity motion.[4][5]

1.5.2 Forces in Quantum Mechanics

Main articles: Quantum mechanics and Pauli principle

The notion "force" keeps its meaning in quantum mechanics, though one is now dealing with operators instead of classical variables and though the physics is now described by the Schrödinger equation instead of Newtonian equations. This has the consequence that the results of a measurement are now sometimes "quantized", i.e. they appear in discrete portions. This is, of course, difficult to imagine in the context of "forces". However, the potentials $V(x,y,z)$ or fields, from which the forces generally can be derived, are treated similar to classical position variables, i.e., $V(x, y, z) \rightarrow \hat{V}(\hat{x}, \hat{y}, \hat{z})$.

This becomes different only in the framework of quantum field theory, where these fields are also quantized.

However, already in quantum mechanics there is one "caveat", namely the particles acting onto each other do not only possess the spatial variable, but also a discrete intrinsic angular momentum-like variable called the "spin", and there is the Pauli principle relating the space and the spin variables. Depending on the value of the spin, identical particles split into two different classes, fermions and bosons. If two identical fermions (e.g. electrons) have a *symmetric* spin function (e.g. parallel spins) the spatial variables must be *antisymmetric* (i.e. they exclude each other from their places much as if there was a repulsive force), and vice versa, i.e. for antiparallel *spins* the *position variables* must be symmetric (i.e. the apparent force must be attractive). Thus in the case of two fermions there is a strictly negative correlation between spatial and spin variables, whereas for two bosons (e.g. quanta of electromagnetic waves, photons) the correlation is strictly positive.

Thus the notion "force" loses already part of its meaning.

1.5.3 Feynman diagrams

Main article: Feynman diagrams
In modern particle physics, forces and the acceleration of

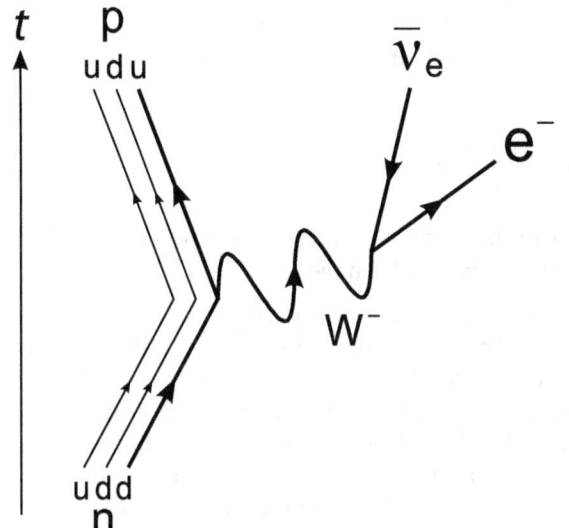

Feynman diagram for the decay of a neutron into a proton. The W boson is between two vertices indicating a repulsion.

particles are explained as a mathematical by-product of exchange of momentum-carrying gauge bosons. With the development of quantum field theory and general relativity, it was realized that force is a redundant concept arising from conservation of momentum (4-momentum in relativity and momentum of virtual particles in quantum electrodynamics). The conservation of momentum can be directly derived from the homogeneity or symmetry of space and so is usually considered more fundamental than the concept of a force. Thus the currently known fundamental forces are considered more accurately to be "fundamental interactions".[6]:199–128 When particle A emits (creates) or absorbs (annihilates) virtual particle B, a momentum conservation results in recoil of particle A making impression of repulsion or attraction between particles A A' exchanging by B. This description applies to all forces arising from fundamental interactions. While sophisticated mathematical descriptions are needed to predict, in full detail, the accurate result of such interactions, there is a conceptually simple way to describe such interactions through the use of Feynman diagrams. In a Feynman diagram, each matter particle is represented as a straight line (see world line) traveling through time, which normally increases up or to the right in the diagram. Matter and anti-matter particles are identical except for their direction of propagation through the Feynman diagram. World lines of particles intersect at interaction vertices, and the Feynman diagram represents any

force arising from an interaction as occurring at the vertex with an associated instantaneous change in the direction of the particle world lines. Gauge bosons are emitted away from the vertex as wavy lines and, in the case of virtual particle exchange, are absorbed at an adjacent vertex.[24]

The utility of Feynman diagrams is that other types of physical phenomena that are part of the general picture of fundamental interactions but are conceptually separate from forces can also be described using the same rules. For example, a Feynman diagram can describe in succinct detail how a neutron decays into an electron, proton, and neutrino, an interaction mediated by the same gauge boson that is responsible for the weak nuclear force.[24]

1.6 Fundamental forces

Main article: Fundamental interaction

All of the forces in the universe are based on four fundamental interactions. The strong and weak forces are nuclear forces that act only at very short distances, and are responsible for the interactions between subatomic particles, including nucleons and compound nuclei. The electromagnetic force acts between electric charges, and the gravitational force acts between masses. All other forces in nature derive from these four fundamental interactions. For example, friction is a manifestation of the electromagnetic force acting between the atoms of two surfaces, and the Pauli exclusion principle,[25] which does not permit atoms to pass through each other. Similarly, the forces in springs, modeled by Hooke's law, are the result of electromagnetic forces and the Exclusion Principle acting together to return an object to its equilibrium position. Centrifugal forces are acceleration forces that arise simply from the acceleration of rotating frames of reference.[4]:12-11 [5]:359

The development of fundamental theories for forces proceeded along the lines of unification of disparate ideas. For example, Isaac Newton unified the force responsible for objects falling at the surface of the Earth with the force responsible for the orbits of celestial mechanics in his universal theory of gravitation. Michael Faraday and James Clerk Maxwell demonstrated that electric and magnetic forces were unified through one consistent theory of electromagnetism. In the 20th century, the development of quantum mechanics led to a modern understanding that the first three fundamental forces (all except gravity) are manifestations of matter (fermions) interacting by exchanging virtual particles called gauge bosons.[26] This standard model of particle physics posits a similarity between the forces and led scientists to predict the unification of the weak and electromagnetic forces in electroweak theory subsequently confirmed by observation. The complete formulation of the standard model predicts an as yet unobserved Higgs mechanism, but observations such as neutrino oscillations indicate that the standard model is incomplete. A Grand Unified Theory allowing for the combination of the electroweak interaction with the strong force is held out as a possibility with candidate theories such as supersymmetry proposed to accommodate some of the outstanding unsolved problems in physics. Physicists are still attempting to develop self-consistent unification models that would combine all four fundamental interactions into a theory of everything. Einstein tried and failed at this endeavor, but currently the most popular approach to answering this question is string theory.[6]:212–219

1.6.1 Gravitational

Main article: Gravity

What we now call gravity was not identified as a universal force until the work of Isaac Newton. Before Newton, the tendency for objects to fall towards the Earth was not understood to be related to the motions of celestial objects. Galileo was instrumental in describing the characteristics of falling objects by determining that the acceleration of every object in free-fall was constant and independent of the mass of the object. Today, this acceleration due to gravity towards the surface of the Earth is usually designated as $\vec{g}$ and has a magnitude of about 9.81 meters per second squared (this measurement is taken from sea level and may vary depending on location), and points toward the center of the Earth.[28] This observation means that the force of gravity on an object at the Earth's surface is directly proportional to the object's mass. Thus an object that has a mass of m will experience a force:

$$\vec{F} = m\vec{g}$$

In free-fall, this force is unopposed and therefore the net force on the object is its weight. For objects not in free-fall, the force of gravity is opposed by the reactions of their supports. For example, a person standing on the ground experiences zero net force, since his weight is balanced by a normal force exerted by the ground.[4][5]

Newton's contribution to gravitational theory was to unify the motions of heavenly bodies, which Aristotle had assumed were in a natural state of constant motion, with falling motion observed on the Earth. He proposed a law of gravity that could account for the celestial motions that had been described earlier using Kepler's laws of planetary motion.[29]

Newton came to realize that the effects of gravity might be observed in different ways at larger distances. In particu-

lar, Newton determined that the acceleration of the Moon around the Earth could be ascribed to the same force of gravity if the acceleration due to gravity decreased as an inverse square law. Further, Newton realized that the acceleration due to gravity is proportional to the mass of the attracting body.[29] Combining these ideas gives a formula that relates the mass ($m_\oplus$) and the radius ($R_\oplus$) of the Earth to the gravitational acceleration:

$$\vec{g} = -\frac{Gm_\oplus}{R_\oplus{}^2}\hat{r}$$

where the vector direction is given by $\hat{r}$, the unit vector directed outward from the center of the Earth.[10]

In this equation, a dimensional constant G is used to describe the relative strength of gravity. This constant has come to be known as Newton's Universal Gravitation Constant,[30] though its value was unknown in Newton's lifetime. Not until 1798 was Henry Cavendish able to make the first measurement of G using a torsion balance; this was widely reported in the press as a measurement of the mass of the Earth since knowing G could allow one to solve for the Earth's mass given the above equation. Newton, however, realized that since all celestial bodies followed the same laws of motion, his law of gravity had to be universal. Succinctly stated, Newton's Law of Gravitation states that the force on a spherical object of mass m_1 due to the gravitational pull of mass m_2 is

$$\vec{F} = -\frac{Gm_1 m_2}{r^2}\hat{r}$$

where r is the distance between the two objects' centers of mass and $\hat{r}$ is the unit vector pointed in the direction away from the center of the first object toward the center of the second object.[10]

This formula was powerful enough to stand as the basis for all subsequent descriptions of motion within the solar system until the 20th century. During that time, sophisticated methods of perturbation analysis[31] were invented to calculate the deviations of orbits due to the influence of multiple bodies on a planet, moon, comet, or asteroid. The formalism was exact enough to allow mathematicians to predict the existence of the planet Neptune before it was observed.[32]

It was only the orbit of the planet Mercury that Newton's Law of Gravitation seemed not to fully explain. Some astrophysicists predicted the existence of another planet (Vulcan) that would explain the discrepancies; however, despite some early indications, no such planet could be found. When Albert Einstein formulated his theory of general relativity (GR) he turned his attention to the problem of Mercury's orbit and found that his theory added a correction, which could account for the discrepancy. This was the first time that Newton's Theory of Gravity had been shown to be less correct than an alternative.[34]

Since then, and so far, general relativity has been acknowledged as the theory that best explains gravity. In GR, gravitation is not viewed as a force, but rather, objects moving freely in gravitational fields travel under their own inertia in straight lines through curved space-time – defined as the shortest space-time path between two space-time events. From the perspective of the object, all motion occurs as if there were no gravitation whatsoever. It is only when observing the motion in a global sense that the curvature of space-time can be observed and the force is inferred from the object's curved path. Thus, the straight line path in space-time is seen as a curved line in space, and it is called the *ballistic trajectory* of the object. For example, a basketball thrown from the ground moves in a parabola, as it is in a uniform gravitational field. Its space-time trajectory (when the extra ct dimension is added) is almost a straight line, slightly curved (with the radius of curvature of the order of few light-years). The time derivative of the changing momentum of the object is what we label as "gravitational force".[5]

1.6.2 Electromagnetic

Main article: Electromagnetic force

The electrostatic force was first described in 1784 by Coulomb as a force that existed intrinsically between two charges.[18]:519 The properties of the electrostatic force were that it varied as an inverse square law directed in the radial direction, was both attractive and repulsive (there was intrinsic polarity), was independent of the mass of the charged objects, and followed the superposition principle. Coulomb's law unifies all these observations into one succinct statement.[35]

Subsequent mathematicians and physicists found the construct of the *electric field* to be useful for determining the electrostatic force on an electric charge at any point in space. The electric field was based on using a hypothetical "test charge" anywhere in space and then using Coulomb's Law to determine the electrostatic force.[36]:4-6 to 4-8 Thus the electric field anywhere in space is defined as

$$\vec{E} = \frac{\vec{F}}{q}$$

where q is the magnitude of the hypothetical test charge.

Meanwhile, the Lorentz force of magnetism was discovered to exist between two electric currents. It has the same math-

ematical character as Coulomb's Law with the proviso that like currents attract and unlike currents repel. Similar to the electric field, the magnetic field can be used to determine the magnetic force on an electric current at any point in space. In this case, the magnitude of the magnetic field was determined to be

$$B = \frac{F}{I\ell}$$

where I is the magnitude of the hypothetical test current and ℓ is the length of hypothetical wire through which the test current flows. The magnetic field exerts a force on all magnets including, for example, those used in compasses. The fact that the Earth's magnetic field is aligned closely with the orientation of the Earth's axis causes compass magnets to become oriented because of the magnetic force pulling on the needle.

Through combining the definition of electric current as the time rate of change of electric charge, a rule of vector multiplication called Lorentz's Law describes the force on a charge moving in a magnetic field.[36] The connection between electricity and magnetism allows for the description of a unified *electromagnetic force* that acts on a charge. This force can be written as a sum of the electrostatic force (due to the electric field) and the magnetic force (due to the magnetic field). Fully stated, this is the law:

$$\vec{F} = q(\vec{E} + \vec{v} \times \vec{B})$$

where $\vec{F}$ is the electromagnetic force, q is the magnitude of the charge of the particle, $\vec{E}$ is the electric field, $\vec{v}$ is the velocity of the particle that is crossed with the magnetic field ($\vec{B}$).

The origin of electric and magnetic fields would not be fully explained until 1864 when James Clerk Maxwell unified a number of earlier theories into a set of 20 scalar equations, which were later reformulated into 4 vector equations by Oliver Heaviside and Josiah Willard Gibbs.[37] These "Maxwell Equations" fully described the sources of the fields as being stationary and moving charges, and the interactions of the fields themselves. This led Maxwell to discover that electric and magnetic fields could be "self-generating" through a wave that traveled at a speed that he calculated to be the speed of light. This insight united the nascent fields of electromagnetic theory with optics and led directly to a complete description of the electromagnetic spectrum.[38]

However, attempting to reconcile electromagnetic theory with two observations, the photoelectric effect, and the nonexistence of the ultraviolet catastrophe, proved troublesome. Through the work of leading theoretical physicists, a new theory of electromagnetism was developed using quantum mechanics. This final modification to electromagnetic theory ultimately led to quantum electrodynamics (or QED), which fully describes all electromagnetic phenomena as being mediated by wave–particles known as photons. In QED, photons are the fundamental exchange particle, which described all interactions relating to electromagnetism including the electromagnetic force.[Note 4]

It is a common misconception to ascribe the stiffness and rigidity of solid matter to the repulsion of like charges under the influence of the electromagnetic force. However, these characteristics actually result from the Pauli exclusion principle. Since electrons are fermions, they cannot occupy the same quantum mechanical state as other electrons. When the electrons in a material are densely packed together, there are not enough lower energy quantum mechanical states for them all, so some of them must be in higher energy states. This means that it takes energy to pack them together. While this effect is manifested macroscopically as a structural force, it is technically only the result of the existence of a finite set of electron states.

1.6.3   Strong nuclear

Main article: Strong interaction

There are two "nuclear forces", which today are usually described as interactions that take place in quantum theories of particle physics. The strong nuclear force[18]:940 is the force responsible for the structural integrity of atomic nuclei while the weak nuclear force[18]:951 is responsible for the decay of certain nucleons into leptons and other types of hadrons.[4][5]

The strong force is today understood to represent the interactions between quarks and gluons as detailed by the theory of quantum chromodynamics (QCD).[39] The strong force is the fundamental force mediated by gluons, acting upon quarks, antiquarks, and the gluons themselves. The (aptly named) strong interaction is the "strongest" of the four fundamental forces.

The strong force only acts *directly* upon elementary particles. However, a residual of the force is observed between hadrons (the best known example being the force that acts between nucleons in atomic nuclei) as the nuclear force. Here the strong force acts indirectly, transmitted as gluons, which form part of the virtual pi and rho mesons, which classically transmit the nuclear force (see this topic for more). The failure of many searches for free quarks has shown that the elementary particles affected are not directly observable. This phenomenon is called color confinement.

1.6.4   Weak nuclear

Main article: Weak interaction

The weak force is due to the exchange of the heavy W and Z bosons. Its most familiar effect is beta decay (of neutrons in atomic nuclei) and the associated radioactivity. The word "weak" derives from the fact that the field strength is some 10^{13} times less than that of the strong force. Still, it is stronger than gravity over short distances. A consistent electroweak theory has also been developed, which shows that electromagnetic forces and the weak force are indistinguishable at a temperatures in excess of approximately 10^{15} kelvins. Such temperatures have been probed in modern particle accelerators and show the conditions of the universe in the early moments of the Big Bang.

1.7   Non-fundamental forces

Some forces are consequences of the fundamental ones. In such situations, idealized models can be utilized to gain physical insight.

1.7.1   Normal force

Main article: Normal force

The normal force is due to repulsive forces of interaction between atoms at close contact. When their electron clouds overlap, Pauli repulsion (due to fermionic nature of electrons) follows resulting in the force that acts in a direction normal to the surface interface between two objects.[18]:93 The normal force, for example, is responsible for the structural integrity of tables and floors as well as being the force that responds whenever an external force pushes on a solid object. An example of the normal force in action is the impact force on an object crashing into an immobile surface.[4][5]

1.7.2   Friction

Main article: Friction

Friction is a surface force that opposes relative motion. The frictional force is directly related to the normal force that acts to keep two solid objects separated at the point of contact. There are two broad classifications of frictional forces: static friction and kinetic friction.

The static friction force (F_{sf}) will exactly oppose forces applied to an object parallel to a surface contact up to the limit specified by the coefficient of static friction (μ_{sf}) multiplied by the normal force (F_N). In other words, the magnitude of the static friction force satisfies the inequality:

$$0 \leq F_{sf} \leq \mu_{sf}F_{N}.$$

The kinetic friction force (F_{kf}) is independent of both the forces applied and the movement of the object. Thus, the magnitude of the force equals:

$$F_{kf} = \mu_{kf}F_{N},$$

where μ_{kf} is the coefficient of kinetic friction. For most surface interfaces, the coefficient of kinetic friction is less than the coefficient of static friction.

1.7.3   Tension

Main article: Tension (physics)

Tension forces can be modeled using ideal strings that are massless, frictionless, unbreakable, and unstretchable. They can be combined with ideal pulleys, which allow ideal strings to switch physical direction. Ideal strings transmit tension forces instantaneously in action-reaction pairs so that if two objects are connected by an ideal string, any force directed along the string by the first object is accompanied by a force directed along the string in the opposite direction by the second object.[40] By connecting the same string multiple times to the same object through the use of a set-up that uses movable pulleys, the tension force on a load can be multiplied. For every string that acts on a load, another factor of the tension force in the string acts on the load. However, even though such machines allow for an increase in force, there is a corresponding increase in the length of string that must be displaced in order to move the load. These tandem effects result ultimately in the conservation of mechanical energy since the work done on the load is the same no matter how complicated the machine.[4][5][41]

1.7.4   Elastic force

Main articles: Elasticity (physics) and Hooke's law
An elastic force acts to return a spring to its natural length. An ideal spring is taken to be massless, frictionless, unbreakable, and infinitely stretchable. Such springs exert forces that push when contracted, or pull when extended, in proportion to the displacement of the spring from its equilibrium position.[42] This linear relationship was described by Robert Hooke in 1676, for whom Hooke's law is named. If Δx is the displacement, the force exerted by an ideal spring equals:

$$\vec{F} = -k\Delta \vec{x}$$

where k is the spring constant (or force constant), which is particular to the spring. The minus sign accounts for the tendency of the force to act in opposition to the applied load.[4][5]

1.7.5   Continuum mechanics

Main articles: Pressure, Drag (physics), and Stress (mechanics)

Newton's laws and Newtonian mechanics in general were first developed to describe how forces affect idealized point particles rather than three-dimensional objects. However, in real life, matter has extended structure and forces that act on one part of an object might affect other parts of an object. For situations where lattice holding together the atoms in an object is able to flow, contract, expand, or otherwise change shape, the theories of continuum mechanics describe the way forces affect the material. For example, in extended fluids, differences in pressure result in forces being directed along the pressure gradients as follows:

$$\frac{\vec{F}}{V} = -\vec{\nabla}P$$

where V is the volume of the object in the fluid and P is the scalar function that describes the pressure at all locations in space. Pressure gradients and differentials result in the buoyant force for fluids suspended in gravitational fields, winds in atmospheric science, and the lift associated with aerodynamics and flight.[4][5]

A specific instance of such a force that is associated with dynamic pressure is fluid resistance: a body force that resists the motion of an object through a fluid due to viscosity. For so-called "Stokes' drag" the force is approximately proportional to the velocity, but opposite in direction:

$$\vec{F}_{d} = -b\vec{v}$$

where:

> b is a constant that depends on the properties of the fluid and the dimensions of the object (usually the cross-sectional area), and
>
> $\vec{v}$ is the velocity of the object.[4][5]

More formally, forces in continuum mechanics are fully described by a stress–tensor with terms that are roughly defined as

$$\sigma = \frac{F}{A}$$

where A is the relevant cross-sectional area for the volume for which the stress-tensor is being calculated. This formalism includes pressure terms associated with forces that act normal to the cross-sectional area (the matrix diagonals of the tensor) as well as shear terms associated with forces that act parallel to the cross-sectional area (the off-diagonal elements). The stress tensor accounts for forces that cause all strains (deformations) including also tensile stresses and compressions.[3][5]:133–134[36]:38-1–38-11

1.7.6 Fictitious forces

Main article: Fictitious forces

There are forces that are frame dependent, meaning that they appear due to the adoption of non-Newtonian (that is, non-inertial) reference frames. Such forces include the centrifugal force and the Coriolis force.[43] These forces are considered fictitious because they do not exist in frames of reference that are not accelerating.[4][5] Because these forces are not genuine they are also referred to as "pseudo forces".[4]:12-11

In general relativity, gravity becomes a fictitious force that arises in situations where spacetime deviates from a flat geometry. As an extension, Kaluza–Klein theory and string theory ascribe electromagnetism and the other fundamental forces respectively to the curvature of differently scaled dimensions, which would ultimately imply that all forces are fictitious.

1.8 Rotations and torque

Main article: Torque

Forces that cause extended objects to rotate are associated with torques. Mathematically, the torque of a force $\vec{F}$ is defined relative to an arbitrary reference point as the cross-product:

$$\vec{\tau} = \vec{r} \times \vec{F}$$

where

$\vec{r}$ is the position vector of the force application point relative to the reference point.

Torque is the rotation equivalent of force in the same way that angle is the rotational equivalent for position, angular velocity for velocity, and angular momentum for momentum. As a consequence of Newton's First Law of Motion, there exists rotational inertia that ensures that all bodies maintain their angular momentum unless acted upon by an unbalanced torque. Likewise, Newton's Second Law of Motion can be used to derive an analogous equation for the instantaneous angular acceleration of the rigid body:

$$\vec{\tau} = I\vec{\alpha}$$

where

I is the moment of inertia of the body

$\vec{\alpha}$ is the angular acceleration of the body.

This provides a definition for the moment of inertia, which is the rotational equivalent for mass. In more advanced treatments of mechanics, where the rotation over a time interval is described, the moment of inertia must be substituted by the tensor that, when properly analyzed, fully determines the characteristics of rotations including precession and nutation.

Equivalently, the differential form of Newton's Second Law provides an alternative definition of torque:

$$\vec{\tau} = \frac{d\vec{L}}{dt}, \quad [44] \text{ where } \vec{L} \text{ is the angular momentum of the particle.}$$

Newton's Third Law of Motion requires that all objects exerting torques themselves experience equal and opposite torques,[45] and therefore also directly implies the conservation of angular momentum for closed systems that experience rotations and revolutions through the action of internal torques.

1.8.1 Centripetal force

Main article: Centripetal force

For an object accelerating in circular motion, the unbalanced force acting on the object equals:[46]

$$\vec{F} = -\frac{mv^2\hat{r}}{r}$$

where m is the mass of the object, v is the velocity of the object and r is the distance to the center of the circular path and $\hat{r}$ is the unit vector pointing in the radial direction outwards from the center. This means that the unbalanced centripetal force felt by any object is always directed toward the center of the curving path. Such forces act perpendicular to the velocity vector associated with the motion of an object, and therefore do not change the speed of the object (magnitude of the velocity), but only the direction of the velocity vector. The unbalanced force that accelerates an object can be resolved into a component that is perpendicular to the path, and one that is tangential to the path. This yields both the tangential force, which accelerates the object by either slowing it down or speeding it up, and the radial (centripetal) force, which changes its direction.[4][5]

1.9   Kinematic integrals

Main articles: Impulse, Mechanical work, and Power (physics)

Forces can be used to define a number of physical concepts by integrating with respect to kinematic variables. For example, integrating with respect to time gives the definition of impulse:[47]

$$\vec{I} = \int_{t_1}^{t_2} \vec{F} \mathrm{d}t,$$

which by Newton's Second Law must be equivalent to the change in momentum (yielding the Impulse momentum theorem).

Similarly, integrating with respect to position gives a definition for the work done by a force:[4]:13-3

$$W = \int_{\vec{x}_1}^{\vec{x}_2} \vec{F} \cdot \mathrm{d}\vec{x},$$

which is equivalent to changes in kinetic energy (yielding the work energy theorem).[4]:13-3

Power P is the rate of change $\mathrm{d}W/\mathrm{d}t$ of the work W, as the trajectory is extended by a position change $\mathrm{d}\vec{x}$ in a time interval $\mathrm{d}t$:[4]:13-2

$$\mathrm{d}W = \frac{\mathrm{d}W}{\mathrm{d}\vec{x}} \cdot \mathrm{d}\vec{x} = \vec{F} \cdot \mathrm{d}\vec{x},$$

$$\text{so}\quad P = \frac{\mathrm{d}W}{\mathrm{d}t} = \frac{\mathrm{d}W}{\mathrm{d}\vec{x}} \cdot \frac{\mathrm{d}\vec{x}}{\mathrm{d}t} = \vec{F} \cdot \vec{v},$$

with $\vec{v} = \mathrm{d}\,\vec{x}/\mathrm{d}t$ the velocity.

1.10   Potential energy

Main article: Potential energy

Instead of a force, often the mathematically related concept of a potential energy field can be used for convenience. For instance, the gravitational force acting upon an object can be seen as the action of the gravitational field that is present at the object's location. Restating mathematically the definition of energy (via the definition of work), a potential scalar field $U(\vec{r})$ is defined as that field whose gradient is equal and opposite to the force produced at every point:

$$\vec{F} = -\vec{\nabla}U.$$

Forces can be classified as conservative or nonconservative. Conservative forces are equivalent to the gradient of a potential while nonconservative forces are not.[4][5]

1.10.1   Conservative forces

Main article: Conservative force

A conservative force that acts on a closed system has an associated mechanical work that allows energy to convert only between kinetic or potential forms. This means that for a closed system, the net mechanical energy is conserved whenever a conservative force acts on the system. The force, therefore, is related directly to the difference in potential energy between two different locations in space,[48] and can be considered to be an artifact of the potential field in the same way that the direction and amount of a flow of water can be considered to be an artifact of the contour map of the elevation of an area.[4][5]

Conservative forces include gravity, the electromagnetic force, and the spring force. Each of these forces has models that are dependent on a position often given as a radial vector $\vec{r}$ emanating from spherically symmetric potentials.[49] Examples of this follow:

For gravity:

$$\vec{F} = -\frac{Gm_1m_2\vec{r}}{r^3}$$

where G is the gravitational constant, and m_n is the mass of object n.

For electrostatic forces:

$$\vec{F} = \frac{q_1q_2\vec{r}}{4\pi\epsilon_0 r^3}$$

where ϵ_0 is electric permittivity of free space, and q_n is the electric charge of object n.

For spring forces:

$$\vec{F} = -k\vec{r}$$

where k is the spring constant.[4][5]

1.10.2 Nonconservative forces

For certain physical scenarios, it is impossible to model forces as being due to gradient of potentials. This is often due to macrophysical considerations that yield forces as arising from a macroscopic statistical average of microstates. For example, friction is caused by the gradients of numerous electrostatic potentials between the atoms, but manifests as a force model that is independent of any macroscale position vector. Nonconservative forces other than friction include other contact forces, tension, compression, and drag. However, for any sufficiently detailed description, all these forces are the results of conservative ones since each of these macroscopic forces are the net results of the gradients of microscopic potentials.[4][5]

The connection between macroscopic nonconservative forces and microscopic conservative forces is described by detailed treatment with statistical mechanics. In macroscopic closed systems, nonconservative forces act to change the internal energies of the system, and are often associated with the transfer of heat. According to the Second law of thermodynamics, nonconservative forces necessarily result in energy transformations within closed systems from ordered to more random conditions as entropy increases.[4][5]

1.11   Units of measurement

The SI unit of force is the newton (symbol N), which is the force required to accelerate a one kilogram mass at a rate of one meter per second squared, or $kg \cdot m \cdot s^{-2}$.[50] The corresponding CGS unit is the dyne, the force required to accelerate a one gram mass by one centimeter per second squared, or $g \cdot cm \cdot s^{-2}$. A newton is thus equal to 100,000 dynes.

The gravitational foot-pound-second English unit of force is the pound-force (lbf), defined as the force exerted by gravity on a pound-mass in the standard gravitational field of 9.80665 $m \cdot s^{-2}$.[50] The pound-force provides an alternative unit of mass: one slug is the mass that will accelerate by one foot per second squared when acted on by one pound-force.[50]

An alternative unit of force in a different foot-pound-second system, the absolute fps system, is the poundal, defined as the force required to accelerate a one-pound mass at a rate of one foot per second squared.[50] The units of slug and poundal are designed to avoid a constant of proportionality in Newton's Second Law.

The pound-force has a metric counterpart, less commonly used than the newton: the kilogram-force (kgf) (sometimes kilopond), is the force exerted by standard gravity on one kilogram of mass.[50] The kilogram-force leads to an alternate, but rarely used unit of mass: the metric slug (sometimes mug or hyl) is that mass that accelerates at 1 $m \cdot s^{-2}$ when subjected to a force of 1 kgf. The kilogram-force is not a part of the modern SI system, and is generally deprecated; however it still sees use for some purposes as expressing aircraft weight, jet thrust, bicycle spoke tension, torque wrench settings and engine output torque. Other arcane units of force include the sthène, which is equivalent to 1000 N, and the kip, which is equivalent to 1000 lbf.

See also Ton-force.

1.12   Force measurement

See force gauge, spring scale, load cell

1.13   See also

- Orders of magnitude (force)

1.14   Notes

[1] Newton's *Principia Mathematica* actually used a finite difference version of this equation based upon *impulse*. See *Impulse*.

[2] "It is important to note that we *cannot* derive a general expression for Newton's second law for variable mass systems by treating the mass in $\mathbf{F} = d\mathbf{P}/dt = d(M\mathbf{v})$ as a *variable*. [...] We *can* use $\mathbf{F} = d\mathbf{P}/dt$ to analyze variable mass systems *only* if we apply it to an *entire system of constant mass* having parts among which there is an interchange of mass." [Emphasis as in the original] (Halliday, Resnick & Krane 2001, p. 199)

[3] "Any single force is only one aspect of a mutual interaction between *two* bodies." (Halliday, Resnick & Krane 2001, pp. 78–79)

[4] For a complete library on quantum mechanics see Quantum mechanics – References

1.15   References

[1] Nave, C. R. (2014). "Force". *Hyperphysics*. Dept. of Physics and Astronomy, Georgia State University. Retrieved 15 August 2014.

[2] Heath, T.L. *"The Works of Archimedes* (1897). The unabridged work in PDF form (19 MB)". Internet Archive. Retrieved 2007-10-14.

[3] *University Physics*, Sears, Young & Zemansky, pp.18–38

[4] Feynman volume 1

[5] Kleppner & Kolenkow 2010

[6] Weinberg, S. (1994). *Dreams of a Final Theory*. Vintage Books USA. ISBN 0-679-74408-8.

[7] Lang, Helen S. (1998). *The order of nature in Aristotle's physics : place and the elements* (1. publ. ed.). Cambridge: Cambridge Univ. Press. ISBN 9780521624534.

[8] Hetherington, Norriss S. (1993). *Cosmology: Historical, Literary, Philosophical, Religious, and Scientific Perspectives*. Garland Reference Library of the Humanities. p. 100. ISBN 0-8153-1085-4.

[9] Drake, Stillman (1978). Galileo At Work. Chicago: University of Chicago Press. ISBN 0-226-16226-5

[10] Newton, Isaac (1999). *The Principia Mathematical Principles of Natural Philosophy*. Berkeley: University of California Press. ISBN 0-520-08817-4. This is a recent translation into English by I. Bernard Cohen and Anne Whitman, with help from Julia Budenz.

[11] DiSalle, Robert (2002-03-30). "Space and Time: Inertial Frames". *Stanford Encyclopedia of Philosophy*. Retrieved 2008-03-24.

[12] Howland, R. A. (2006). *Intermediate dynamics a linear algebraic approach* (Online-Ausg. ed.). New York: Springer. pp. 255–256. ISBN 9780387280592.

[13] One exception to this rule is: Landau, L. D.; Akhiezer, A. I.; Lifshitz, A. M. (196). *General Physics; mechanics and molecular physics* (First English ed.). Oxford: Pergamon Press. ISBN 0-08-003304-0. Translated by: J. B. Sykes, A. D. Petford, and C. L. Petford. Library of Congress Catalog Number 67-30260. In section 7, pages 12–14, this book defines force as *dp/dt*.

[14] Jammer, Max (1999). *Concepts of force : a study in the foundations of dynamics* (Facsim. ed.). Mineola, N.Y.: Dover Publications. pp. 220–222. ISBN 9780486406893.

[15] Noll, Walter (April 2007). "On the Concept of Force" (pdf). Carnegie Mellon University. Retrieved 28 October 2013.

[16] C. Hellingman (1992). "Newton's third law revisited". *Phys. Educ.* **27** (2): 112–115. Bibcode:1992PhyEd..27..112H. doi:10.1088/0031-9120/27/2/011. Quoting Newton in the *Principia*: It is not one action by which the Sun attracts Jupiter, and another by which Jupiter attracts the Sun; but it is one action by which the Sun and Jupiter mutually endeavour to come nearer together.

[17] Dr. Nikitin (2007). "Dynamics of translational motion". Retrieved 2008-01-04.

[18] Cutnell & Johnson 2003

[19] "Seminar: Visualizing Special Relativity". *The Relativistic Raytracer*. Retrieved 2008-01-04.

[20] Wilson, John B. "Four-Vectors (4-Vectors) of Special Relativity: A Study of Elegant Physics". *The Science Realm: John's Virtual Sci-Tech Universe*. Archived from the original on 26 June 2009. Retrieved 2008-01-04.

[21] "Introduction to Free Body Diagrams". *Physics Tutorial Menu*. University of Guelph. Retrieved 2008-01-02.

[22] Henderson, Tom (2004). "The Physics Classroom". *The Physics Classroom and Mathsoft Engineering & Education, Inc.* Retrieved 2008-01-02.

[23] "Static Equilibrium". *Physics Static Equilibrium (forces and torques)*. University of the Virgin Islands. Archived from the original on October 19, 2007. Retrieved 2008-01-02.

[24] Shifman, Mikhail (1999). *ITEP lectures on particle physics and field theory*. World Scientific. ISBN 981-02-2639-X.

[25] Nave, Carl Rod. "Pauli Exclusion Principle". *HyperPhysics*. University of Guelph. Retrieved 2013-10-28.

[26] "Fermions & Bosons". *The Particle Adventure*. Retrieved 2008-01-04.

[27] http://www.pha.jhu.edu/~{}dfehling/particle.gif

[28] Cook, A. H. (1965). "A New Absolute Determination of the Acceleration due to Gravity at the National Physical Laboratory". *Nature*. **208** (5007): 279. Bibcode:1965Natur.208..279C. doi:10.1038/208279a0.

[29] Young, Hugh; Freedman, Roger; Sears, Francis and Zemansky, Mark (1949) *University Physics*. Pearson Education. pp. 59–82

[30] "Sir Isaac Newton: The Universal Law of Gravitation". *Astronomy 161 The Solar System*. Retrieved 2008-01-04.

[31] Watkins, Thayer. "Perturbation Analysis, Regular and Singular". *Department of Economics*. San José State University.

[32] Kollerstrom, Nick (2001). "Neptune's Discovery. The British Case for Co-Prediction.". University College London. Archived from the original on 2005-11-11. Retrieved 2007-03-19.

[33] "Powerful New Black Hole Probe Arrives at Paranal". Retrieved 13 August 2015.

[34] Einstein, Albert (1916). "The Foundation of the General Theory of Relativity" (PDF). *Annalen der Physik*. **49** (7): 769–822. Bibcode:1916AnP...354..769E. doi:10.1002/andp.19163540702. Retrieved 2006-09-03.

[35] Coulomb, Charles (1784). "Recherches théoriques et expérimentales sur la force de torsion et sur l'élasticité des fils de metal". *Histoire de l'Académie Royale des Sciences*: 229–269.

[36] Feynman volume 2

[37] Scharf, Toralf (2007). *Polarized light in liquid crystals and polymers*. John Wiley and Sons. p. 19. ISBN 0-471-74064-0., Chapter 2, p. 19

[38] Duffin, William (1980). *Electricity and Magnetism, 3rd Ed*. McGraw-Hill. pp. 364–383. ISBN 0-07-084111-X.

[39] Stevens, Tab (10 July 2003). "Quantum-Chromodynamics: A Definition – Science Articles". Archived from the original on 2011-10-16. Retrieved 2008-01-04.

[40] "Tension Force". *Non-Calculus Based Physics I*. Retrieved 2008-01-04.

[41] Fitzpatrick, Richard (2006-02-02). "Strings, pulleys, and inclines". Retrieved 2008-01-04.

[42] Nave, Carl Rod. "Elasticity". *HyperPhysics*. University of Guelph. Retrieved 2013-10-28.

[43] Mallette, Vincent (1982–2008). "Inwit Publishing, Inc. and Inwit, LLC – Writings, Links and Software Distributions – The Coriolis Force". *Publications in Science and Mathematics, Computing and the Humanities*. Inwit Publishing, Inc. Retrieved 2008-01-04.

[44] Nave, Carl Rod. "Newton's 2nd Law: Rotation". *HyperPhysics*. University of Guelph. Retrieved 2013-10-28.

[45] Fitzpatrick, Richard (2007-01-07). "Newton's third law of motion". Retrieved 2008-01-04.

[46] Nave, Carl Rod. "Centripetal Force". *HyperPhysics*. University of Guelph. Retrieved 2013-10-28.

[47] Hibbeler, Russell C. (2010). *Engineering Mechanics, 12th edition*. Pearson Prentice Hall. p. 222. ISBN 0-13-607791-9.

[48] Singh, Sunil Kumar (2007-08-25). "Conservative force". *Connexions*. Retrieved 2008-01-04.

[49] Davis, Doug. "Conservation of Energy". *General physics*. Retrieved 2008-01-04.

[50] Wandmacher, Cornelius; Johnson, Arnold (1995). *Metric Units in Engineering*. ASCE Publications. p. 15. ISBN 0-7844-0070-9.

1.16 Further reading

- Corben, H.C.; Philip Stehle (1994). *Classical Mechanics*. New York: Dover publications. pp. 28–31. ISBN 0-486-68063-0.

- Cutnell, John D.; Johnson, Kenneth W. (2003). *Physics, Sixth Edition*. Hoboken, New Jersey: John Wiley & Sons Inc. ISBN 0471151831.

- Feynman, Richard P.; Leighton; Sands, Matthew (2010). *The Feynman lectures on physics. Vol. I: Mainly mechanics, radiation and heat* (New millennium ed.). New York: BasicBooks. ISBN 978-0465024933.

- Feynman, Richard P.; Leighton, Robert B.; Sands, Matthew (2010). *The Feynman lectures on physics. Vol. II: Mainly electromagnetism and matter* (New millennium ed.). New York: BasicBooks. ISBN 978-0465024940.

- Halliday, David; Resnick, Robert; Krane, Kenneth S. (2001). *Physics v. 1*. New York: John Wiley & Sons. ISBN 0-471-32057-9.

- Kleppner, Daniel; Kolenkow, Robert J. (2010). *An introduction to mechanics* (3. print ed.). Cambridge: Cambridge University Press. ISBN 0521198216.

- Parker, Sybil (1993). "force". *Encyclopedia of Physics*. Ohio: McGraw-Hill. p. 107,. ISBN 0-07-051400-3.

- Sears F., Zemansky M. & Young H. (1982). *University Physics*. Reading, Massachusetts: Addison-Wesley. ISBN 0-201-07199-1.

- Serway, Raymond A. (2003). *Physics for Scientists and Engineers*. Philadelphia: Saunders College Publishing. ISBN 0-534-40842-7.

- Tipler, Paul (2004). *Physics for Scientists and Engineers: Mechanics, Oscillations and Waves, Thermodynamics* (5th ed.). W. H. Freeman. ISBN 0-7167-0809-4.

- Verma, H.C. (2004). *Concepts of Physics Vol 1*. (2004 Reprint ed.). Bharti Bhavan. ISBN 8177091875.

1.17 External links

- Video lecture on Newton's three laws by Walter Lewin from MIT OpenCourseWare

- A Java simulation on vector addition of forces

- Force demonstrated as any influence on an object that changes the object's shape or motion (video)

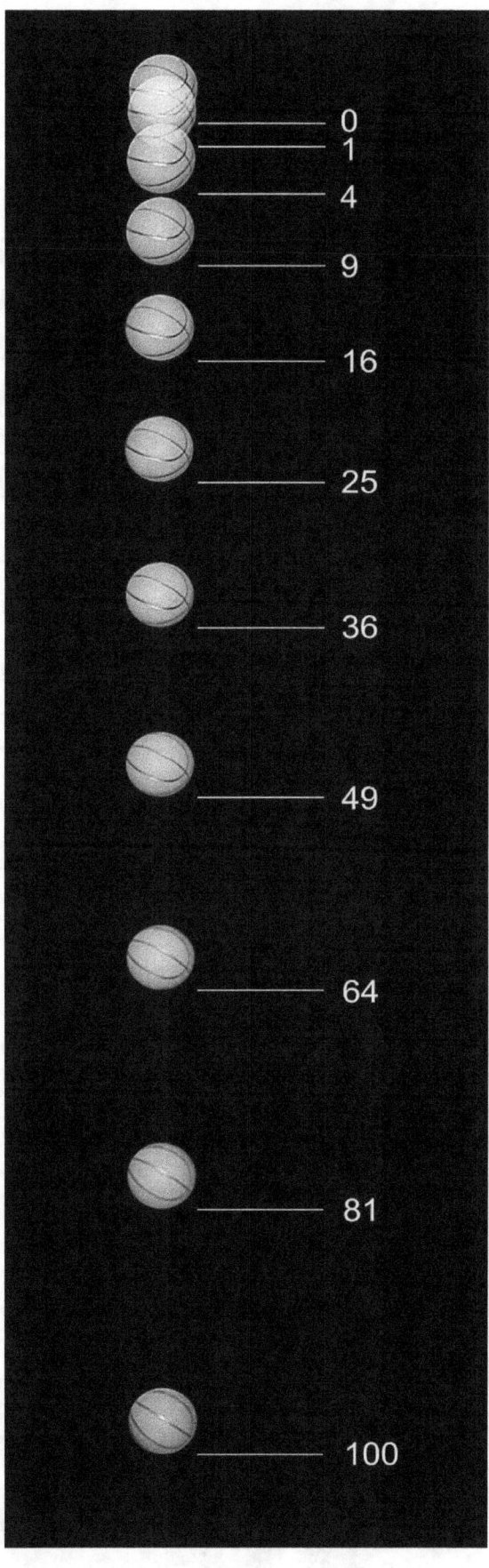

Instruments like GRAVITY provide a powerful probe for gravity force detection.[33]

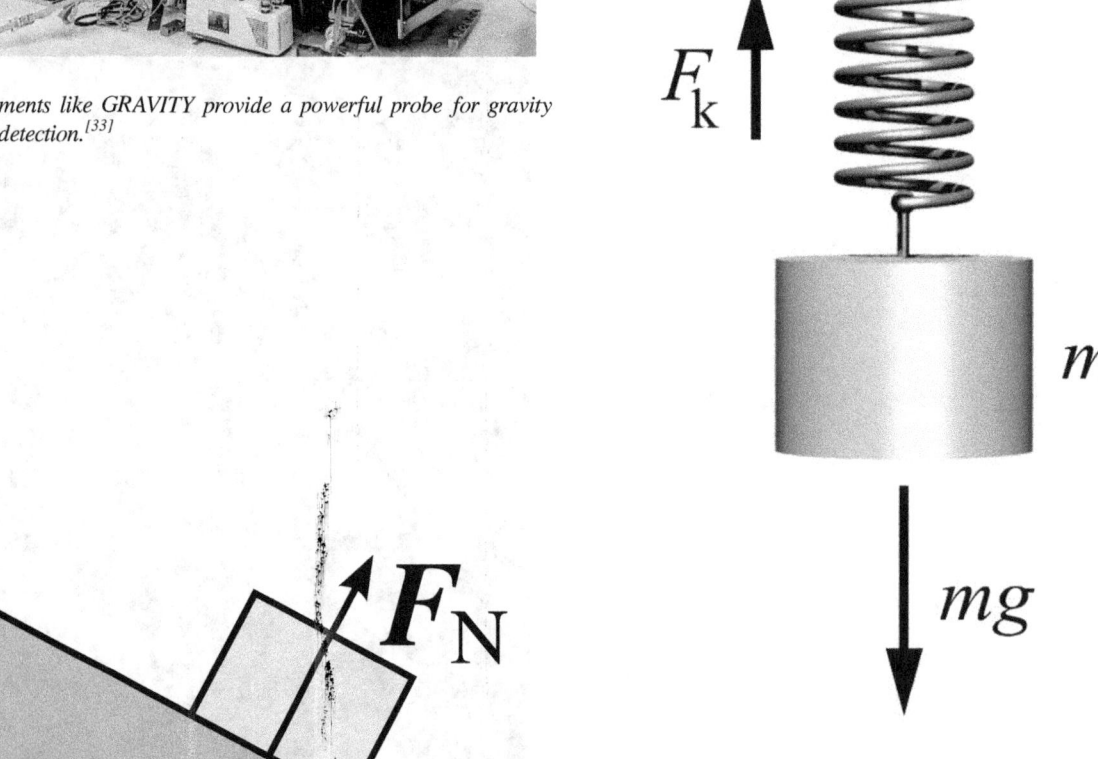

F_k *is the force that responds to the load on the spring*

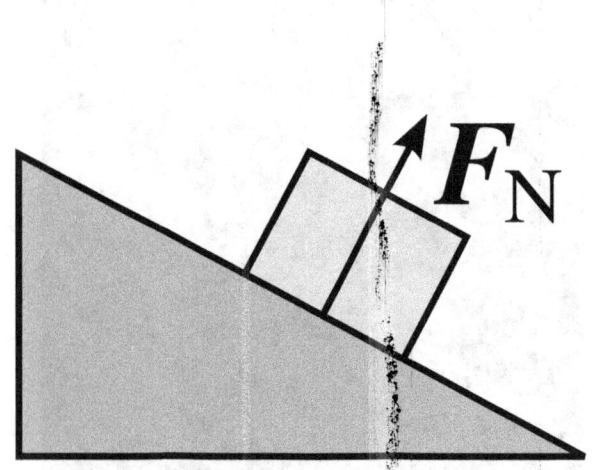

FN *represents the normal force exerted on the object.*

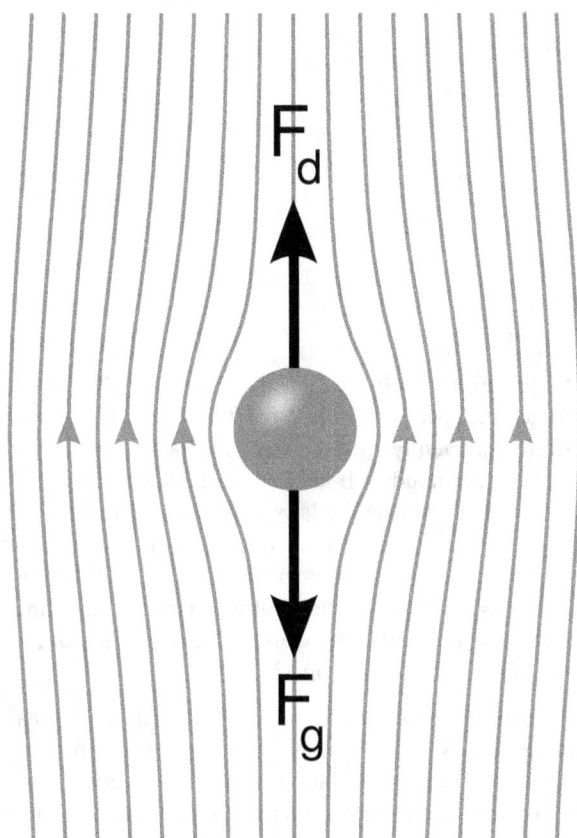

When the drag force (F_d) associated with air resistance becomes equal in magnitude to the force of gravity on a falling object (F_g), the object reaches a state of dynamic equilibrium at terminal velocity.

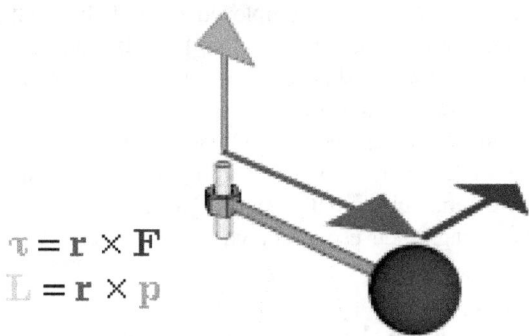

Relationship between force (F), torque (τ), and momentum vectors (p and L) in a rotating system.

Chapter 2

Classical mechanics

For the textbooks, see Classical Mechanics (Goldstein book) and Classical Mechanics (Kibble and Berkshire book).

In physics, **classical mechanics** is one of the two major

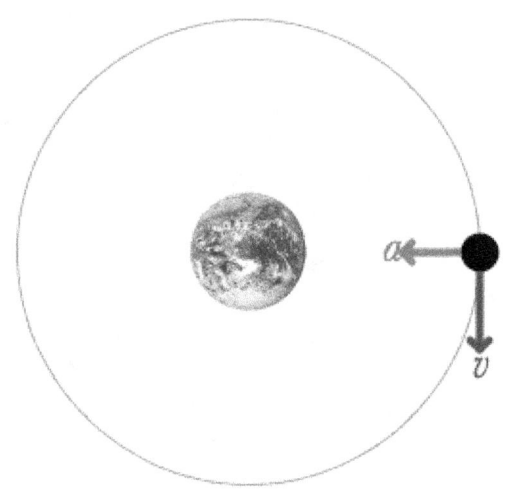

Diagram of orbital motion of a satellite around the earth, showing perpendicular velocity and acceleration (force) vectors.

sub-fields of mechanics, along with quantum mechanics. Classical mechanics is concerned with the set of physical laws describing the motion of bodies under the influence of a system of forces. The study of the motion of bodies is an ancient one, making classical mechanics one of the oldest and largest subjects in science, engineering and technology. It is also widely known as **Newtonian mechanics**.

Classical mechanics describes the motion of macroscopic objects, from projectiles to parts of machinery, as well as astronomical objects, such as spacecraft, planets, stars, and galaxies. Within classical mechanics are fields of study that describe the behavior of solids, liquids and gases and other specific sub-topics. Classical mechanics also provides extremely accurate results as long as the domain of study is

restricted to large objects and the speeds involved do not approach the speed of light. When the objects being examined are sufficiently small, it becomes necessary to introduce the other major sub-field of mechanics, quantum mechanics, which adjusts the laws of physics of macroscopic objects for the atomic nature of matter by including the wave–particle duality of atoms and molecules. When both quantum mechanics and classical mechanics cannot apply, such as at the quantum level with high speeds, quantum field theory (QFT) becomes applicable.

The term *classical mechanics* was coined in the early 20th century to describe the system of physics begun by Isaac Newton and many contemporary 17th century natural philosophers, and is built upon the earlier astronomical theories of Johannes Kepler, which in turn were based on the precise observations of Tycho Brahe and the studies of terrestrial projectile motion of Galileo. Since these aspects of physics were developed long before the emergence of quantum physics and relativity, some sources exclude Einstein's theory of relativity from this category. However, a number of modern sources *do* include relativistic mechanics, which in their view represents *classical mechanics* in its most developed and most accurate form.[note 1]

The earliest development of classical mechanics is often referred to as Newtonian mechanics, and is associated with the physical concepts employed by and the mathematical methods invented by Newton, Leibniz, and others. Later, more abstract and general methods were developed, leading to reformulations of classical mechanics known as Lagrangian mechanics and Hamiltonian mechanics. These advances were largely made in the 18th and 19th centuries, and they extend substantially beyond Newton's work, particularly through their use of analytical mechanics.

2.1 Description of the theory

The following introduces the basic concepts of classical mechanics. For simplicity, it often models real-world objects as point particles (objects with negligible size). The mo-

21

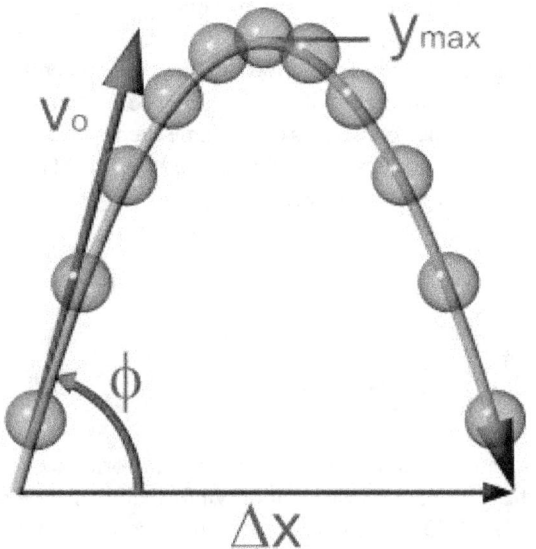

The analysis of projectile motion is a part of classical mechanics.

tion of a point particle is characterized by a small number of parameters: its position, mass, and the forces applied to it. Each of these parameters is discussed in turn.

In reality, the kind of objects that classical mechanics can describe always have a non-zero size. (The physics of *very* small particles, such as the electron, is more accurately described by quantum mechanics.) Objects with non-zero size have more complicated behavior than hypothetical point particles, because of the additional degrees of freedom: a baseball can spin while it is moving, for example. However, the results for point particles can be used to study such objects by treating them as composite objects, made of a large number of collectively acting point particles. The center of mass of a composite object behaves like a point particle.

Classical mechanics uses common-sense notions of how matter and forces exist and interact. It assumes that matter and energy have definite, knowable attributes such as where an object is in space and its speed. It also assumes that objects may be directly influenced only by their immediate surroundings, known as the principle of locality.

2.1.1   Position and its derivatives

Main article: Kinematics

The *position* of a point particle is defined with respect to an arbitrary fixed reference point in space called the origin **O**, in space, to which is attached a coordinate system. A simple coordinate system might describe the position of a point **P** by means of an arrow designated as **r** (in classical

mechanics called a vector), pointing from the origin **O** to the point particle. In general, the point particle need not be stationary relative to **O**, such that **r** is a function of t, the time. In pre-Einstein relativity (known as Galilean relativity), time is considered an absolute, i.e., the time interval that is observed to elapse between any given pair of events is the same for all observers.[1] In addition to relying on absolute time, classical mechanics assumes Euclidean geometry for the structure of space.[2]

Velocity and speed

Main articles: Velocity and speed

The *velocity*, or the rate of change of position with time, is defined as the derivative of the position with respect to time:

$$\mathbf{v} = \frac{d\mathbf{r}}{dt}$$

In classical mechanics, velocities are directly additive and subtractive. For example, if one car traveling east at 60 km/h passes another car traveling east at 50 km/h, then from the perspective of the slower car, the faster car is traveling east at 60 − 50 = 10 km/h. Whereas, from the perspective of the faster car, the slower car is moving 10 km/h to the west. Velocities are directly additive as vector quantities; they must be dealt with using vector analysis.

Mathematically, if the velocity of the first object in the previous discussion is denoted by the vector $\mathbf{u} = u\mathbf{d}$ and the velocity of the second object by the vector $\mathbf{v} = v\mathbf{e}$, where u is the speed of the first object, v is the speed of the second object, and $\mathbf{d}$ and $\mathbf{e}$ are unit vectors in the directions of motion of each particle respectively, then the velocity of the first object as seen by the second object is

$$\mathbf{u}' = \mathbf{u} - \mathbf{v}.$$

Similarly,

$$\mathbf{v}' = \mathbf{v} - \mathbf{u}.$$

When both objects are moving in the same direction, this equation can be simplified to

$$\mathbf{u}' = (u - v)\mathbf{d}.$$

Or, by ignoring direction, the difference can be given in terms of speed only:

$$u' = u - v.$$

Acceleration

Main article: Acceleration

The *acceleration*, or rate of change of velocity, is the derivative of the velocity with respect to time (the second derivative of the position with respect to time):

$$\mathbf{a} = \frac{d\mathbf{v}}{dt} = \frac{d^2\mathbf{r}}{dt^2}.$$

Acceleration represents the velocity's change over time: either of the velocity's magnitude or direction, or both. If only the magnitude v of the velocity decreases, this is sometimes referred to as *deceleration*, but generally any change in the velocity with time, including deceleration, is simply referred to as acceleration.

Frames of reference

Main articles: Inertial frame of reference and Galilean transformation

While the position, velocity and acceleration of a particle can be described with respect to any observer in any state of motion, classical mechanics assumes the existence of a special family of reference frames in terms of which the mechanical laws of nature take a comparatively simple form. These special reference frames are called inertial frames. An inertial frame is one that when an object that has no force interactions (an idealized situation) is viewed from that frame, it appears either to be at rest or in a state of uniform motion in a straight line. This is the fundamental definition of an inertial frame. They are characterized by the requirement that all forces entering the observer's physical laws originate from identifiable sources caused by fields, such as electro-static field (caused by static electrical charges), electro-magnetic field (caused by moving charges), gravitational field (caused by mass), and so forth. A non-inertial reference frame is one that is accelerating with respect to an inertial frame, and in such a non-inertial frame a particle appears to be acted on by other forces not explained by existing fields. Such other forces are called variously fictitious forces, inertia forces, pseudo-forces. Hence, there appears to be other forces that enter the equations of motion solely as a result of the observed accelerations. A key concept of inertial frames is the method for identifying them. For practical purposes, reference frames that are unaccelerated with respect to the distant stars (an extremely distant point) are regarded as good approximations to inertial frames.

Consider two reference frames S and S'. For observers in each of the reference frames an event has space-time coordinates of (x,y,z,t) in frame S and (x',y',z',t') in frame S'. Assuming time is measured the same in all reference frames, and if we require $x = x'$ when $t = 0$, then the relation between the space-time coordinates of the same event observed from the reference frames S' and S, which are moving at a relative velocity of u in the x direction is:

$$x' = x - u \cdot t$$
$$y' = y$$
$$z' = z$$
$$t' = t.$$

This set of formulas defines a group transformation known as the Galilean transformation (informally, the *Galilean transform*). This group is a limiting case of the Poincaré group used in special relativity. The limiting case applies when the velocity u is very small compared to c, the speed of light.

The transformations have the following consequences:

- $\mathbf{v}' = \mathbf{v} - \mathbf{u}$ (the velocity $\mathbf{v}'$ of a particle from the perspective of S' is slower by $\mathbf{u}$ than its velocity $\mathbf{v}$ from the perspective of S)

- $\mathbf{a}' = \mathbf{a}$ (the acceleration of a particle is the same in any inertial reference frame)

- $\mathbf{F}' = \mathbf{F}$ (the force on a particle is the same in any inertial reference frame)

- the speed of light is not a constant in classical mechanics, nor does the special position given to the speed of light in relativistic mechanics have a counterpart in classical mechanics.

For some problems, it is convenient to use rotating coordinates (reference frames). Thereby one can either keep a mapping to a convenient inertial frame, or introduce additionally a fictitious centrifugal force and Coriolis force.

2.1.2 Forces; Newton's second law

Main articles: Force and Newton's laws of motion

Newton was the first to mathematically express the relationship between force and momentum. Some physicists interpret Newton's second law of motion as a definition of force and mass, while others consider it a fundamental postulate, a law of nature. Either interpretation has the same mathematical consequences, historically known as "Newton's Second Law":

$$\mathbf{F} = \frac{d\mathbf{p}}{dt} = \frac{d(m\mathbf{v})}{dt}.$$

The quantity $m\mathbf{v}$ is called the (canonical) momentum. The net force on a particle is thus equal to the rate of change of the momentum of the particle with time. Since the definition of acceleration is $\mathbf{a} = d\mathbf{v}/dt$, the second law can be written in the simplified and more familiar form:

$$\mathbf{F} = m\mathbf{a}.$$

So long as the force acting on a particle is known, Newton's second law is sufficient to describe the motion of a particle. Once independent relations for each force acting on a particle are available, they can be substituted into Newton's second law to obtain an ordinary differential equation, which is called the *equation of motion*.

As an example, assume that friction is the only force acting on the particle, and that it may be modeled as a function of the velocity of the particle, for example:

$$\mathbf{F}_{\mathrm{R}} = -\lambda\mathbf{v},$$

where λ is a positive constant, the negative sign states that the force is opposite the sense of the velocity. Then the equation of motion is

$$-\lambda\mathbf{v} = m\mathbf{a} = m\frac{d\mathbf{v}}{dt}.$$

This can be integrated to obtain

$$\mathbf{v} = \mathbf{v}_0 e^{\frac{-\lambda t}{m}}$$

where $\mathbf{v}_0$ is the initial velocity. This means that the velocity of this particle decays exponentially to zero as time progresses. In this case, an equivalent viewpoint is that the kinetic energy of the particle is absorbed by friction (which converts it to heat energy in accordance with the conservation of energy), and the particle is slowing down. This expression can be further integrated to obtain the position $\mathbf{r}$ of the particle as a function of time.

Important forces include the gravitational force and the Lorentz force for electromagnetism. In addition, Newton's third law can sometimes be used to deduce the forces acting on a particle: if it is known that particle A exerts a force $\mathbf{F}$ on another particle B, it follows that B must exert an equal and opposite *reaction force*, $-\mathbf{F}$, on A. The strong form of Newton's third law requires that $\mathbf{F}$ and $-\mathbf{F}$ act along the line connecting A and B, while the weak form does not. Illustrations of the weak form of Newton's third law are often found for magnetic forces.

2.1.3 Work and energy

Main articles: Work (physics), kinetic energy, and potential energy

If a constant force $\mathbf{F}$ is applied to a particle that makes a displacement $\Delta\mathbf{r}$,[note 2] the *work done* by the force is defined as the scalar product of the force and displacement vectors:

$$W = \mathbf{F} \cdot \Delta\mathbf{r}.$$

More generally, if the force varies as a function of position as the particle moves from $\mathbf{r}_1$ to $\mathbf{r}_2$ along a path C, the work done on the particle is given by the line integral

$$W = \int_C \mathbf{F}(\mathbf{r}) \cdot d\mathbf{r}.$$

If the work done in moving the particle from $\mathbf{r}_1$ to $\mathbf{r}_2$ is the same no matter what path is taken, the force is said to be conservative. Gravity is a conservative force, as is the force due to an idealized spring, as given by Hooke's law. The force due to friction is non-conservative.

The kinetic energy E_{k} of a particle of mass m travelling at speed v is given by

$$E_{\mathrm{k}} = \tfrac{1}{2}mv^2.$$

For extended objects composed of many particles, the kinetic energy of the composite body is the sum of the kinetic energies of the particles.

The work–energy theorem states that for a particle of constant mass m, the total work W done on the particle as it moves from position $\mathbf{r}_1$ to $\mathbf{r}_2$ is equal to the change in kinetic energy E_{k} of the particle:

$$W = \Delta E_{\mathrm{k}} = E_{\mathrm{k},2} - E_{\mathrm{k},1} = \tfrac{1}{2}m\left(v_2^2 - v_1^2\right).$$

Conservative forces can be expressed as the gradient of a scalar function, known as the potential energy and denoted E_{p}:

$$\mathbf{F} = -\nabla E_{\mathrm{p}}.$$

If all the forces acting on a particle are conservative, and E_{p} is the total potential energy (which is defined as a work of involved forces to rearrange mutual positions of bodies), obtained by summing the potential energies corresponding to each force

$$\mathbf{F} \cdot \Delta \mathbf{r} = -\nabla E_{\mathrm{p}} \cdot \Delta \mathbf{r} = -\Delta E_{\mathrm{p}} \,.$$

The decrease in the potential energy is equal to the increase in the kinetic energy

$$-\Delta E_{\mathrm{p}} = \Delta E_{\mathrm{k}} \Rightarrow \Delta(E_{\mathrm{k}} + E_{\mathrm{p}}) = 0 \,.$$

This result is known as *conservation of energy* and states that the total energy,

$$\sum E = E_{\mathrm{k}} + E_{\mathrm{p}} \,,$$

is constant in time. It is often useful, because many commonly encountered forces are conservative.

2.1.4 Beyond Newton's laws

Classical mechanics also describes the more complex motions of extended non-pointlike objects. Euler's laws provide extensions to Newton's laws in this area. The concepts of angular momentum rely on the same calculus used to describe one-dimensional motion. The rocket equation extends the notion of rate of change of an object's momentum to include the effects of an object "losing mass".

There are two important alternative formulations of classical mechanics: Lagrangian mechanics and Hamiltonian mechanics. These, and other modern formulations, usually bypass the concept of "force", instead referring to other physical quantities, such as energy, speed and momentum, for describing mechanical systems in generalized coordinates.

The expressions given above for momentum and kinetic energy are only valid when there is no significant electromagnetic contribution. In electromagnetism, Newton's second law for current-carrying wires breaks down unless one includes the electromagnetic field contribution to the momentum of the system as expressed by the Poynting vector divided by c^2, where c is the speed of light in free space.

2.2 Limits of validity

Many branches of classical mechanics are simplifications or approximations of more accurate forms; two of the most accurate being general relativity and relativistic statistical mechanics. Geometric optics is an approximation to the quantum theory of light, and does not have a superior "classical" form.

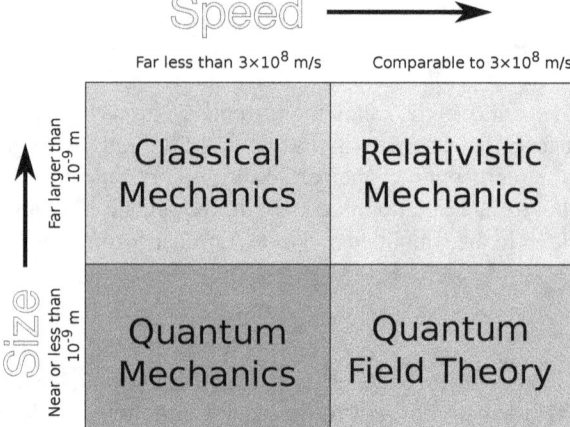

Domain of validity for Classical Mechanics

When both quantum mechanics and classical mechanics cannot apply, such as at the quantum level with many degrees of freedom, quantum field theory (QFT) is of use. QFT deals with small distances and large speeds with many degrees of freedom as well as the possibility of any change in the number of particles throughout the interaction. When treating large degrees of freedom at the macroscopic level, statistical mechanics becomes useful. Statistical mechanics describes the behavior of large (but countable) numbers of particles and their interactions as a whole at the macroscopic level. Statistical mechanics is mainly used in thermodynamics for systems that lie outside the bounds of the assumptions of classical thermodynamics. In the case of high velocity objects approaching the speed of light, classical mechanics is enhanced by special relativity. General relativity unifies special relativity with Newton's law of universal gravitation, allowing physicists to handle gravitation at a deeper level.

2.2.1 The Newtonian approximation to special relativity

In special relativity, the momentum of a particle is given by

$$\mathbf{p} = \frac{m\mathbf{v}}{\sqrt{1 - v^2/c^2}} \,,$$

where m is the particle's rest mass, $\mathbf{v}$ its velocity, and c is the speed of light.

If v is very small compared to c, v^2/c^2 is approximately zero, and so

$$\mathbf{p} \approx m\mathbf{v} \,.$$

Thus the Newtonian equation $\mathbf{p} = m\mathbf{v}$ is an approximation of the relativistic equation for bodies moving with low speeds compared to the speed of light.

For example, the relativistic cyclotron frequency of a cyclotron, gyrotron, or high voltage magnetron is given by

$$f = f_\mathrm{c} \frac{m_0}{m_0 + T/c^2},$$

where f_c is the classical frequency of an electron (or other charged particle) with kinetic energy T and (rest) mass m_0 circling in a magnetic field. The (rest) mass of an electron is 511 keV. So the frequency correction is 1% for a magnetic vacuum tube with a 5.11 kV direct current accelerating voltage.

2.2.2 The classical approximation to quantum mechanics

The ray approximation of classical mechanics breaks down when the de Broglie wavelength is not much smaller than other dimensions of the system. For non-relativistic particles, this wavelength is

$$\lambda = \frac{h}{p}$$

where h is Planck's constant and p is the momentum.

Again, this happens with electrons before it happens with heavier particles. For example, the electrons used by Clinton Davisson and Lester Germer in 1927, accelerated by 54 volts, had a wavelength of 0.167 nm, which was long enough to exhibit a single diffraction side lobe when reflecting from the face of a nickel crystal with atomic spacing of 0.215 nm. With a larger vacuum chamber, it would seem relatively easy to increase the angular resolution from around a radian to a milliradian and see quantum diffraction from the periodic patterns of integrated circuit computer memory.

More practical examples of the failure of classical mechanics on an engineering scale are conduction by quantum tunneling in tunnel diodes and very narrow transistor gates in integrated circuits.

Classical mechanics is the same extreme high frequency approximation as geometric optics. It is more often accurate because it describes particles and bodies with rest mass. These have more momentum and therefore shorter De Broglie wavelengths than massless particles, such as light, with the same kinetic energies.

2.3 History

Main article: History of classical mechanics
See also: Timeline of classical mechanics

Some Greek philosophers of antiquity, among them Aristotle, founder of Aristotelian physics, may have been the first to maintain the idea that "everything happens for a reason" and that theoretical principles can assist in the understanding of nature. While to a modern reader, many of these preserved ideas come forth as eminently reasonable, there is a conspicuous lack of both mathematical theory and controlled experiment, as we know it. These latter became decisive factors in forming modern science, and their early application came to be known as classical mechanics.

In his *Elementa super demonstrationem ponderum*, medieval mathematician Jordanus de Nemore introduced the concept of "positional gravity" and the use of component forces.

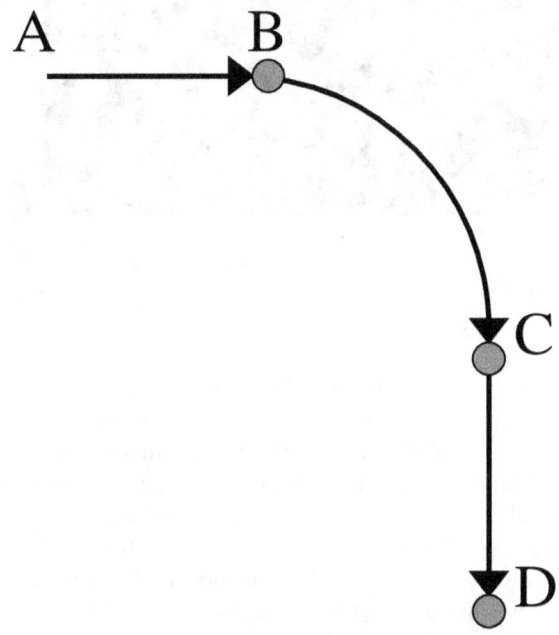

Three stage Theory of impetus according to Albert of Saxony.

The first published causal explanation of the motions of planets was Johannes Kepler's *Astronomia nova* published in 1609. He concluded, based on Tycho Brahe's observations of the orbit of Mars, that the orbits were ellipses. This break with ancient thought was happening around the same time that Galileo was proposing abstract mathematical laws for the motion of objects. He may (or may not) have performed the famous experiment of dropping two cannonballs of different weights from the tower of Pisa, showing that they both hit the ground at the same time. The reality of this experiment is disputed, but, more importantly, he did carry out quantitative experiments by rolling balls on an inclined plane. His theory of accelerated motion derived from the results of such experiments, and forms a cornerstone of classical mechanics.

Sir Isaac Newton (1643–1727), an influential figure in the history of physics and whose three laws of motion form the basis of classical mechanics

Newton founded his principles of natural philosophy on three proposed laws of motion: the law of inertia, his second law of acceleration (mentioned above), and the law of action and reaction; and hence laid the foundations for classical mechanics. Both Newton's second and third laws were given the proper scientific and mathematical treatment in Newton's *Philosophiæ Naturalis Principia Mathematica*, which distinguishes them from earlier attempts at explaining similar phenomena, which were either incomplete, incorrect, or given little accurate mathematical expression. Newton also enunciated the principles of conservation of momentum and angular momentum. In mechanics, Newton was also the first to provide the first correct scientific and mathematical formulation of gravity in Newton's law of universal gravitation. The combination of Newton's laws of motion and gravitation provide the fullest and most accurate description of classical mechanics. He demonstrated that these laws apply to everyday objects as well as to celestial objects. In particular, he obtained a theoretical explanation of Kepler's laws of motion of the planets.

Newton had previously invented the calculus, of mathematics, and used it to perform the mathematical calculations.

For acceptability, his book, the *Principia*, was formulated entirely in terms of the long-established geometric methods, which were soon eclipsed by his calculus. However, it was Leibniz who developed the notation of the derivative and integral preferred[3] today.

Hamilton's greatest contribution is perhaps the reformulation of Newtonian mechanics, now called Hamiltonian mechanics.

Newton, and most of his contemporaries, with the notable exception of Huygens, worked on the assumption that classical mechanics would be able to explain all phenomena, including light, in the form of geometric optics. Even when discovering the so-called Newton's rings (a wave interference phenomenon) he maintained his own corpuscular theory of light.

After Newton, classical mechanics became a principal field of study in mathematics as well as physics. Several reformulations progressively allowed finding solutions to a far greater number of problems. The first notable reformulation was in 1788 by Joseph Louis Lagrange. Lagrangian mechanics was in turn re-formulated in 1833 by William Rowan Hamilton.

Some difficulties were discovered in the late 19th century that could only be resolved by more modern physics. Some of these difficulties related to compatibility with electromagnetic theory, and the famous Michelson–Morley experiment. The resolution of these problems led to the special theory of relativity, often included in the term classical mechanics.

A second set of difficulties were related to thermodynamics. When combined with thermodynamics, classical mechanics leads to the Gibbs paradox of classical statistical mechanics, in which entropy is not a well-defined quantity. Black-body radiation was not explained without the introduction of quanta. As experiments reached the atomic level, classical mechanics failed to explain, even approximately, such basic things as the energy levels and sizes of atoms and the photo-electric effect. The effort at resolving these problems led to the development of quantum mechanics.

Since the end of the 20th century, the place of classical mechanics in physics has been no longer that of an independent theory. Instead, classical mechanics is now considered an approximate theory to the more general quantum mechanics. Emphasis has shifted to understanding the fundamental forces of nature as in the Standard model and its more modern extensions into a unified theory of everything.[4] Classical mechanics is a theory useful for the study of the motion of non-quantum mechanical, low-energy particles in weak gravitational fields. In the 21st century classical mechanics has been extended into the complex domain and complex classical mechanics exhibits behaviors very similar to quantum mechanics.[5]

2.4   Branches

Classical mechanics was traditionally divided into three main branches:

- Statics, the study of equilibrium and its relation to forces

- Dynamics, the study of motion and its relation to forces

- Kinematics, dealing with the implications of observed motions without regard for circumstances causing them

Another division is based on the choice of mathematical formalism:

- Newtonian mechanics

- Lagrangian mechanics

- Hamiltonian mechanics

Alternatively, a division can be made by region of application:

- Celestial mechanics, relating to stars, planets and other celestial bodies

- Continuum mechanics, for materials modelled as a continuum, e.g., solids and fluids (i.e., liquids and gases).

- Relativistic mechanics (i.e. including the special and general theories of relativity), for bodies whose speed is close to the speed of light.

- Statistical mechanics, which provides a framework for relating the microscopic properties of individual atoms and molecules to the macroscopic or bulk thermodynamic properties of materials.

2.5   See also

- Dynamical systems

- History of classical mechanics

- List of equations in classical mechanics

- List of publications in classical mechanics

- Molecular dynamics

- Newton's laws of motion

- Special theory of relativity

- Quantum Mechanics

- Quantum Field Theory

2.6   Notes

[1] The notion of "classical" may be somewhat confusing, insofar as this term usually refers to the era of classical antiquity in European history. While many discoveries within the mathematics of that period are applicable today and of great use, much of the science that emerged then has since been superseded by more accurate models. This in no way detracts from the science of that time because most of modern physics is built directly upon those developments. The emergence of classical mechanics was a decisive stage in the development of science, in the modern sense of the term. Above all, what characterizes it is its insistence that the descriptions of the behavior of bodies be placed on a more exacting basis that could only be provided by a mathematical treatment and its reliance on experiment, rather than speculation. Classical mechanics established the means of predicting in a quantitative manner the behavior of objects, and how to test them by carefully designed measurement. The emerging globally cooperative endeavor increasingly provided for much closer scrutiny and testing, both of theory and experiment. This was, and remains, a key factor in establishing certain knowledge, and in bringing it to the service of society. History shows how closely

the health and wealth of a society depends on nurturing this investigative and critical approach.

[2] The displacement $\Delta \mathbf{r}$ is the difference of the particle's initial and final positions: $\Delta \mathbf{r} = \mathbf{r}\text{fi}_{\text{nal}} - \mathbf{r}_{\text{initial}}$.

2.7 References

[1] Knudsen, Jens M.; Hjorth, Poul (2012). *Elements of Newtonian Mechanics* (illustrated ed.). Springer Science & Business Media. p. 30. ISBN 978-3-642-97599-8. Extract of page 30

[2] MIT physics 8.01 lecture notes (page 12) (PDF)

[3] Jesseph, Douglas M. (1998). "Leibniz on the Foundations of the Calculus: The Question of the Reality of Infinitesimal Magnitudes". Perspectives on Science. 6.1&2: 6–40. Retrieved 31 December 2011.

[4] Page 2-10 of the *Feynman Lectures on Physics* says "For already in classical mechanics there was indeterminability from a practical point of view." The past tense here implies that classical physics is no longer fundamental.

[5] Complex Elliptic Pendulum, Carl M. Bender, Daniel W. Hook, Karta Kooner in Asymptotics in Dynamics, Geometry and PDEs; Generalized Borel Summation vol. I

2.8 Further reading

- Alonso, M.; Finn, J. (1992). *Fundamental University Physics*. Addison-Wesley.

- Feynman, Richard (1999). *The Feynman Lectures on Physics*. Perseus Publishing. ISBN 0-7382-0092-1.

- Feynman, Richard; Phillips, Richard (1998). *Six Easy Pieces*. Perseus Publishing. ISBN 0-201-32841-0.

- Goldstein, Herbert; Charles P. Poole; John L. Safko (2002). *Classical Mechanics* (3rd ed.). Addison Wesley. ISBN 0-201-65702-3.

- Kibble, Tom W.B.; Berkshire, Frank H. (2004). *Classical Mechanics (5th ed.)*. Imperial College Press. ISBN 978-1-86094-424-6.

- Kleppner, D.; Kolenkow, R. J. (1973). *An Introduction to Mechanics*. McGraw-Hill. ISBN 0-07-035048-5.

- Landau, L.D.; Lifshitz, E.M. (1972). *Course of Theoretical Physics, Vol. 1—Mechanics*. Franklin Book Company. ISBN 0-08-016739-X.

- Morin, David (2008). *Introduction to Classical Mechanics: With Problems and Solutions* (1st ed.). Cambridge, UK: Cambridge University Press. ISBN 978-0-521-87622-3.*Gerald Jay Sussman; Jack Wisdom (2001). *Structure and Interpretation of Classical Mechanics*. MIT Press. ISBN 0-262-19455-4.

- O'Donnell, Peter J. (2015). *Essential Dynamics and Relativity*. CRC Press. ISBN 978-1-466-58839-4.

- Thornton, Stephen T.; Marion, Jerry B. (2003). *Classical Dynamics of Particles and Systems (5th ed.)*. Brooks Cole. ISBN 0-534-40896-6.

2.9 External links

- Crowell, Benjamin. Newtonian Physics (an introductory text, uses algebra with optional sections involving calculus)

- Fitzpatrick, Richard. Classical Mechanics (uses calculus)

- Hoiland, Paul (2004). Preferred Frames of Reference & Relativity

- Horbatsch, Marko, "*Classical Mechanics Course Notes*".

- Rosu, Haret C., "*Classical Mechanics*". Physics Education. 1999. [arxiv.org : physics/9909035]

- Shapiro, Joel A. (2003). Classical Mechanics

- Sussman, Gerald Jay & Wisdom, Jack & Mayer,Meinhard E. (2001). Structure and Interpretation of Classical Mechanics

- Tong, David. Classical Dynamics (Cambridge lecture notes on Lagrangian and Hamiltonian formalism)

- Kinematic Models for Design Digital Library (KMODDL)
  Movies and photos of hundreds of working mechanical-systems models at Cornell University. Also includes an e-book library of classic texts on mechanical design and engineering.

- MIT OpenCourseWare 8.01: Classical Mechanics Free videos of actual course lectures with links to lecture notes, assignments and exams.

- Alejandro A. Torassa, On Classical Mechanics

Chapter 3

Newton's laws of motion

For other uses, see Laws of motion.

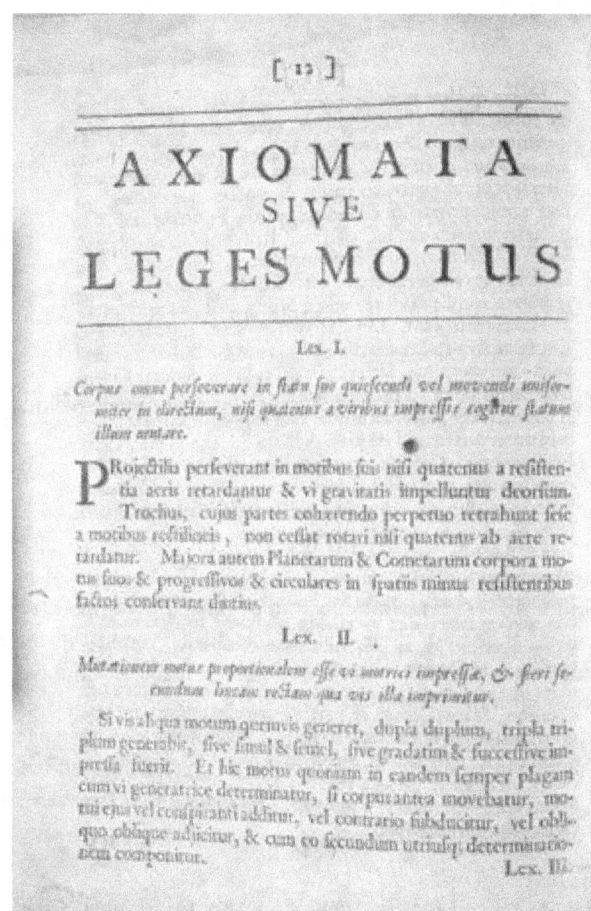

Newton's First and Second laws, in Latin, from the original 1687 Principia Mathematica.

Newton's laws of motion are three physical laws that, together, laid the foundation for classical mechanics. They describe the relationship between a body and the forces acting upon it, and its motion in response to those forces. They have been expressed in several different ways, over nearly three centuries,[1] and can be summarised as follows.

The three laws of motion were first compiled by Isaac Newton in his *Philosophiæ Naturalis Principia Mathematica* (*Mathematical Principles of Natural Philosophy*), first published in 1687.[4] Newton used them to explain and investigate the motion of many physical objects and systems.[5] For example, in the third volume of the text, Newton showed that these laws of motion, combined with his law of universal gravitation, explained Kepler's laws of planetary motion.

3.1 Overview

Isaac Newton (1643–1727), the physicist who formulated the laws

Newton's laws are applied to objects which are idealised as single point masses,[6] in the sense that the size and shape of the object's body are neglected to focus on its motion more easily. This can be done when the object is small compared to the distances involved in its analysis, or the deformation and rotation of the body are of no importance. In this way, even a planet can be idealised as a particle for analysis of its orbital motion around a star.

In their original form, Newton's laws of motion are not adequate to characterise the motion of rigid bodies and deformable bodies. Leonhard Euler in 1750 introduced a generalisation of Newton's laws of motion for rigid bodies called Euler's laws of motion, later applied as well for deformable bodies assumed as a continuum. If a body is represented as an assemblage of discrete particles, each governed by Newton's laws of motion, then Euler's laws can be derived from Newton's laws. Euler's laws can, however, be taken as axioms describing the laws of motion for extended bodies, independently of any particle structure.[7]

Newton's laws hold only with respect to a certain set of frames of reference called Newtonian or inertial reference frames. Some authors interpret the first law as defining what an inertial reference frame is; from this point of view, the second law only holds when the observation is made from an inertial reference frame, and therefore the first law cannot be proved as a special case of the second. Other authors do treat the first law as a corollary of the second.[8][9] The explicit concept of an inertial frame of reference was not developed until long after Newton's death.

In the given interpretation mass, acceleration, momentum, and (most importantly) force are assumed to be externally defined quantities. This is the most common, but not the only interpretation of the way one can consider the laws to be a definition of these quantities.

Newtonian mechanics has been superseded by special relativity, but it is still useful as an approximation when the speeds involved are much slower than the speed of light.[10]

3.2 Newton's first law

Main article: Inertia

The first law states that if the net force (the vector sum of all forces acting on an object) is zero, then the velocity of the object is constant. Velocity is a vector quantity which expresses both the object's speed and the direction of its motion; therefore, the statement that the object's velocity is constant is a statement that both its speed and the direction of its motion are constant.

The first law can be stated mathematically when the mass is a non-zero constant, as,

Explanation of Newton's first law and reference frames. (MIT Course 8.01)[11]

$$\sum \mathbf{F} = 0 \;\Leftrightarrow\; \frac{d\mathbf{v}}{dt} = 0.$$

Consequently,

- An object that is at rest will stay at rest unless a force acts upon it.

- An object that is in motion will not change its velocity unless a force acts upon it.

This is known as *uniform motion*. An object *continues* to do whatever it happens to be doing unless a force is exerted upon it. If it is at rest, it continues in a state of rest (demonstrated when a tablecloth is skilfully whipped from under dishes on a tabletop and the dishes remain in their initial state of rest). If an object is moving, it continues to move without turning or changing its speed. This is evident in space probes that continually move in outer space. Changes in motion must be imposed against the tendency of an object to retain its state of motion. In the absence of net forces, a moving object tends to move along a straight line path indefinitely.

Newton placed the first law of motion to establish frames of reference for which the other laws are applicable. The first law of motion postulates the existence of at least one frame of reference called a Newtonian or inertial reference frame, relative to which the motion of a particle not subject to forces is a straight line at a constant speed.[8][12] Newton's first law is often referred to as the *law of inertia*. Thus, a condition necessary for the uniform motion of a particle relative to an inertial reference frame is that the total net force acting on it is zero. In this sense, the first law can be restated as:

In every material universe, the motion of a particle in a preferential reference frame Φ is determined by the action of forces whose total vanished for all times when and only when the velocity of the particle is constant in Φ. That is, a particle initially at rest or in uniform motion in the preferential frame Φ continues in that state unless compelled by forces to change it.[13]

Newton's laws are valid only in an inertial reference frame. Any reference frame that is in uniform motion with respect to an inertial frame is also an inertial frame, i.e. Galilean invariance or the principle of Newtonian relativity.[14]

3.3 Newton's second law

Explanation of Newton's second law, using gravity as an example. (MIT OCW)[15]

The second law states that the rate of change of momentum of a body, is directly proportional to the force applied and this change in momentum takes place in the direction of the applied force.

$$\mathbf{F} = \frac{d\mathbf{p}}{dt} = \frac{d(m\mathbf{v})}{dt}.$$

The second law can also be stated in terms of an object's acceleration. Since Newton's second law is only valid for constant-mass systems,[16][17][18] mass can be taken outside the differentiation operator by the constant factor rule in differentiation. Thus,

$$\mathbf{F} = m\,\frac{d\mathbf{v}}{dt} = m\mathbf{a},$$

where **F** is the net force applied, m is the mass of the body, and **a** is the body's acceleration. Thus, the net force applied

to a body produces a proportional acceleration. In other words, if a body is accelerating, then there is a force on it.

Consistent with the first law, the time derivative of the momentum is non-zero when the momentum changes direction, even if there is no change in its magnitude; such is the case with uniform circular motion. The relationship also implies the conservation of momentum: when the net force on the body is zero, the momentum of the body is constant. Any net force is equal to the rate of change of the momentum.

Any mass that is gained or lost by the system will cause a change in momentum that is not the result of an external force. A different equation is necessary for variable-mass systems (see below).

Newton's second law requires modification if the effects of special relativity are to be taken into account, because at high speeds the approximation that momentum is the product of rest mass and velocity is not accurate.

3.3.1 Impulse

An impulse **J** occurs when a force **F** acts over an interval of time Δt, and it is given by[19][20]

$$\mathbf{J} = \int_{\Delta t} \mathbf{F}\,dt.$$

Since force is the time derivative of momentum, it follows that

$$\mathbf{J} = \Delta \mathbf{p} = m\Delta \mathbf{v}.$$

This relation between impulse and momentum is closer to Newton's wording of the second law.[21]

Impulse is a concept frequently used in the analysis of collisions and impacts.[22]

3.3.2 Variable-mass systems

Main article: Variable-mass system

Variable-mass systems, like a rocket burning fuel and ejecting spent gases, are not closed and cannot be directly treated by making mass a function of time in the second law;[17] that is, the following formula is wrong:[18]

$$\mathbf{F}_{\text{net}} = \frac{d}{dt}\big[m(t)\mathbf{v}(t)\big] = m(t)\frac{d\mathbf{v}}{dt} + \mathbf{v}(t)\frac{dm}{dt}. \qquad \text{(wrong)}$$

The falsehood of this formula can be seen by noting that it does not respect Galilean invariance: a variable-mass object with $\mathbf{F} = 0$ in one frame will be seen to have $\mathbf{F} \neq 0$ in another frame.[16] The correct equation of motion for a body whose mass m varies with time by either ejecting or accreting mass is obtained by applying the second law to the entire, constant-mass system consisting of the body and its ejected/accreted mass; the result is[16]

$$\mathbf{F} + \mathbf{u}\frac{\mathrm{d}m}{\mathrm{d}t} = m\frac{\mathrm{d}\mathbf{v}}{\mathrm{d}t}$$

where $\mathbf{u}$ is the velocity of the escaping or incoming mass relative to the body. From this equation one can derive the equation of motion for a varying mass system, for example, the Tsiolkovsky rocket equation. Under some conventions, the quantity $\mathbf{u}\,\mathrm{d}m/\mathrm{d}t$ on the left-hand side, which represents the advection of momentum, is defined as a force (the force exerted on the body by the changing mass, such as rocket exhaust) and is included in the quantity $\mathbf{F}$. Then, by substituting the definition of acceleration, the equation becomes $\mathbf{F} = m\mathbf{a}$.

3.4 Newton's third law

An illustration of Newton's third law in which two skaters push against each other. The first skater on the left exerts a normal force N_{12} on the second skater directed towards the right, and the second skater exerts a normal force N_{21} on the first skater directed towards the left.
The magnitudes of both forces are equal, but they have opposite directions, as dictated by Newton's third law.

The third law states that all forces between two objects exist in equal magnitude and opposite direction: if one object A exerts a force $\mathbf{F}A$ on a second object B, then B simultaneously exerts a force $\mathbf{F}B$ on A, and the two forces are equal

A description of Newton's third law and contact forces[23]

and opposite: $\mathbf{F}A = -\mathbf{F}B$.[24] The third law means that all forces are *interactions* between different bodies,[25][26] and thus that there is no such thing as a unidirectional force or a force that acts on only one body. This law is sometimes referred to as the *action-reaction law*, with $\mathbf{F}A$ called the "action" and $\mathbf{F}B$ the "reaction". The action and the reaction are simultaneous, and it does not matter which is called the *action* and which is called *reaction*; both forces are part of a single interaction, and neither force exists without the other.[24]

The two forces in Newton's third law are of the same type (e.g., if the road exerts a forward frictional force on an accelerating car's tires, then it is also a frictional force that Newton's third law predicts for the tires pushing backward on the road).

From a conceptual standpoint, Newton's third law is seen when a person walks: they push against the floor, and the floor pushes against the person. Similarly, the tires of a car push against the road while the road pushes back on the tires—the tires and road simultaneously push against each other. In swimming, a person interacts with the water, pushing the water backward, while the water simultaneously pushes the person forward—both the person and the water push against each other. The reaction forces account for the motion in these examples. These forces depend on friction; a person or car on ice, for example, may be unable to exert the action force to produce the needed reaction force.[27]

3.5 History

3.5.1 Newton's 1st Law

From the original Latin of Newton's *Principia*:

Translated to English, this reads:

The ancient Greek philosopher Aristotle had the view that all objects have a natural place in the universe: that heavy objects (such as rocks) wanted to be at rest on the Earth and that light objects like smoke wanted to be at rest in the sky and the stars wanted to remain in the heavens. He thought that a body was in its natural state when it was at rest, and for the body to move in a straight line at a constant speed an external agent was needed continually to propel it, otherwise it would stop moving. Galileo Galilei, however, realised that a force is necessary to change the velocity of a body, i.e., acceleration, but no force is needed to maintain its velocity. In other words, Galileo stated that, in the *absence* of a force, a moving object will continue moving. The tendency of objects to resist changes in motion was what Galileo called *inertia*. This insight was refined by Newton, who made it into his first law, also known as the "law of inertia"—no force means no acceleration, and hence the body will maintain its velocity. As Newton's first law is a restatement of the law of inertia which Galileo had already described, Newton appropriately gave credit to Galileo.

The law of inertia apparently occurred to several different natural philosophers and scientists independently, including Thomas Hobbes in his *Leviathan*.[29] The 17th century philosopher and mathematician René Descartes also formulated the law, although he did not perform any experiments to confirm it.[30][31]

3.5.2   Newton's 2nd Law

Newton's original Latin reads:

This was translated quite closely in Motte's 1729 translation as:

According to modern ideas of how Newton was using his terminology,[32] this is understood, in modern terms, as an equivalent of:

> *The change of momentum of a body is proportional to the impulse impressed on the body, and happens along the straight line on which that impulse is impressed.*

This may be expressed by the formula $F = p'$, where p' is the time derivative of the momentum p. This equation can be seen clearly in the Wren Library of Trinity College, Cambridge, in a glass case in which Newton's manuscript is open to the relevant page.

Motte's 1729 translation of Newton's Latin continued with Newton's commentary on the second law of motion, reading:

> *If a force generates a motion, a double force will generate double the motion, a triple force triple the motion, whether that force be impressed altogether and at once, or gradually and successively. And this motion (being always directed the same way with the generating force), if the body moved before, is added to or subtracted from the former motion, according as they directly conspire with or are directly contrary to each other; or obliquely joined, when they are oblique, so as to produce a new motion compounded from the determination of both.*

The sense or senses in which Newton used his terminology, and how he understood the second law and intended it to be understood, have been extensively discussed by historians of science, along with the relations between Newton's formulation and modern formulations.[33]

3.5.3   Newton's 3rd Law

Translated to English, this reads:

Newton's Scholium (explanatory comment) to this law:

> Whatever draws or presses another is as much drawn or pressed by that other. If you press a stone with your finger, the finger is also pressed by the stone. If a horse draws a stone tied to a rope, the horse (if I may so say) will be equally drawn back towards the stone: for the distended rope, by the same endeavour to relax or unbend itself, will draw the horse as much towards the stone, as it does the stone towards the horse, and will obstruct the progress of the one as much as it advances that of the other. If a body impinges upon another, and by its force changes the motion of the other, that body also (because of the equality of the mutual pressure) will undergo an equal change, in its own motion, toward the contrary part. The changes made by these actions are equal, not in the velocities but in the motions of the bodies; that is to say, if the bodies are not hindered by any other impediments. For, as the motions are equally changed, the changes of the velocities made toward contrary parts are reciprocally proportional to the bodies. This law takes place also in attractions, as will be proved in the next scholium.[34]

In the above, as usual, *motion* is Newton's name for momentum, hence his careful distinction between motion and velocity.

Newton used the third law to derive the law of conservation of momentum;[35] from a deeper perspective, however, conservation of momentum is the more fundamental idea (derived via Noether's theorem from Galilean invariance), and holds in cases where Newton's third law appears to fail, for instance when force fields as well as particles carry momentum, and in quantum mechanics.

3.6 Importance and range of validity

Newton's laws were verified by experiment and observation for over 200 years, and they are excellent approximations at the scales and speeds of everyday life. Newton's laws of motion, together with his law of universal gravitation and the mathematical techniques of calculus, provided for the first time a unified quantitative explanation for a wide range of physical phenomena.

These three laws hold to a good approximation for macroscopic objects under everyday conditions. However, Newton's laws (combined with universal gravitation and classical electrodynamics) are inappropriate for use in certain circumstances, most notably at very small scales, very high speeds (in special relativity, the Lorentz factor must be included in the expression for momentum along with the rest mass and velocity) or very strong gravitational fields. Therefore, the laws cannot be used to explain phenomena such as conduction of electricity in a semiconductor, optical properties of substances, errors in non-relativistically corrected GPS systems and superconductivity. Explanation of these phenomena requires more sophisticated physical theories, including general relativity and quantum field theory.

In quantum mechanics, concepts such as force, momentum, and position are defined by linear operators that operate on the quantum state; at speeds that are much lower than the speed of light, Newton's laws are just as exact for these operators as they are for classical objects. At speeds comparable to the speed of light, the second law holds in the original form $\mathbf{F} = d\mathbf{p}/dt$, where $\mathbf{F}$ and $\mathbf{p}$ are four-vectors.

3.7 Relationship to the conservation laws

In modern physics, the laws of conservation of momentum, energy, and angular momentum are of more general validity than Newton's laws, since they apply to both light and matter, and to both classical and non-classical physics.

This can be stated simply, "Momentum, energy and angular momentum cannot be created or destroyed."

Because force is the time derivative of momentum, the concept of force is redundant and subordinate to the conservation of momentum, and is not used in fundamental theories (e.g., quantum mechanics, quantum electrodynamics, general relativity, etc.). The standard model explains in detail how the three fundamental forces known as gauge forces originate out of exchange by virtual particles. Other forces, such as gravity and fermionic degeneracy pressure, also arise from the momentum conservation. Indeed, the conservation of 4-momentum in inertial motion via curved space-time results in what we call gravitational force in general relativity theory. The application of the space derivative (which is a momentum operator in quantum mechanics) to the overlapping wave functions of a pair of fermions (particles with half-integer spin) results in shifts of maxima of compound wavefunction away from each other, which is observable as the "repulsion" of the fermions.

Newton stated the third law within a world-view that assumed instantaneous action at a distance between material particles. However, he was prepared for philosophical criticism of this action at a distance, and it was in this context that he stated the famous phrase "I feign no hypotheses". In modern physics, action at a distance has been completely eliminated, except for subtle effects involving quantum entanglement. However, in modern engineering, in all practical applications involving the motion of vehicles and satellites, the concept of action at a distance is used extensively.

The discovery of the second law of thermodynamics by Carnot in the 19th century showed that every physical quantity is not conserved over time, thus disproving the validity of inducing the opposite metaphysical view from Newton's laws. Hence, a "steady-state" worldview based solely on Newton's laws and the conservation laws does not take entropy into account.

3.8 See also

- Euler's laws
- Hamiltonian mechanics
- Lagrangian mechanics
- List of scientific laws named after people
- Mercury, orbit of
- Modified Newtonian dynamics

- Newton's law of universal gravitation

- Principle of least action

- Reaction (physics)

- Principle of relativity

3.9   References and notes

[1] For explanations of Newton's laws of motion by Newton in the early 18th century, by the physicist William Thomson (Lord Kelvin) in the mid-19th century, and by a modern text of the early 21st century, see:-

- Newton's "Axioms or Laws of Motion" starting on page 19 of volume 1 of the 1729 translation of the "Principia";

- Section 242, *Newton's laws of motion* in Thomson, W (Lord Kelvin), and Tait, P G, (1867), *Treatise on natural philosophy*, volume 1; and

- Benjamin Crowell (2000), *Newtonian Physics*.

[2] Browne, Michael E. (July 1999). *Schaum's outline of theory and problems of physics for engineering and science* (Series: Schaum's Outline Series). McGraw-Hill Companies. p. 58. ISBN 978-0-07-008498-8.

[3] Holzner, Steven (December 2005). *Physics for Dummies.* Wiley, John & Sons, Incorporated. p. 64. ISBN 978-0-7645-5433-9.

[4] See the *Principia* on line at Andrew Motte Translation

[5] Andrew Motte translation of Newton's *Principia* (1687) *Axioms or Laws of Motion*

[6] *[...]while Newton had used the word 'body' vaguely and in at least three different meanings, Euler realized that the statements of Newton are generally correct only when applied to masses concentrated at isolated points;*Truesdell, Clifford A.; Becchi, Antonio; Benvenuto, Edoardo (2003). *Essays on the history of mechanics: in memory of Clifford Ambrose Truesdell and Edoardo Benvenuto.* New York: Birkhäuser. p. 207. ISBN 3-7643-1476-1.

[7] Lubliner, Jacob (2008). *Plasticity Theory (Revised Edition)* (PDF). Dover Publications. ISBN 0-486-46290-0.

[8] Galili, I.; Tseitlin, M. (2003). "Newton's First Law: Text, Translations, Interpretations and Physics Education". *Science & Education.* **12** (1): 45–73. Bibcode:2003Sc&Ed..12...45G. doi:10.1023/A:1022632600805.

[9] Benjamin Crowell. "4. Force and Motion". *Newtonian Physics.* ISBN 0-9704670-1-X.

[10] In making a modern adjustment of the second law for (some of) the effects of relativity, *m* would be treated as the relativistic mass, producing the relativistic expression for momentum, and the third law might be modified if possible to allow for the finite signal propagation speed between distant interacting particles.

[11] Walter Lewin (20 September 1999). *Newton's First, Second, and Third Laws. MIT Course 8.01: Classical Mechanics, Lecture 6.* (ogg) (videotape). Cambridge, MA USA: MIT OCW. Event occurs at 0:00–6:53. Retrieved 23 December 2010.

[12] NMJ Woodhouse (2003). *Special relativity.* London/Berlin: Springer. p. 6. ISBN 1-85233-426-6.

[13] Beatty, Millard F. (2006). *Principles of engineering mechanics Volume 2 of Principles of Engineering Mechanics: Dynamics-The Analysis of Motion,.* Springer. p. 24. ISBN 0-387-23704-6.

[14] Thornton, Marion (2004). *Classical dynamics of particles and systems* (5th ed.). Brooks/Cole. p. 53. ISBN 0-534-40896-6.

[15] Lewin, Newton's First, Second, and Third Laws, Lecture 6. (6:53–11:06)

[16] Plastino, Angel R.; Muzzio, Juan C. (1992). "On the use and abuse of Newton's second law for variable mass problems". *Celestial Mechanics and Dynamical Astronomy.* Netherlands: Kluwer Academic Publishers. **53** (3): 227–232. Bibcode:1992CeMDA..53..227P. doi:10.1007/BF00052611. ISSN 0923-2958. "We may conclude emphasizing that Newton's second law is valid for constant mass only. When the mass varies due to accretion or ablation, [an alternate equation explicitly accounting for the changing mass] should be used."

[17] Halliday; Resnick. *Physics.* **1**. p. 199. ISBN 0-471-03710-9. It is important to note that we *cannot* derive a general expression for Newton's second law for variable mass systems by treating the mass in $\mathbf{F} = d\mathbf{P}/dt = d(M\mathbf{v})$ as a *variable.* [...] We *can* use $\mathbf{F} = d\mathbf{P}/dt$ to analyze variable mass systems *only* if we apply it to an *entire system of constant mass* having parts among which there is an interchange of mass. [Emphasis as in the original]

[18] Kleppner, Daniel; Robert Kolenkow (1973). *An Introduction to Mechanics.* McGraw-Hill. pp. 133–134. ISBN 0-07-035048-5. Recall that $\mathbf{F} = d\mathbf{P}/dt$ was established for a system composed of a certain set of particles[. ... I]t is essential to deal with the same set of particles throughout the time interval[. ...] Consequently, the mass of the system can not change during the time of interest.

[19] Hannah, J, Hillier, M J, *Applied Mechanics*, p221, Pitman Paperbacks, 1971

[20] Raymond A. Serway; Jerry S. Faughn (2006). *College Physics.* Pacific Grove CA: Thompson-Brooks/Cole. p. 161. ISBN 0-534-99724-4.

[21] I Bernard Cohen (Peter M. Harman & Alan E. Shapiro, Eds) (2002). *The investigation of difficult things: essays on Newton and the history of the exact sciences in honour of D.T. Whiteside.* Cambridge UK: Cambridge University Press. p. 353. ISBN 0-521-89266-X.

[22] WJ Stronge (2004). *Impact mechanics.* Cambridge UK: Cambridge University Press. p. 12 ff. ISBN 0-521-60289-0.

[23] Lewin, Newton's First, Second, and Third Laws, Lecture 6. (14:11–16:00)

[24] Resnick; Halliday; Krane (1992). *Physics, Volume 1* (4th ed.). p. 83.

[25] C Hellingman (1992). "Newton's third law revisited". *Phys. Educ.* **27** (2): 112–115. Bibcode:1992PhyEd..27..112H. doi:10.1088/0031-9120/27/2/011. Quoting Newton in the *Principia*: It is not one action by which the Sun attracts Jupiter, and another by which Jupiter attracts the Sun; but it is one action by which the Sun and Jupiter mutually endeavour to come nearer together.

[26] Resnick & Halliday (1977). *Physics* (Third ed.). John Wiley & Sons. pp. 78–79. Any single force is only one aspect of a mutual interaction between *two* bodies.

[27] Hewitt (2006), p. 75

[28] Isaac Newton, *The Principia*, A new translation by I.B. Cohen and A. Whitman, University of California press, Berkeley 1999.

[29] Thomas Hobbes wrote in *Leviathan*:

> That when a thing lies still, unless somewhat else stir it, it will lie still forever, is a truth that no man doubts. But [the proposition] that when a thing is in motion it will eternally be in motion unless somewhat else stay it, though the reason be the same (namely that nothing can change itself), is not so easily assented to. For men measure not only other men but all other things by themselves. And because they find themselves subject after motion to pain and lassitude, [they] think every thing else grows weary of motion and seeks repose of its own accord, little considering whether it be not some other motion wherein that desire of rest they find in themselves, consists.

[30] Cohen, I. B. (1995). *Science and the Founding Fathers: Science in the Political Thought of Jefferson, Franklin, Adams and Madison.* New York: W.W. Norton. p. 117. ISBN 978-0393315103.

[31] Cohen, I. B. (1980). *The Newtonian Revolution: With Illustrations of the Transformation of Scientific Ideas.* Cambridge, England: Cambridge University Press. p. 183-4. ISBN 978-0521273800.

[32] According to Maxwell in *Matter and Motion*, Newton meant by *motion* "the quantity of matter moved as well as the rate at which it travels" and by impressed force *he meant "the time during which the force acts as well as the intensity of the force"*. See Harman and Shapiro, cited below.

[33] See for example (1) I Bernard Cohen, "Newton's Second Law and the Concept of Force in the Principia", in "The Annus Mirabilis of Sir Isaac Newton 1666–1966" (Cambridge, Massachusetts: The MIT Press, 1967), pages 143–185; (2) Stuart Pierson, "'Corpore cadente. . .': Historians Discuss Newton's Second Law", Perspectives on Science, 1 (1993), pages 627–658; and (3) Bruce Pourciau, "Newton's Interpretation of Newton's Second Law", Archive for History of Exact Sciences, vol.60 (2006), pages 157–207; also an online discussion by G E Smith, in 5. Newton's Laws of Motion, s.5 of "Newton's Philosophiae Naturalis Principia Mathematica" in (online) Stanford Encyclopedia of Philosophy, 2007.

[34] This translation of the third law and the commentary following it can be found in the "Principia" on page 20 of volume 1 of the 1729 translation.

[35] Newton, *Principia*, Corollary III to the laws of motion

3.10 Further reading and works cited

- Crowell, Benjamin (2011), *Light and Matter* (2011, Light and Matter), especially at Section *4.2, Newton's First Law*, Section *4.3, Newton's Second Law*, and Section *5.1, Newton's Third Law*.

- Feynman, R. P.; Leighton, R. B.; Sands, M. (2005). *The Feynman Lectures on Physics.* Vol. 1 (2nd ed.). Pearson/Addison-Wesley. ISBN 0-8053-9049-9.

- Fowles, G. R.; Cassiday, G. L. (1999). *Analytical Mechanics* (6th ed.). Saunders College Publishing. ISBN 0-03-022317-2.

- Likins, Peter W. (1973). *Elements of Engineering Mechanics.* McGraw-Hill Book Company. ISBN 0-07-037852-5.

- Marion, Jerry; Thornton, Stephen (1995). *Classical Dynamics of Particles and Systems.* Harcourt College Publishers. ISBN 0-03-097302-3.

- NMJ Woodhouse (2003). *Special Relativity.* London/Berlin: Springer. p. 6. ISBN 1-85233-426-6.

Historical

- Newton, Isaac, "Mathematical Principles of Natural Philosophy", 1729 English translation based on 3rd Latin edition (1726), volume 1, containing Book 1, especially at the section *Axioms or Laws of Motion*, starting page 19.

- Newton, Isaac, "Mathematical Principles of Natural Philosophy", 1729 English translation based on 3rd Latin edition (1726), volume 2, containing Books 2 & 3.

- Thomson, W (Lord Kelvin), and Tait, P G, (1867), *Treatise on natural philosophy*, volume 1, especially at Section 242, *Newton's laws of motion*.

3.11   External links

- MIT Physics video lecture on Newton's three laws

- Light and Matter – an on-line textbook

- Simulation on Newton's first law of motion

- "Newton's Second Law" by Enrique Zeleny, Wolfram Demonstrations Project.

- Newton's 3rd Law demonstrated in a vacuum on YouTube

Chapter 4

Four-force

In the special theory of relativity, **four-force** is a four-vector that replaces the classical force.

4.1   In Special Relativity

The four-force is the four-vector defined as the change in four-momentum over the particle's own time:

$$\mathbf{F} = \frac{d\mathbf{P}}{d\tau}$$

For a particle of constant invariant mass $m > 0$, $\mathbf{P} = m\mathbf{U}$ where $\mathbf{U} = \gamma(c, \mathbf{u})$ is the four-velocity, so we can relate the four-force with the four-acceleration $\mathbf{A}$ as in Newton's second law:

$$\mathbf{F} = m\mathbf{A} = \left(\gamma \frac{\mathbf{f} \cdot \mathbf{u}}{c}, \gamma\mathbf{f} \right)$$

Here

$$\mathbf{f} = \frac{d}{dt}\left(\gamma m\mathbf{u} \right) = \frac{d\mathbf{p}}{dt}$$

and

$$\mathbf{f} \cdot \mathbf{u} = \frac{d}{dt}\left(\gamma mc^2 \right) = \frac{dE}{dt}$$

where $\mathbf{u}$, $\mathbf{p}$ and $\mathbf{f}$ are 3-vectors describing the velocity and the momentum of the particle and the force acting on it respectively.

4.2   In General Relativity

In general relativity the relation between four-force, and four-acceleration remains the same, but the elements of the four-force are related to the elements of the four-momentum through a covariant derivative with respect to proper time.

$$F^\lambda := \frac{DP^\lambda}{d\tau} = \frac{dP^\lambda}{d\tau} + \Gamma^\lambda{}_{\mu\nu}U^\mu P^\nu$$

In addition, we can formulate force using the concept of coordinate transformations between different coordinate systems. Assume that we know the correct expression for force in a coordinate system at which the particle is momentarily at rest. Then we can perform a transformation to another system to get the corresponding expression of force.[1] In special relativity the transformation will be a Lorentz transformation between coordinate systems moving with a relative constant velocity whereas in general relativity it will be a general coordinate transformation.

Consider the four-force $F^\mu = (F^0, \mathbf{F})$ acting on a particle of mass m which is momentarily at rest in a coordinate system. The relativistic force f^μ in another coordinate system moving with constant velocity v, relative to the other one, is obtained using a Lorentz transformation:

$$\mathbf{f} = \mathbf{F} + (\gamma - 1)\mathbf{v}\frac{\mathbf{v} \cdot \mathbf{F}}{v^2},$$
$$f^0 = \gamma\boldsymbol{\beta} \cdot \mathbf{F} = \boldsymbol{\beta} \cdot \mathbf{f}.$$

where $\boldsymbol{\beta} = \mathbf{v}/c$.

In general relativity, the expression for force becomes

$$f^\mu = m\frac{DU^\mu}{d\tau}$$

with covariant derivative $D/d\tau$. The equation of motion becomes

$$m\frac{d^2x^\mu}{d\tau^2} = f^\mu - m\Gamma^\mu{}_{\nu\lambda}\frac{dx^\nu}{d\tau}\frac{dx^\lambda}{d\tau},$$

where $\Gamma^\mu{}_{\nu\lambda}$ is the Christoffel symbol. If there is no external force, this becomes the equation for geodesics in the curved space-time. The second term in the above equation, plays the role of a gravitational force. If f_f^α is the correct expression for force in a freely falling frame ξ^α, we can use the then the equivalence principle to write the four-force in an arbitrary coordinate x^μ :

$$f^\mu = \frac{\partial x^\mu}{\partial \xi^\alpha} f_f^\alpha.$$

4.3 Examples

In special relativity, Lorentz 4-force (4-force acting to charged particle situated in electromagnetic field) can be expressed as:

$$F_\mu = q F_{\mu\nu} U^\nu$$

where

- $F_{\mu\nu}$ is electromagnetic tensor,
- U^ν is 4-velocity, and
- q - electric charge.

4.4 See also

- four-vector
- four-velocity
- four-acceleration
- four-momentum

4.5 References

[1] Steven, Weinberg (1972). *Gravitation and Cosmology: Principles and Applications of the General Theory of Relativity.* John Wiley & Sons, Inc. ISBN 0-471-92567-5.

- Rindler, Wolfgang (1991). *Introduction to Special Relativity* (2nd ed.). Oxford: Oxford University Press. ISBN 0-19-853953-3.

Chapter 5

Net force

For other uses, see Net force (disambiguation).

In physics, **net force** is the *overall* force acting on an object. In order to calculate the net force, the body is isolated and interactions with the environment or other constraints are represented as forces and torques in a free-body diagram.

The net force does not have the same effect on the movement of the object as the original system forces, unless the point of application of the net force and an associated torque are determined so that they form the resultant force and torque. It is always possible to determine the torque associated with a point of application of a net force so that it maintains the movement of the object under the original system of forces.

With its associated torque, the net force becomes the *resultant force* and has the same effect on the rotational motion of the object as all actual forces taken together.[1] It is possible for a system of forces to define a torque-free resultant force. In this case, the net force when applied at the proper line of action has the same effect on the body as all of the forces at their points of application. It is not always possible to find a torque-free resultant force.

5.1 Total force

The sum of forces acting on a particle is called the total force or the net force. The net force is a single force that replaces the effect of the original forces on the particle's motion. It gives the particle the same acceleration as all those actual forces together as described by the Newton's second law of motion.

Force is a vector quantity, which means that it has a magnitude and a direction, and it is usually denoted using boldface such as $\mathbf{F}$ or by using an arrow over the symbol, such as $\vec{F}$.

Graphically, a force is represented as line segment from its point of application A to a point B which defines its direction and magnitude. The length of the segment AB represents

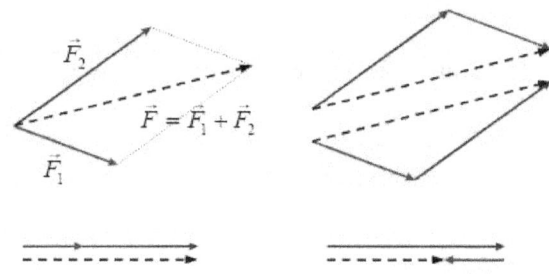

A diagrammatic method for the addition of forces.

the magnitude of the force.

Vector calculus was developed in the late 1800s and early 1900s. The parallelogram rule used for the addition of forces, however, dates from antiquity and is noted explicitly by Galileo and Newton.[2]

The diagram shows the addition of the forces $\vec{F}_1$ and $\vec{F}_2$. The sum $\vec{F}$ of the two forces is drawn as the diagonal of a parallelogram defined by the two forces.

Forces applied to an extended body can have different points of application. Forces are bound vectors and can be added only if they are applied at the same point. The net force obtained from all the forces acting on a body will not preserve its motion unless they are applied at the same point and the appropriate torque associated with the new point of application is determined. The net force on a body applied at a single point with the appropriate torque is known as the resultant force and torque.

5.2 Parallelogram rule for the addition of forces

A force is known as a bound vector which means it has a direction and magnitude and a point of application. A convenient way to define a force is by a line segment from a point A to a point B. If we denote the coordinates of these

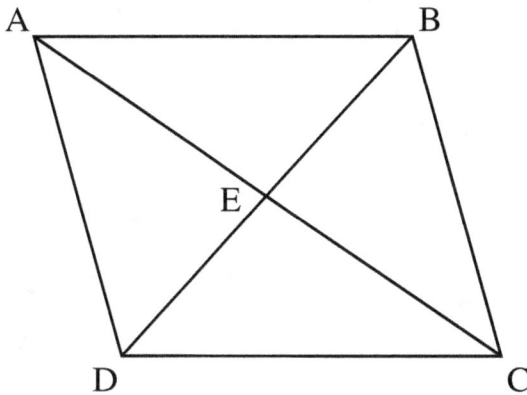

Parallelogram ABCD

points as $\mathbf{A}=(A_x, A_y, A_z)$ and $\mathbf{B}=(B_x, B_y, B_z)$, then the force vector applied at A is given by

$$\mathbf{F} = \mathbf{B} - \mathbf{A} = (B_x - A_x, B_y - A_y, B_z - A_z).$$

The length of the vector $\mathbf{B}$-$\mathbf{A}$ defines the magnitude of $\mathbf{F}$ and is given by

$$|\mathbf{F}| = \sqrt{(B_x - A_x)^2 + (B_y - A_y)^2 + (B_z - A_z)^2}.$$

The sum of two forces $\mathbf{F}_1$ and $\mathbf{F}_2$ applied at A can be computed from the sum of the segments that define them. Let $\mathbf{F}_1$=$\mathbf{B}$-$\mathbf{A}$ and $\mathbf{F}_2$=$\mathbf{D}$-$\mathbf{A}$, then the sum of these two vectors is

$$\mathbf{F} = \mathbf{F}_1 + \mathbf{F}_2 = \mathbf{B} - \mathbf{A} + \mathbf{D} - \mathbf{A},$$

which can be written as

$$\mathbf{F} = \mathbf{F}_1 + \mathbf{F}_2 = 2(\frac{\mathbf{B} + \mathbf{D}}{2} - \mathbf{A}) = 2(\mathbf{E} - \mathbf{A}),$$

where $\mathbf{E}$ is the midpoint of the segment $\mathbf{BD}$ that joins the points B and D.

Thus, the sum of the forces $\mathbf{F}_1$ and $\mathbf{F}_2$ is twice the segment joining A to the midpoint E of the segment joining the endpoints B and D of the two forces. The doubling of this length is easily achieved by defining a segments $\mathbf{BC}$ and $\mathbf{DC}$ parallel to $\mathbf{AD}$ and $\mathbf{AB}$, respectively, to complete the parallelogram $ABCD$. The diagonal $\mathbf{AC}$ of this parallelogram is the sum of the two force vectors. This is known as the parallelogram rule for the addition of forces.

5.3 Translation and rotation due to a force

5.3.1 Point forces

When a force acts on a particle, it is applied to a single point (the particle volume is negligible): this is a point force and the particle is its application point. But an external force on an extended body (object) can be applied to a number of its constituent particles, i.e. can be "spread" over some volume or surface of the body. However, in order to determine its rotational effect on the body, it is necessary to specify its point of application (actually, the line of application, as explained below). The problem is usually resolved in the following ways:

- Often the volume or surface on which the force acts is relatively small compared to the size of the body, so that it can be approximated by a point. It is usually not difficult to determine whether the error caused by such approximation is acceptable.

- If it is not acceptable (obviously e.g. in the case of gravitational force), such "volume/surface" force should be described as a system of forces (components), each acting on a single particle, and then the calculation should be done for each of them separately. Such a calculation is typically simplified by the use of differential elements of the body volume/surface, and the integral calculus. In a number of cases, though, it can be shown that such a system of forces may be replaced by a single point force without the actual calculation (as in the case of uniform gravitational force).

In any case, the analysis of the rigid body motion begins with the point force model. And when a force acting on a body is shown graphically, the oriented line segment representing the force is usually drawn so as to "begin" (or "end") at the application point.

5.3.2 Rigid bodies

In the example shown in the diagram opposite, a single force $\vec{F}$ acts at the application point $\mathbf{H}$ on a free rigid body. The body has the mass m and its center of mass is the point $\mathbf{C}$. In the constant mass approximation, the force causes changes in the body motion described by the following expressions:

$\vec{a} = \frac{\vec{F}}{m}$ is the center of mass acceleration; and

$\vec{\alpha} = \frac{\vec{\tau}}{I}$ is the angular acceleration of the body.

= 7,5 rad/s^2, and to its center of mass it gives the linear acceleration a = F/m = 4 m/s^2.

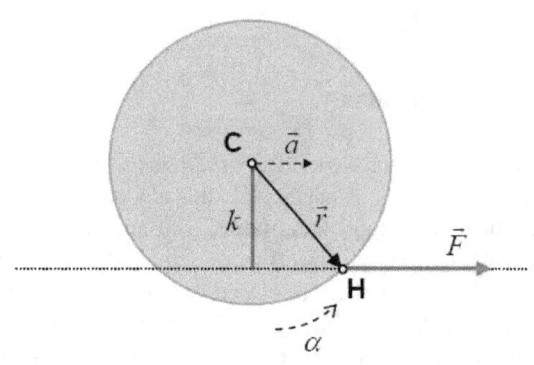

How a force accelerates a body.

5.4 Resultant force

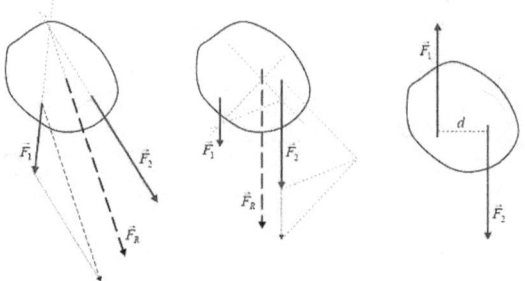

Graphical placing of the resultant force.

In the second expression, $\vec{\tau}$ is the torque or moment of force, whereas I is the moment of inertia of the body. A torque caused by a force $\vec{F}$ is a vector quantity defined with respect to some reference point:

$\vec{\tau} = \vec{r} \times \vec{F}$ is the torque vector, and

$\tau = Fk$ is the amount of torque.

The vector $\vec{r}$ is the position vector of the force application point, and in this example it is drawn from the center of mass as the reference point of(see diagram). The straight line segment k is the lever arm of the force $\vec{F}$ with respect to the center of mass. As the illustration suggests, the torque does not change (the same lever arm) if the application point is moved along the line of the application of the force (dotted black line). More formally, this follows from the properties of the vector product, and shows that rotational effect of the force depends only on the position of its line of application, and not on the particular choice of the point of application along that line.

The torque vector is perpendicular to the plane defined by the force and the vector $\vec{r}$, and in this example it is directed towards the observer; the angular acceleration vector has the same direction. The right hand rule relates this direction to the clockwise or counter-clockwise rotation in the plane of the drawing.

The moment of inertia I is calculated with respect to the axis through the center of mass that is parallel with the torque. If the body shown in the illustration is a homogeneous disc, this moment of inertia is $I=mr^2/2$. If the disc has the mass 0,5 kg and the radius 0,8 m, the moment of inertia is 0,16 kgm^2. If the amount of force is 2 N, and the lever arm 0,6 m, the amount of torque is 1,2 Nm. At the instant shown, the force gives to the disc the angular acceleration $\alpha = \tau/I$

Resultant force and torque replaces the effects of a system of forces acting on the movement of a rigid body. An interesting special case is a torque-free resultant which can be found as follows:

1. Vector addition is used to find the net force;

2. Use the equation to determine the point of application with zero torque:

$$\vec{r} \times \vec{F}_R = \sum_{i=1}^{N} (\vec{r}_i \times \vec{F}_i)$$

where $\vec{F}_R$ is the net force, $\vec{r}$ locates its application point, and individual forces are $\vec{F}_i$ with application points $\vec{r}_i$. It may be that there is no point of application that yields a torque-free resultant.

The diagram opposite illustrates simple graphical methods for finding the line of application of the resultant force of simple planar systems:

1. Lines of application of the actual forces $\vec{F}_1$ and $\vec{F}_2$ on the leftmost illustration intersect. After vector addition is performed "at the location of $\vec{F}_1$ ", the net force obtained is translated so that its line of application passes through the common intersection point. With respect to that point all torques are zero, so the torque of the resultant force $\vec{F}_R$ is equal to the sum of the torques of the actual forces.

2. The illustration in the middle of the diagram shows two parallel actual forces. After vector addition "at the location of $\vec{F}_2$ ", the net force is translated to the appropriate line of application, where it becomes the

resultant force $\vec{F}_R$. The procedure is based on decomposition of all forces into components for which the lines of application (pale dotted lines) intersect at one point (the so-called pole, arbitrarily set at the right side of the illustration). Then the arguments from the previous case are applied to the forces and their components to demonstrate the torque relationships.

3. The rightmost illustration shows a couple, two equal but opposite forces for which the amount of the net force is zero, but they produce the net torque $\tau = Fd$ where d is the distance between their lines of application. Since there is no resultant force, this torque can be [is?] described as "pure" torque.

5.5   Usage

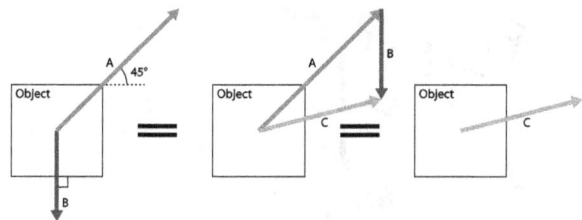

Vector diagram for addition of non-parallel forces.

In general, a system of forces acting on a rigid body can always be replaced by one force plus one pure (see previous section) torque. The force is the net force, but in order to calculate the additional torque, the net force must be assigned the line of action. The line of action can be selected arbitrarily, but the additional pure torque will depend on this choice. In a special case it is possible to find such line of action that this additional torque is zero.

The resultant force and torque can be determined for any configuration of forces. However, an interesting special case is a torque-free resultant which it is useful both conceptually and practically, because the body moves without rotating as if it was a particle.

Some authors do not distinguish the resultant force from the net force and use the terms as synonyms.[3]

5.6   See also

- Screw theory

- Center of mass

- Centers of gravity in non-uniform fields

5.7   References

[1] Symon, Keith R. (1964), Mechanics, Addison-Wesley, LCCN 60-5164

[2] Michael J. Crowe (1967). *A History of Vector Analysis : The Evolution of the Idea of a Vectorial System.* Dover Publications (reprint edition; ISBN 0-486-67910-1).

[3] Resnick, Robert and Halliday, David (1966), Physics, (Vol I and II, Combined edition), Wiley International Edition, Library of Congress Catalog Card No. 66-11527

Chapter 6

Normal force

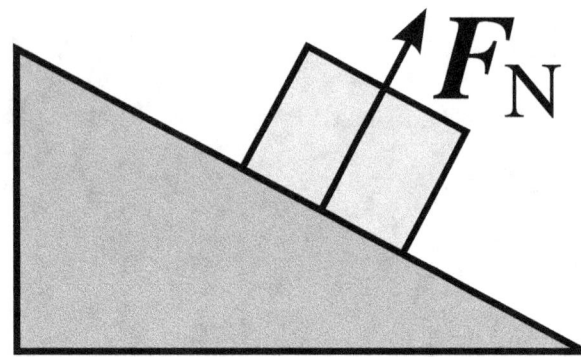

FN *represents the **normal force***

In mechanics, the **normal force** F_n is the component, perpendicular to the surface (surface being a plane) of contact, of the contact force exerted on an object by, for example, the surface of a floor or wall, preventing the object from falling. Here "normal" refers to the geometry terminology for being perpendicular, as opposed the common language use of "normal" meaning common or expected. For example, consider a person standing still on the ground, in which case the ground reaction force reduces to the normal force. In another common situation, if an object hits a surface with some speed, and the surface can withstand it, the normal force provides for a rapid deceleration, which will depend on the flexibility of the surface.

6.1 Equations

In a simple case such as an object resting upon a table, the normal force on the object is equal but in opposite direction to the gravitational force applied on the object (or the weight of the object), that is, $N = mg$, where m is mass, and g is the gravitational field strength (about 9.81 m/s^2 on Earth). The normal force here represents the force applied by the table against the object that prevents it from sinking through the table, and requires that the table is sturdy enough to deliver this normal force without breaking.

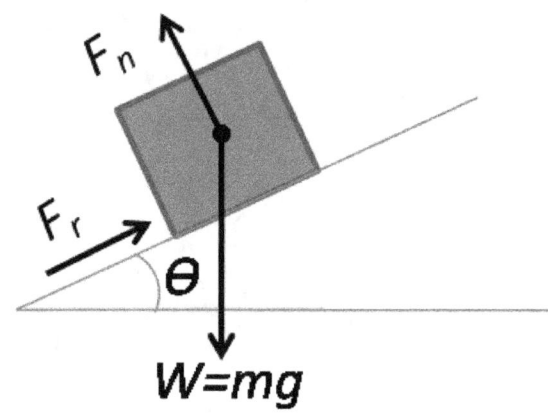

Weight (W), *the frictional force* (F$_r$), *and the normal force* (F$_n$) *impacting a cube. Weight is mass* (m) *multiplied by gravity* (g).

Where an object rests on an incline, the normal force is perpendicular to the plane the object rests on. Still, the normal force will be as large as necessary to prevent sinking through the surface, presuming the surface is sturdy enough. The strength of the force can be calculated as:

$$N = mg \cos(\theta)$$

where N is the normal force, m is the mass of the object, g is the gravitational field strength, and θ is the angle of the inclined surface measured from the horizontal.

The normal force is one of several forces which act on the object. In the simple situations so far considered, the most important other forces acting on it are friction and the force of gravity.

6.1.1 Using vectors

In general, the magnitude of the normal force, N, is the projection of the net surface interaction force, T, in the normal direction, $\mathbf{n}$, and so the normal force vector can be found by scaling the normal direction by the net surface interaction force. The surface interaction force, in turn, is equal to the dot product of the unit normal with the Cauchy stress tensor describing the stress state of the surface. That is,

$$\mathbf{N} = \mathbf{n}\,N = \mathbf{n}\,(\mathbf{T} \cdot \mathbf{n}) = \mathbf{n}\,(\mathbf{n} \cdot \tau \cdot \mathbf{n}).$$

Or, in indicial notation,

$$N_i = n_i N = n_i T_j n_j = n_i n_k \tau_{jk} n_j.$$

The parallel shear component of the contact force is known as the frictional force ($F_f r$).

The static coefficient of friction for an object on an inclined plane can be calculated as follows:[1]

$$\mu_s = \tan(\theta)$$

for an object on the point of sliding where θ is the angle between the slope and the horizontal.

6.2   Real-world applications

For a person standing in an elevator either stationary or moving at constant velocity, the normal force on the person's feet balances the person's weight. In an elevator that is accelerating upward, the normal force is greater than the person's ground weight and so the person's perceived weight increases (making the person feel heavier). In an elevator that is accelerating downward, the normal force is less than the person's ground weight and so a passenger's perceived weight decreases. If a passenger were to stand on a "weighing scale", such as a conventional bathroom scale, while riding the elevator, the scale will be reading the normal force it delivers to the passenger's feet, and will be different than the person's ground weight if the elevator cab is *accelerating* up or down. The weighing scale measures normal force (which varies as the elevator cab accelerates), not gravitational force (which does not vary as the cab accelerates). It is impossible to measure true gravitational force without knowledge of the motion of one's immediate environment.

When we define upward to be the positive direction, constructing Newton's second law and solving for the normal force on a passenger yields the following equation:

$$N = m(g + a)$$

6.3   References

[1] Nichols, Edward Leamington; Franklin, William Suddards (1898). *The Elements of Physics* **1**. Macmillan. p. 101.

Chapter 7

Equilibrant Force

Equilibrant Force as defined by Prof. Samuel udofia is a force which brings equilibrium[1] state. It is considered to be the equal and opposite of the resultant force. Equilibrant force is the force, which keeps any object motionless and acts on virtually every object in the world that is not moving. This term was first used in the 1800s to indicate counterbalancing a force system.[2]

7.1   Equilibrant Force

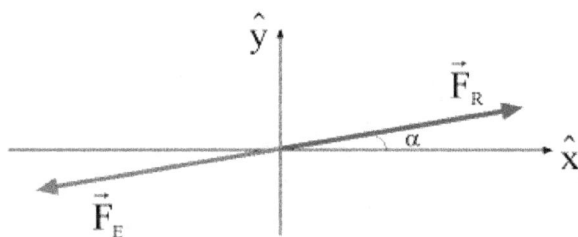

Equilibrant Force

A force equal to, but opposite of, the resultant sum of vector forces; that force which balances other forces, thus bringing an object to equilibrium.

For forces (F_1, F_2, ...), equilibrium (FE) will be opposite to the resultant force (FR).

So, : $\mathbf{F}_1 + \mathbf{F}_2 = \mathbf{F}_R = \mathbf{F}_E$ (Opposite Direction)

7.2   Example for Equilibrant Force

Let us consider the following example:

Two forces are pushing an object along the ground. One force is 10 N [W] and the other is 8.0 N [S]. For Calculating the equilibrant for the given situation.

By using Pythagoras' theorem, $c^2 = a^2 + b^2$

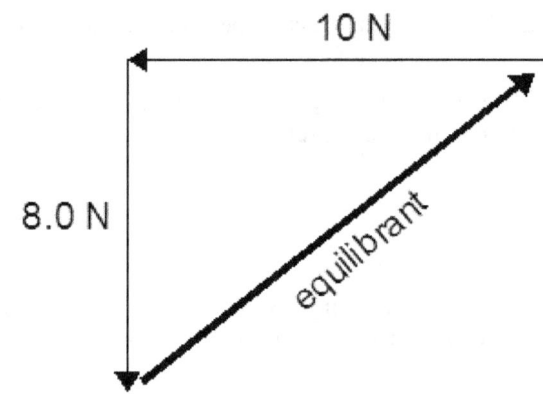

Equilibrant Force Example

$c^2 = 10{\wedge}2 + 8{\wedge}2$
So, c = 13 N

For Calculating the angle where the 8.0 N force and the equilibrant touch,

$\tan\Theta = opp/adj = 10/8$
Hence, $\Theta = 51°$

The equilibrant is **13 N [N51°E]**.

7.3   References

[1] "Physics" (PDF). Retrieved 28 May 2014.

[2] "Equilibrant Force". Retrieved 28 May 2014.

7.4   External links

- Equilibrium

Chapter 8

Line of action

For the board game, see Lines of action.

In physics, the **line of action** of a force F is a geometric representation of how the force is applied. It is the line through the point at which the force is applied in the same direction as the vector $F\rightarrow$.[1]

The concept is essential, for instance, for understanding the net effect of multiple forces applied to a body. As an example, if two forces of equal magnitude act upon a rigid body along the same line of action but in opposite directions, then they have no net effect—loosely speaking, they cancel one another out. But if, instead, their lines of action are not identical, but merely parallel, then their effect is to create a moment on the body, which tends to rotate it.

8.1 References

[1]   • Kane, Thomas R.; Levinson, David A. (1985), *Dynamics: Theory and Application*, McGraw-Hill Series in Mechanical Engineering, McGraw-Hill, Inc., ISBN 0-07-037846-0

Chapter 9

Resultant force

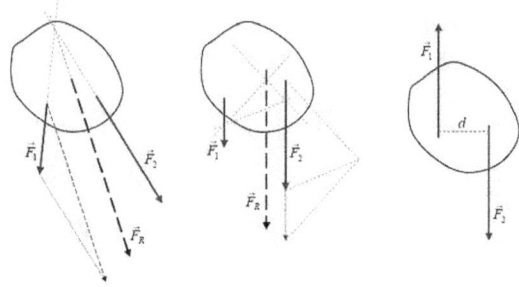

Graphical placing of the resultant force

A **resultant force** is the single force and associated torque obtained by combining a system of forces and torques acting on a rigid body. The defining feature of a resultant force, or resultant force-torque, is that it has the same effect on the rigid body as the original system of forces.[1]

The point of application of the resultant force determines its associated torque. The term *resultant force* should be understood to refer to both the forces and torques acting on a rigid body, which is why some use the term **resultant force-torque**.

9.1 Illustration

The diagram illustrates simple graphical methods for finding the line of application of the resultant force of simple planar systems.

1. Lines of application of the actual forces $\vec{F}_1$ and $\vec{F}_2$ on the leftmost illustration intersect. After vector addition is performed "at the location of $\vec{F}_1$", the net force obtained is translated so that its line of application passes through the common intersection point. With respect to that point all torques are zero, so the torque of the resultant force $\vec{F}_R$ is equal to the sum of the torques of the actual forces.

2. Illustration in the middle of the diagram shows two parallel actual forces. After vector addition "at the location of $\vec{F}_2$", the net force is translated to the appropriate line of application, whereof it becomes the resultant force $\vec{F}_R$. The procedure is based on decomposition of all forces into components for which the lines of application (pale dotted lines) intersect at one point (the so-called pole, arbitrarily set at the right side of the illustration). Then the arguments from the previous case are applied to the forces and their components to demonstrate the torque relationships.

3. The rightmost illustration shows a couple, two equal but opposite forces for which the amount of the net force is zero, but they produce the net torque $\tau = Fd$ where d is the distance between their lines of application. This is "pure" torque, since there is no resultant force.

9.2 Bound vector

A force applied to a body has a point of application. The effect the force is different for different points of application. For this reason a force is called a *bound vector*, which means that it is bound to its point of application.

Forces applied at the same point can be added together to obtain the same effect on the body. However, forces with different points of application cannot be added together and maintain the same effect on the body.

It is a simple matter to change the point of application of a force by introducing equal and opposite forces at two different points of application that produce a pure torque on the body. In this way, all of the forces acting on a body can be moved to the same point of application with associated torques.

A system of forces on a rigid body are combined by moving the forces to the same point of application and computing the associated torques. The sum of these forces and torques yields the resultant force-torque.

9.3   Associated torque

If a point $\mathbf{R}$ is selected as the point of application of the resultant force $\mathbf{F}$ of a system of n forces $\mathbf{F}_i$ then the associated torque $\mathbf{T}$ is determined from the formulas

$$\mathbf{F} = \sum_{i=1}^{n} \mathbf{F}_i,$$

and

$$\mathbf{T} = \sum_{i=1}^{n} (\mathbf{R}_i - \mathbf{R}) \times \mathbf{F}_i.$$

It is useful to note that the point of application $\mathbf{R}$ of the resultant force may be anywhere along the line of action of $\mathbf{F}$ without changing the value of the associated torque. To see this add the vector $k\mathbf{F}$ to the point of application $\mathbf{R}$ in the calculation of the associated torque,

$$\mathbf{T} = \sum_{i=1}^{n} (\mathbf{R}_i - (\mathbf{R} + k\mathbf{F})) \times \mathbf{F}_i.$$

The right side of this equation can be separated into the original;formula for $\mathbf{T}$ plus the additional term including $k\mathbf{F}$,

$$\mathbf{T} = \sum_{i=1}^{n} (\mathbf{R}_i - \mathbf{R}) \times \mathbf{F}_i - \sum_{i=1}^{n} k\mathbf{F} \times \mathbf{F}_i = \sum_{i=1}^{n} (\mathbf{R}_i - \mathbf{R}) \times \mathbf{F}_i,$$

because the second term is zero. To see this notice that $\mathbf{F}$ is the sum of the vectors $\mathbf{F}_i$ which yields

$$\sum_{i=1}^{n} k\mathbf{F} \times \mathbf{F}_i = k\mathbf{F} \times (\sum_{i=1}^{n} \mathbf{F}_i) = 0,$$

thus the value of the associated torque is unchanged.

9.4   Torque-free resultant

It is useful to consider whether there is a point of application $\mathbf{R}$ such that the associated torque is zero. This point is defined by the property

$$\mathbf{R} \times \mathbf{F} = \sum_{i=1}^{n} \mathbf{R}_i \times \mathbf{F}_i,$$

where $\mathbf{F}$ is resultant force and $\mathbf{F}_i$ form the system of forces.

Notice that this equation for $\mathbf{R}$ has a solution only if the sum of the individual torques on the right side yield a vector that is perpendicular to $\mathbf{F}$. Thus, the condition that a system of forces has a torque-free resultant can be written as

$$\mathbf{F} \cdot (\sum_{i=1}^{n} \mathbf{R}_i \times \mathbf{F}_i) = 0.$$

If this condition is satisfied then there is a point of application for the resultant which results in a pure force. If this condition is not satisfied, then the system of forces includes a pure torque for every point of application.

9.5   Wrench

The forces and torques acting on a rigid body can be assembled into the pair of vectors called a *wrench*.[2]If a system of forces and torques has a net resultant force $\mathbf{F}$ and a net resultant torque $\mathbf{T}$, then the entire system can be replaced by a force $\mathbf{F}$ and an arbitrarily located couple that yields a torque of $\mathbf{T}$. In general, if $\mathbf{F}$ and $\mathbf{T}$ are orthogonal, it is possible to derive a radial vector $\mathbf{R}$ such that $\mathbf{R} \times \mathbf{F} = \mathbf{T}$, meaning that the single force $\mathbf{F}$, acting at displacement $\mathbf{R}$, can replace the system. If the system is zero-force (torque only), it is termed a *screw* and is mathematically formulated as screw theory.[3][4]

The resultant force and torque on a rigid body obtained from a system of forces $\mathbf{F}_i$ i=1,...,n, is simply the sum of the individual wrenches $\mathbf{W}_i$, that is

$$\mathbf{W} = \sum_{i=1}^{n} \mathbf{W}_i = \sum_{i=1}^{n} (\mathbf{F}_i, \mathbf{R}_i \times \mathbf{F}_i).$$

Notice that the case of two equal but opposite forces $\mathbf{F}$ and -$\mathbf{F}$ acting at points $\mathbf{A}$ and $\mathbf{B}$ respectively, yields the resultant $\mathbf{W}=(\mathbf{F}-\mathbf{F}, \mathbf{A}\times\mathbf{F} - \mathbf{B}\times \mathbf{F}) = (0, (\mathbf{A}-\mathbf{B})\times\mathbf{F})$. This shows that wrenches of the form $\mathbf{W}=(0, \mathbf{T})$ can be interpreted as pure torques.

9.6   References

[1] H. Dadourian, *Analytical Mechanics for Students of Physics and Engineering,* Van Nostrand Co., Boston, MA 1913

[2] R. M. Murray, Z. Li, and S. Sastry, *A Mathematical Introduction to Robotic Manipulation,* CRC Press, 1994

[3] R. S. Ball, *The Theory of Screws: A study in the dynamics of a rigid body*, Hodges, Foster & Co., 1876

[4] J. M. McCarthy and G. S. Soh, *Geometric Design of Linkages*. 2nd Edition, Springer 2010

Chapter 10

Force field (physics)

For other uses, see Force field (disambiguation).
In physics a **force field** is a vector field that describes a

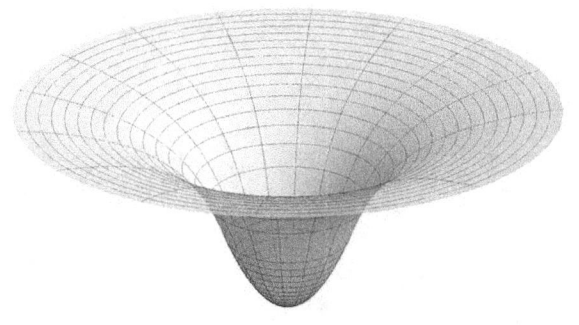

Plot of a two-dimensional slice of the gravitational potential in and around a uniform spherical body. The inflection points of the cross-section are at the surface of the body.

non-contact force acting on a particle at various positions in space. Specifically, a force field is a vector field $\vec{F}$, where $\vec{F}(\vec{x})$ is the force that a particle would feel if it were at the point $\vec{x}$.[1]

10.1 Examples of force fields

- In Newtonian gravity, a particle of mass M creates a gravitational field $\vec{g} = \frac{-GM}{r^2}\hat{r}$, where the radial unit vector $\hat{r}$ points away from the particle. The gravitational force experienced by a particle of mass m is given by $\vec{F} = m\vec{g}$. [2][3]

- An electric field $\vec{E}$ is a vector field. It exerts a force on a point charge q given by $\vec{F} = q\vec{E}$. [4]

- a **gravitational force field** is a model used to explain the influence that a massive body extends into the space around itself, producing a force on another massive body.,[5]

10.2 Restriction to position-dependent forces

Some forces, including friction, air drag, and the magnetic force on a charged particle, depend on the particle's velocity as well as its position. Therefore these forces are not characterized by a force field.

10.3 Work done by a force field

As a particle moves through a force field along a path C, the work done by the force is a line integral

$$W = \int_C \vec{F} \cdot d\vec{r}$$

This value is independent of the velocity/momentum that the particle travels along the path. For a conservative force field, it is also independent of the path itself, but depends only on the starting and ending points. Therefore, if the starting and ending points are the same, the work is zero for a conservative field:

$$\oint_C \vec{F} \cdot d\vec{r} = 0$$

If the field is conservative, the work done can be more easily evaluated by realizing that a conservative vector field can be written as the gradient of some scalar potential function:

$$\vec{F} = \nabla\phi$$

The work done is then simply the difference in the value of this potential in the starting and end points of the path. If these points are given by x = a and x = b, respectively:

$$W = \phi(b) - \phi(a)$$

10.4   See also

- Field line

- Force

10.5   References

[1] Mathematical methods in chemical engineering, by V. G. Jenson and G. V. Jeffreys, p211

[2] Vector calculus, by Marsden and Tromba, p288

[3] Engineering mechanics, by Kumar, p104

[4] Calculus: Early Transcendental Functions, by Larson, Hostetler, Edwards, p1055

[5] Geroch, Robert (1981). *General relativity from A to B*. University of Chicago Press. p. 181. ISBN 0-226-28864-1., Chapter 7, page 181

Chapter 11

Fundamental interaction

Fundamental interactions, also known as **fundamental forces**, are the interactions in physical systems that do not appear to be reducible to more basic interactions. There are four conventionally accepted fundamental interactions—gravitational, electromagnetic, strong nuclear, and weak nuclear. Each one is understood as the dynamics of a *field*. The gravitational force is modelled as a continuous classical field. The other three are each modelled as discrete quantum fields, and exhibit a measurable unit or *elementary particle*.

The two nuclear interactions produce strong forces at minuscule, subatomic distances. The strong nuclear interaction is responsible for the binding of atomic nuclei. The weak nuclear interaction also acts on the nucleus, mediating radioactive decay. Electromagnetism and gravity produce significant forces at macroscopic scales where the effects can be seen directly in everyday life. Electrical and magnetic fields tend to cancel each other out when large collections of objects are considered, so over the largest distances (on the scale of planets and galaxies), gravity tends to be the dominant force.

Theoretical physicists working beyond the Standard Model seek to quantize the gravitational field toward predictions that particle physicists can experimentally confirm, thus yielding acceptance to a theory of quantum gravity (QG). (Phenomena suitable to model as a fifth force—perhaps an added gravitational effect—remain widely disputed.) Other theorists seek to unite the electroweak and strong fields within a Grand Unified Theory (GUT). While all four fundamental interactions are widely thought to align on a highly minuscule scale, particle accelerators cannot produce the massive energy levels required to experimentally probe at that Planck scale (which would experimentally confirm such theories.) Yet some theories, such as the string theory, seek both QG and GUT within one framework, unifying all four fundamental interactions along with mass generation within a theory of everything (ToE).

11.1 History

11.1.1 Classical theory

In his 1687 theory, Isaac Newton postulated space as an infinite and unalterable physical structure existing before, within, and around all objects while their states and relations unfold at a constant pace everywhere, thus absolute space and time. Inferring that all objects bearing mass approach at a constant rate, but collide by impact proportional to their masses, Newton inferred that matter exhibits an attractive force. His law of universal gravitation mathematically stated it to span the entire universe instantly (despite absolute time), or, if not actually a force, to be instant interaction among all objects (despite absolute space.) As conventionally interpreted, Newton's theory of motion modelled a *central force* without a communicating medium.[1] Thus Newton's theory violated the first principle of mechanical philosophy, as stated by Descartes, *No action at a distance*. Conversely, during the 1820s, when explaining magnetism, Michael Faraday inferred a *field* filling space and transmitting that force. Faraday conjectured that ultimately, all forces unified into one.

In the early 1870s, James Clerk Maxwell unified electricity and magnetism as effects of an electromagnetic field whose third consequence was light, travelling at constant speed in a vacuum. The electromagnetic field theory contradicted predictions of Newton's theory of motion, unless physical states of the luminiferous aether—presumed to fill all space whether within matter or in a vacuum and to manifest the electromagnetic field—aligned all phenomena and thereby held valid the Newtonian principle relativity or invariance.

11.1.2 The Standard Model

Main article: Standard Model
See also: Lambda-CDM model

The Standard Model of particle physics was developed throughout the latter half of the 20th century. In the Stan-

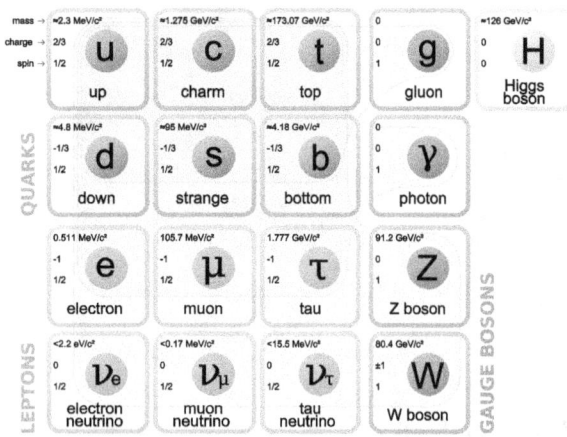

The Standard Model of elementary particles, with the fermions in the first three columns, the gauge bosons in the fourth column, and the Higgs boson in the fifth column

dard Model, the electromagnetic, strong, and weak interactions associate with elementary particles, whose behaviours are modelled in quantum mechanics (QM). For predictive success with QM's probabilistic outcomes, particle physics conventionally models QM events across a field set to special relativity, altogether relativistic quantum field theory (QFT).[2] Force particles, called gauge bosons—*force carriers* or *messenger particles* of underlying fields—interact with matter particles, called fermions. Everyday matter is atoms, composed of three fermion types: up-quarks and down-quarks constituting, as well as electrons orbiting, the atom's nucleus. Atoms interact, form molecules, and manifest further properties through electromagnetic interactions among their electrons absorbing and emitting photons, the electromagnetic field's force carrier, which if unimpeded traverse potentially infinite distance. Electromagnetism's QFT is quantum electrodynamics (QED).

The electromagnetic interaction was modelled with the weak interaction, whose force carriers are W and Z bosons, traversing the minuscule distance, in electroweak theory (EWT). Electroweak interaction would operate at such high temperatures as soon after the presumed Big Bang, but, as the early universe cooled, split into electromagnetic and weak interactions. The strong interaction, whose force carrier is the gluon, traversing minuscule distance among quarks, is modeled in quantum chromodynamics (QCD). EWT, QCD, and the Higgs mechanism, whereby the Higgs field manifests Higgs bosons that interact with some quantum particles and thereby endow those particles with mass comprise particle physics' Standard Model (SM). Predictions are usually made using calculational approximation methods, although such perturbation theory is inadequate to model some experimental observations (for instance bound states and solitons.) Still, physicists widely accept the Stan-

dard Model as science's most experimentally confirmed theory.

Beyond the Standard Model, some theorists work to unite the electroweak and strong interactions within a Grand Unified Theory (GUT). Some attempts at GUTs hypothesize "shadow" particles, such that every known matter particle associates with an undiscovered force particle, and vice versa, altogether supersymmetry (SUSY). Other theorists seek to quantize the gravitational field by the modelling behaviour of its hypothetical force carrier, the graviton and achieve quantum gravity (QG). One approach to QG is loop quantum gravity (LQG). Still other theorists seek both QG and GUT within one framework, reducing all four fundamental interactions to a Theory of Everything (ToE). The most prevalent aim at a ToE is string theory, although to model matter particles, it added SUSY to force particles—and so, strictly speaking, became superstring theory. Multiple, seemingly disparate superstring theories were unified on a backbone, M-theory. Theories beyond the Standard Model remain highly speculative, lacking great experimental support.

11.2 Overview of the fundamental interactions

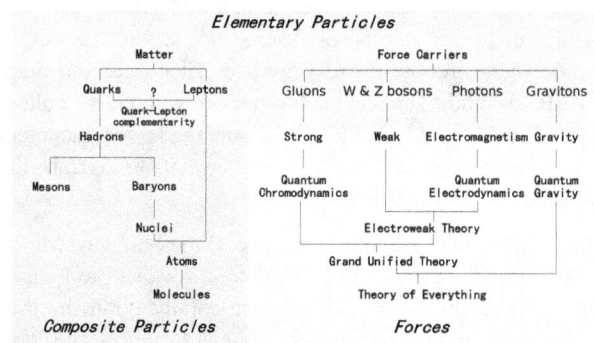

An overview of the various families of elementary and composite particles, and the theories describing their interactions. Fermions are on the left, and Bosons are on the right.

In the conceptual model of fundamental interactions, matter consists of fermions, which carry properties called charges and spin $\pm\frac{1}{2}$ (intrinsic angular momentum $\pm\hbar/2$, where $\hbar$ is the reduced Planck constant). They attract or repel each other by exchanging bosons.

The interaction of any pair of fermions in perturbation theory can then be modelled thus:

Two fermions go in → *interaction* by boson exchange → Two changed fermions go out.

The exchange of bosons always carries energy and momentum between the fermions, thereby changing their speed and direction. The exchange may also transport a charge between the fermions, changing the charges of the fermions in the process (e.g., turn them from one type of fermion to another). Since bosons carry one unit of angular momentum, the fermion's spin direction will flip from $+\frac{1}{2}$ to $-\frac{1}{2}$ (or vice versa) during such an exchange (in units of the reduced Planck's constant).

Because an interaction results in fermions attracting and repelling each other, an older term for "interaction" is force.

According to the present understanding, there are four fundamental interactions or forces: gravitation, electromagnetism, the weak interaction, and the strong interaction. Their magnitude and behaviour vary greatly, as described in the table below. Modern physics attempts to explain every observed physical phenomenon by these fundamental interactions. Moreover, reducing the number of different interaction types is seen as desirable. Two cases in point are the unification of:

- Electric and magnetic force into electromagnetism;
- The electromagnetic interaction and the weak interaction into the electroweak interaction; see below.

Both magnitude ("relative strength") and "range", as given in the table, are meaningful only within a rather complex theoretical framework. It should also be noted that the table below lists properties of a conceptual scheme that is still the subject of ongoing research.

The modern (perturbative) quantum mechanical view of the fundamental forces other than gravity is that particles of matter (fermions) do not directly interact with each other, but rather carry a charge, and exchange virtual particles (gauge bosons), which are the interaction carriers or force mediators. For example, photons mediate the interaction of electric charges, and gluons mediate the interaction of color charges.

11.3   The interactions

11.3.1   Gravity

Main article: Gravity

Gravitation is by far the weakest of the four interactions. The weakness of gravity can easily be demonstrated by suspending a pin using a simple magnet (such as a refrigerator magnet). The magnet is able to hold the pin against the gravitational pull of the entire Earth.

Yet gravitation is very important for macroscopic objects and over macroscopic distances for the following reasons. Gravitation:

- Is the only interaction that acts on all particles having mass, energy and/or momentum
- Has an infinite range, like electromagnetism but unlike strong and weak interaction
- Cannot be absorbed, transformed, or shielded against
- Always attracts and never repels

Even though electromagnetism is far stronger than gravitation, electrostatic attraction is not relevant for large celestial bodies, such as planets, stars, and galaxies, simply because such bodies contain equal numbers of protons and electrons and so have a net electric charge of zero. Nothing "cancels" gravity, since it is only attractive, unlike electric forces which can be attractive or repulsive. On the other hand, all objects having mass are subject to the gravitational force, which only attracts. Therefore, only gravitation matters on the large-scale structure of the universe.

The long range of gravitation makes it responsible for such large-scale phenomena as the structure of galaxies and black holes and it retards the expansion of the universe. Gravitation also explains astronomical phenomena on more modest scales, such as planetary orbits, as well as everyday experience: objects fall; heavy objects act as if they were glued to the ground, and animals can only jump so high.

Gravitation was the first interaction to be described mathematically. In ancient times, Aristotle hypothesized that objects of different masses fall at different rates. During the Scientific Revolution, Galileo Galilei experimentally determined that this was not the case — neglecting the friction due to air resistance, and buoyancy forces if an atmosphere is present (e.g. the case of a dropped air-filled balloon vs a water-filled balloon) all objects accelerate toward the Earth at the same rate. Isaac Newton's law of Universal Gravitation (1687) was a good approximation of the behaviour of gravitation. Our present-day understanding of gravitation stems from Albert Einstein's General Theory of Relativity of 1915, a more accurate (especially for cosmological masses and distances) description of gravitation in terms of the geometry of spacetime.

Merging general relativity and quantum mechanics (or quantum field theory) into a more general theory of quantum gravity is an area of active research. It is hypothesized that gravitation is mediated by a massless spin-2 particle called the graviton.

Although general relativity has been experimentally confirmed (at least for weak fields) on all but the smallest scales, there are rival theories of gravitation. Those taken seriously by [citation needed] the physics community all reduce to general relativity in some limit, and the focus of observational work is to establish limitations on what deviations from general relativity are possible.

Proposed extra dimensions could explain why the gravity force is so weak.[5]

11.3.2 Electroweak interaction

Main article: Electroweak interaction

Electromagnetism and weak interaction appear to be very different at everyday low energies. They can be modeled using two different theories. However, above unification energy, on the order of 100 GeV, they would merge into a single electroweak force.

Electroweak theory is very important for modern cosmology, particularly on how the universe evolved. This is because shortly after the Big Bang, the temperature was approximately above 10^{15} K. Electromagnetic force and weak force were merged into a combined electroweak force.

For contributions to the unification of the weak and electromagnetic interaction between elementary particles, Abdus Salam, Sheldon Glashow and Steven Weinberg were awarded the Nobel Prize in Physics in 1979.[6][7]

Electromagnetism

Main article: Electromagnetism

Electromagnetism is the force that acts between electrically charged particles. This phenomenon includes the electrostatic force acting between charged particles at rest, and the combined effect of electric and magnetic forces acting between charged particles moving relative to each other.

Electromagnetism is infinite-ranged like gravity, but vastly stronger, and therefore describes a number of macroscopic phenomena of everyday experience such as friction, rainbows, lightning, and all human-made devices using electric current, such as television, lasers, and computers. Electromagnetism fundamentally determines all macroscopic, and many atomic levels, properties of the chemical elements, including all chemical bonding.

In a four kilogram (~1 gallon) jug of water there are

$$4000 \text{ g } H_2O \cdot \frac{1 \text{ mol } H_2O}{18 \text{ g } H_2O} \cdot \frac{10 \text{ mol } e^-}{1 \text{ mol } H_2O} \cdot \frac{96{,}000 \text{ C}}{1 \text{ mol } e^-} = 2.1 \times 10^8 C$$

of total electron charge. Thus, if we place two such jugs a meter apart, the electrons in one of the jugs repel those in the other jug with a force of

$$\frac{1}{4\pi\varepsilon_0} \frac{(2.1\times 10^8 C)^2}{(1m)^2} = 4.1 \times 10^{26} N.$$

This is larger than the planet Earth would weigh if weighed on another Earth. The atomic nuclei in one jug also repel those in the other with the same force. However, these repulsive forces are canceled by the attraction of the electrons in jug A with the nuclei in jug B and the attraction of the nuclei in jug A with the electrons in jug B, resulting in no net force. Electromagnetic forces are tremendously stronger than gravity but cancel out so that for large bodies gravity dominates.

Electrical and magnetic phenomena have been observed since ancient times, but it was only in the 19th century that it was discovered that electricity and magnetism are two aspects of the same fundamental interaction. By 1864, Maxwell's equations had rigorously quantified this unified interaction. Maxwell's theory, restated using vector calculus, is the classical theory of electromagnetism, suitable for most technological purposes.

The constant speed of light in a vacuum (customarily described with the letter "c") can be derived from Maxwell's equations, which are consistent with the theory of special relativity. Einstein's 1905 theory of special relativity, however, which flows from the observation that the speed of light is constant no matter how fast the observer is moving, showed that the theoretical result implied by Maxwell's equations has profound implications far beyond electromagnetism on the very nature of time and space.

In another work that departed from classical electromagnetism, Einstein also explained the photoelectric effect by hypothesizing that light was transmitted in quanta, which we now call photons. Starting around 1927, Paul Dirac combined quantum mechanics with the relativistic theory of electromagnetism. Further work in the 1940s, by Richard Feynman, Freeman Dyson, Julian Schwinger, and Sin-Itiro Tomonaga, completed this theory, which is now called quantum electrodynamics, the revised theory of electromagnetism. Quantum electrodynamics and quantum mechanics provide a theoretical basis for electromagnetic behavior such as quantum tunneling, in which a certain percentage of electrically charged particles move in ways that would be impossible under the classical electromagnetic theory, that is necessary for everyday electronic devices such as transistors to function.

Weak interaction

Main article: Weak interaction

The *weak interaction* or *weak nuclear force* is responsible for some nuclear phenomena such as beta decay. Electromagnetism and the weak force are now understood to be two aspects of a unified electroweak interaction — this discovery was the first step toward the unified theory known as the Standard Model. In the theory of the electroweak interaction, the carriers of the weak force are the massive gauge bosons called the W and Z bosons. The weak interaction is the only known interaction which does not conserve parity; it is left-right asymmetric. The weak interaction even violates CP symmetry but does conserve CPT.

11.3.3   Strong interaction

Main article: Strong interaction

The *strong interaction*, or *strong nuclear force*, is the most complicated interaction, mainly because of the way it varies with distance. At distances greater than 10 femtometers, the strong force is practically unobservable. Moreover, it holds only inside the atomic nucleus.

After the nucleus was discovered in 1908, it was clear that a new force was needed to overcome the electrostatic repulsion, a manifestation of electromagnetism, of the positively charged protons. Otherwise, the nucleus could not exist. Moreover, the force had to be strong enough to squeeze the protons into a volume that is 10^{-15} of that of the entire atom. From the short range of this force, Hideki Yukawa predicted that it was associated with a massive particle, whose mass is approximately 100 MeV.

The 1947 discovery of the pion ushered in the modern era of particle physics. Hundreds of hadrons were discovered from the 1940s to 1960s, and an extremely complicated theory of hadrons as strongly interacting particles was developed. Most notably:

- The pions were understood to be oscillations of vacuum condensates;

- Jun John Sakurai proposed the rho and omega vector bosons to be force carrying particles for approximate symmetries of isospin and hypercharge;

- Geoffrey Chew, Edward K. Burdett and Steven Frautschi grouped the heavier hadrons into families that could be understood as vibrational and rotational excitations of strings.

While each of these approaches offered deep insights, no approach led directly to a fundamental theory.

Murray Gell-Mann along with George Zweig first proposed fractionally charged quarks in 1961. Throughout the 1960s, different authors considered theories similar to the modern fundamental theory of quantum chromodynamics (QCD) as simple models for the interactions of quarks. The first to hypothesize the gluons of QCD were Moo-Young Han and Yoichiro Nambu, who introduced the quark color charge and hypothesized that it might be associated with a force-carrying field. At that time, however, it was difficult to see how such a model could permanently confine quarks. Han and Nambu also assigned each quark color an integer electrical charge, so that the quarks were fractionally charged only on average, and they did not expect the quarks in their model to be permanently confined.

In 1971, Murray Gell-Mann and Harald Fritzsch proposed that the Han/Nambu color gauge field was the correct theory of the short-distance interactions of fractionally charged quarks. A little later, David Gross, Frank Wilczek, and David Politzer discovered that this theory had the property of asymptotic freedom, allowing them to make contact with experimental evidence. They concluded that QCD was the complete theory of the strong interactions, correct at all distance scales. The discovery of asymptotic freedom led most physicists to accept QCD since it became clear that even the long-distance properties of the strong interactions could be consistent with experiment if the quarks are permanently confined.

Assuming that quarks are confined, Mikhail Shifman, Arkady Vainshtein, and Valentine Zakharov were able to compute the properties of many low-lying hadrons directly from QCD, with only a few extra parameters to describe the vacuum. In 1980, Kenneth G. Wilson published computer calculations based on the first principles of QCD, establishing, to a level of confidence tantamount to certainty, that QCD will confine quarks. Since then, QCD has been the established theory of the strong interactions.

QCD is a theory of fractionally charged quarks interacting by means of 8 photon-like particles called gluons. The gluons interact with each other, not just with the quarks, and at long distances the lines of force collimate into strings. In this way, the mathematical theory of QCD not only explains how quarks interact over short distances but also the string-like behavior, discovered by Chew and Frautschi, which they manifest over longer distances.

11.3.4   Beyond the Standard Model

Main article: Physics beyond the Standard Model
See also: Elementary particle § Beyond the Standard Model

Numerous theoretical efforts have been made to systematize the existing four fundamental interactions on the model of electroweak unification.

Grand Unified Theories (GUTs) are proposals to show that all of the fundamental interactions, other than gravity, arise from a single interaction with symmetries that break down at low energy levels. GUTs predict relationships among constants of nature that are unrelated in the SM. GUTs also predict gauge coupling unification for the relative strengths of the electromagnetic, weak, and strong forces, a prediction verified at the Large Electron–Positron Collider in 1991 for supersymmetric theories.

Theories of everything, which integrate GUTs with a quantum gravity theory face a greater barrier, because no quantum gravity theories, which include string theory, loop quantum gravity, and twistor theory, have secured wide acceptance. Some theories look for a graviton to complete the Standard Model list of force-carrying particles, while others, like loop quantum gravity, emphasize the possibility that time-space itself may have a quantum aspect to it.

Some theories beyond the Standard Model include a hypothetical fifth force, and the search for such a force is an ongoing line of experimental research in physics. In supersymmetric theories, there are particles that acquire their masses only through supersymmetry breaking effects and these particles, known as moduli can mediate new forces. Another reason to look for new forces is the recent discovery that the expansion of the universe is accelerating (also known as dark energy), giving rise to a need to explain a nonzero cosmological constant, and possibly to other modifications of general relativity. Fifth forces have also been suggested to explain phenomena such as CP violations, dark matter, and dark flow.

In December 2015, two observations in the ATLAS and CMS detectors at the Large Hadron Collider hinted at the existence of a new particle six times heavier than the Higgs Boson.[8] However, after obtaining more experimental data, the anomaly appeared not be significant.[9]

11.4   See also

- Standard Model
  - Strong interaction
  - Electroweak interaction
  - Weak interaction
  - Gravity
    - Quantum gravity
    - String Theory

- Theory of Everything
- Grand Unified Theory
  - Gauge coupling unification
  - Unified Field Theory
- Quintessence, a hypothesized fifth force.
- *People*: Isaac Newton, James Clerk Maxwell, Albert Einstein, Richard Feynman, Sheldon Glashow, Abdus Salam, Steven Weinberg, Gerardus 't Hooft, David Gross, Edward Witten, Howard Georgi.

11.5   References

[1] Newton's absolute space was a medium, but not one transmitting gravitation.

[2] Meinard Kuhlmann, "Physicists debate whether the world is made of particles or fields—or something else entirely", *Scientific American*, 24 Jul 2013.

[3] "Standard Model of Particles and Interactions". *jhu.edu*. Johns Hopkins University. Retrieved August 18, 2016. .gif

[4] Approximate. See Coupling constant for more exact strengths, depending on the particles and energies involved.

[5] CERN (20 January 2012). "Extra dimensions, gravitons, and tiny black holes".

[6] Bais, Sander (2005), *The Equations. Icons of knowledge*, ISBN 0-674-01967-9 p.84

[7] "The Nobel Prize in Physics 1979". The Nobel Foundation. Retrieved 2008-12-16.

[8] "Is the Large Hadron Collider set to reveal secrets of the universe?". Retrieved 2016-07-12.

[9] Elizabeth Gibney (August 5, 2016). "Hopes for revolutionary new LHC particle dashed". *Nature*. Retrieved August 18, 2016.

Bibliography General:

- Davies, Paul (1986), *The Forces of Nature*, Cambridge Univ. Press 2nd ed.

- Feynman, Richard (1967), *The Character of Physical Law*, MIT Press, ISBN 0-262-56003-8

- Schumm, Bruce A. (2004), *Deep Down Things*, Johns Hopkins University Press While all interactions are discussed, discussion is especially thorough on the weak.

- Weinberg, Steven (1993), *The First Three Minutes: A Modern View of the Origin of the Universe*, Basic Books, ISBN 0-465-02437-8

- Weinberg, Steven (1994), *Dreams of a Final Theory*, Basic Books, ISBN 0-679-74408-8

Texts:

- Padmanabhan, T. (1998), *After The First Three Minutes: The Story of Our Universe*, Cambridge Univ. Press, ISBN 0-521-62972-1

- Perkins, Donald H. (2000), *Introduction to High Energy Physics*, Cambridge Univ. Press, ISBN 0-521-62196-8

- Riazuddin (December 29, 2009). "Non-standard interactions" (PDF). *NCP 5th Particle Physics Sypnoisis*. Islamabad: Riazuddin, Head of High-Energy Theory Group at National Center for Physics. **1** (1): 1–25. Retrieved March 19, 2011.

Chapter 12

Coulomb's law

Coulomb's law or Coulomb's inverse-square law, is a law of physics that describes force interacting between static electrically charged particles. In its scalar form the law is:

$$F = k_e \frac{q_1 q_2}{r^2}$$

where k_e is Coulomb's constant ($k_e = 8.99 \times 10^9$ N m^2 C^{-2}), q_1 and q_2 are the signed magnitudes of the charges, and the scalar r is the distance between the charges. The force of interaction between the charges is attractive if the charges have opposite signs (i.e. F is negative) and repulsive if like-signed (i.e. F is positive).

The law was first published in 1784 by French physicist Charles Augustin de Coulomb and was essential to the development of the theory of electromagnetism. It is analogous to Isaac Newton's inverse-square law of universal gravitation. Coulomb's law can be used to derive Gauss's law, and vice versa. The law has been tested heavily, and all observations have upheld the law's principle.

Charles-Augustin de Coulomb

12.1 History

Ancient cultures around the Mediterranean knew that certain objects, such as rods of amber, could be rubbed with cat's fur to attract light objects like feathers. Thales of Miletus made a series of observations on static electricity around 600 BC, from which he believed that friction rendered amber magnetic, in contrast to minerals such as magnetite, which needed no rubbing.[1][2] Thales was incorrect in believing the attraction was due to a magnetic effect, but later science would prove a link between magnetism and electricity. Electricity would remain little more than an intellectual curiosity for millennia until 1600, when the English scientist William Gilbert made a careful study of electricity and magnetism, distinguishing the lodestone effect from static electricity produced by rubbing amber.[1] He coined the New Latin word *electricus* ("of amber" or "like amber", from ἤλεκτρον [*elektron*], the Greek word

for "amber") to refer to the property of attracting small objects after being rubbed.[3] This association gave rise to the English words "electric" and "electricity", which made their first appearance in print in Thomas Browne's *Pseudodoxia Epidemica* of 1646.[4]

Early investigators of the 18th century who suspected that the electrical force diminished with distance as the force of gravity did (i.e., as the inverse square of the distance) included Daniel Bernoulli[5] and Alessandro Volta, both of whom measured the force between plates of a capacitor, and Franz Aepinus who supposed the inverse-square law in 1758.[6]

Based on experiments with electrically charged spheres, Joseph Priestley of England was among the first to propose that electrical force followed an inverse-square law, similar to Newton's law of universal gravitation. However, he did not generalize or elaborate on this.[7] In 1767, he conjectured that the force between charges varied as the inverse square of the distance.[8][9]

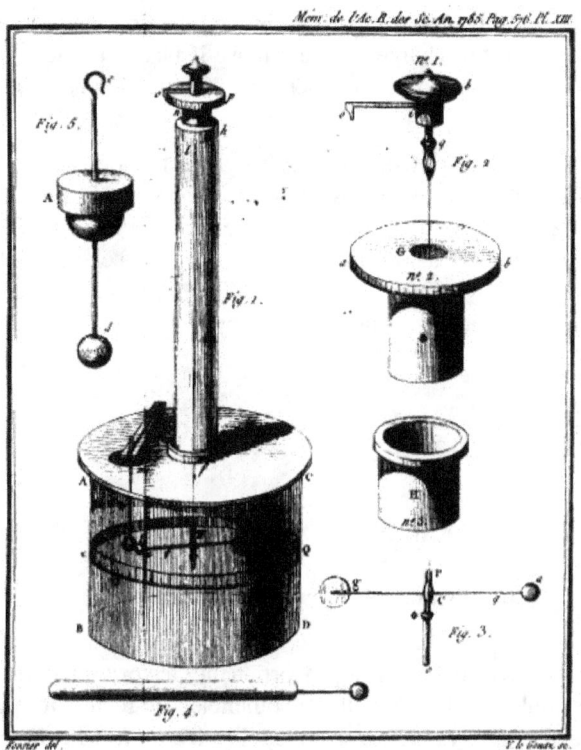

Coulomb's torsion balance

In 1769, Scottish physicist John Robison announced that, according to his measurements, the force of repulsion between two spheres with charges of the same sign varied as $x^{-2.06}$.[10]

In the early 1770s, the dependence of the force between charged bodies upon both distance and charge had already been discovered, but not published, by Henry Cavendish of England.[11]

Finally, in 1785, the French physicist Charles-Augustin de Coulomb published his first three reports of electricity and magnetism where he stated his law. This publication was essential to the development of the theory of electromagnetism.[12] He used a torsion balance to study the repulsion and attraction forces of charged particles, and determined that the magnitude of the electric force between two point charges is directly proportional to the product of the charges and inversely proportional to the square of the distance between them.

The torsion balance consists of a bar suspended from its middle by a thin fiber. The fiber acts as a very weak torsion spring. In Coulomb's experiment, the torsion balance was an insulating rod with a metal-coated ball attached to one end, suspended by a silk thread. The ball was charged with a known charge of static electricity, and a second charged ball of the same polarity was brought near it. The two charged balls repelled one another, twisting the fiber through a certain angle, which could be read from a scale on the instrument. By knowing how much force it took to twist the fiber through a given angle, Coulomb was able to calculate the force between the balls and derive his inverse-square proportionality law.

12.2 The law

Coulomb's law states that:

> The magnitude of the electrostatic force of interaction between two point charges is directly proportional to the scalar multiplication of the magnitudes of charges and inversely proportional to the square of the distance between them.[12]
>
> The force is along the straight line joining them. If the two charges have the same sign, the electrostatic force between them is repulsive; if they have different signs, the force between them is attractive.

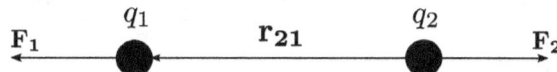

A graphical representation of Coulomb's law

Coulomb's law can also be stated as a simple mathematical expression. The scalar and vector forms of the mathematical equation are

$$|\mathbf{F}| = k_e \frac{|q_1 q_2|}{r^2} \quad \text{and} \quad \mathbf{F}_1 = k_e \frac{q_1 q_2}{|\mathbf{r}_{12}|^2}\hat{\mathbf{r}}_{21}, \quad \text{respectively,}$$

where k_e is Coulomb's constant ($k_e = 8.9875517873681764 \times 10^9$ N m^2 C^{-2}), q_1 and q_2 are the signed magnitudes of the charges, the scalar r is the distance between the charges, the vector $\mathbf{r}_{21} = \mathbf{r}_1 - \mathbf{r}_2$ is the vectorial distance between the charges, and $\hat{\mathbf{r}}_{21} = \mathbf{r}_{21}/|\mathbf{r}_{21}|$ (a unit vector pointing from q_2 to q_1). The vector form of the equation calculates the force $\mathbf{F}_1$ applied on q_1 by q_2. If $\mathbf{r}_{12}$ is used instead, then the effect on q_2 can be found. It can be also calculated using Newton's third law: $\mathbf{F}_2 = -\mathbf{F}_1$.

12.2.1 Units

Electromagnetic theory is usually expressed using the standard SI units. Force is measured in newtons, charge in coulombs, and distance in metres. Coulomb's constant is given by $ke = 1/4\pi\varepsilon_0$. The constant ε_0 is the permittivity of free space in $C^2 \, m^{-2} \, N^{-1}$. And ε is the relative permittivity of the material in which the charges are immersed, and is dimensionless.

The SI derived units for the electric field are volts per meter, newtons per coulomb, or tesla meters per second.

Coulomb's law and Coulomb's constant can also be interpreted in various terms:

- Atomic units. In atomic units the force is expressed in hartrees per Bohr radius, the charge in terms of the elementary charge, and the distances in terms of the *Bohr radius*.

- Electrostatic units or Gaussian units. In electrostatic units and Gaussian units, the unit charge (*esu* or statcoulomb) is defined in such a way that the Coulomb constant k disappears because it has the value of one and becomes dimensionless.

12.2.2 Electric field

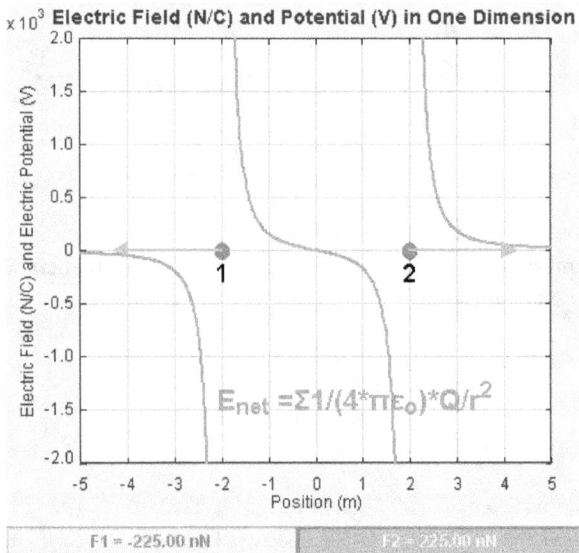

If the two charges have the same sign, the electrostatic force between them is repulsive; if they have different sign, the force between them is attractive.

An electric field is a vector field that associates to each point in space the Coulomb force experienced by a test charge. In the simplest case, the field is considered to be generated solely by a single source point charge. The strength and direction of the Coulomb force **F** on a test charge q_t depends on the electric field **E** that it finds itself in, such that $\mathbf{F} = q_t\mathbf{E}$. If the field is generated by a positive source point charge q, the direction of the electric field points along lines directed radially outwards from it, i.e. in the direction that a positive point test charge q_t would move if placed in the field. For a negative point source charge, the direction is radially inwards.

The magnitude of the electric field **E** can be derived from Coulomb's law. By choosing one of the point charges to be the source, and the other to be the test charge, it follows from Coulomb's law that the magnitude of the electric field **E** created by a single source point charge q at a certain distance from it r in vacuum is given by:

$$|E| = \frac{1}{4\pi\varepsilon_0}\frac{|q|}{r^2}$$

12.2.3 Coulomb's constant

Main article: Coulomb's constant

Coulomb's constant is a proportionality factor that appears in Coulomb's law as well as in other electric-related formulas. Denoted ke, it is also called the electric force constant or electrostatic constant, hence the subscript e.

The exact value of Coulomb's constant is:

$$k_e = \frac{1}{4\pi\varepsilon_0} = \frac{c_0^2\mu_0}{4\pi} = c_0^2 \times 10^{-7} \, \mathrm{H \cdot m^{-1}}$$
$$= 8.987\,551\,787\,368\,176\,4 \times 10^9 \, \mathrm{N \cdot m^2 \cdot C^{-2}}$$

12.2.4 Conditions for validity

There are two conditions to be fulfilled for the validity of Coulomb's law:

1. **The charges considered must be point charges.**

2. **They should be stationary with respect to each other.**

12.3 Scalar form

When it is only of interest to know the magnitude of the electrostatic force (and not its direction), it may be easiest to consider a scalar version of the law. The scalar form of

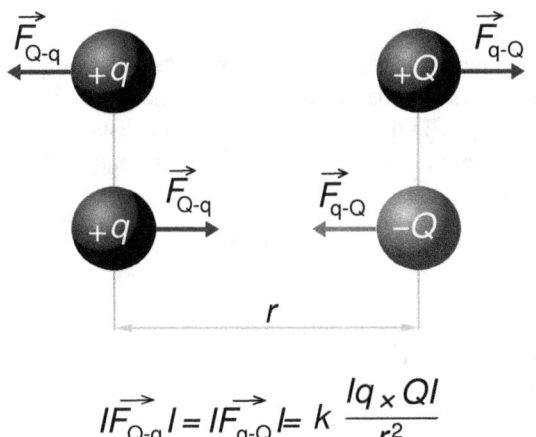

$$|\vec{F}_{Q\text{-}q}| = |\vec{F}_{q\text{-}Q}| = k\,\frac{|q \times Q|}{r^2}$$

*The absolute value of the force **F** between two point charges q and Q relates to the distance between the point charges and to the simple product of their charges. The diagram shows that like charges repel each other, and opposite charges attract each other.*

Coulomb's Law relates the magnitude and sign of the electrostatic force **F** acting simultaneously on two point charges q_1 and q_2 as follows:

$$|\boldsymbol{F}| = k_e \frac{|q_1 q_2|}{r^2}$$

where r is the separation distance and *ke* is Coulomb's constant. If the product $q_1 q_2$ is positive, the force between the two charges is repulsive; if the product is negative, the force between them is attractive.[13]

12.4   Vector form

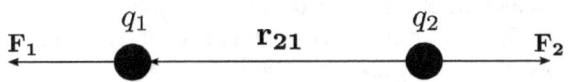

In the image, the vector $\boldsymbol{F}_1$ is the force experienced by q_1, and the vector $\boldsymbol{F}_2$ is the force experienced by q_2. When $q_1 q_2 > 0$ the forces are repulsive (as in the image) and when $q_1 q_2 < 0$ the forces are attractive (opposite to the image). The magnitude of the forces will always be equal.

Coulomb's law states that the electrostatic force $\boldsymbol{F}_1$ experienced by a charge, q_1 at position $\boldsymbol{r}_1$, in the vicinity of another charge, q_2 at position $\boldsymbol{r}_2$, in a vacuum is equal to:

$$\boldsymbol{F}_1 = \frac{q_1 q_2}{4\pi\varepsilon_0} \frac{(\boldsymbol{r}_1 - \boldsymbol{r}_2)}{|\boldsymbol{r}_1 - \boldsymbol{r}_2|^2} = \frac{q_1 q_2}{4\pi\varepsilon_0} \frac{\hat{\boldsymbol{r}}_{21}}{|\boldsymbol{r}_{21}|^2},$$

where $\mathbf{r}_{21} = \mathbf{r}_1 - \mathbf{r}_2$, the unit vector $\hat{\mathbf{r}}_{21} = \mathbf{r}_{21}/|\mathbf{r}_{21}|$, and ε_0 is the electric constant.

The vector form of Coulomb's law is simply the scalar definition of the law with the direction given by the unit vector, $\hat{\mathbf{r}}_{21}$, parallel with the line *from* charge q_2 *to* charge q_1.[14] If both charges have the same sign (like charges) then the product $q_1 q_2$ is positive and the direction of the force on q_1 is given by $\hat{\mathbf{r}}_{21}$; the charges repel each other. If the charges have opposite signs then the product $q_1 q_2$ is negative and the direction of the force on q_1 is given by $-\hat{\mathbf{r}}_{21} = \hat{\mathbf{r}}_{12}$; the charges attract each other.

The electrostatic force $\mathbf{F}_2$ experienced by q_2, according to Newton's third law, is $\mathbf{F}_2 = -\mathbf{F}_1$.

12.4.1   System of discrete charges

The law of superposition allows Coulomb's law to be extended to include any number of point charges. The force acting on a point charge due to a system of point charges is simply the vector addition of the individual forces acting alone on that point charge due to each one of the charges. The resulting force vector is parallel to the electric field vector at that point, with that point charge removed.

The force $\mathbf{F}$ on a small charge q at position $\mathbf{r}$, due to a system of N discrete charges in vacuum is:

$$\boldsymbol{F}(\boldsymbol{r}) = \frac{q}{4\pi\varepsilon_0} \sum_{i=1}^{N} q_i \frac{\boldsymbol{r} - \boldsymbol{r}_i}{|\boldsymbol{r} - \boldsymbol{r}_i|^2} = \frac{q}{4\pi\varepsilon_0} \sum_{i=1}^{N} q_i \frac{\widehat{\boldsymbol{R}_i}}{|\boldsymbol{R}_i|^2},$$

where qi and $\mathbf{r}i$ are the magnitude and position respectively of the ith charge, $\hat{\mathbf{R}}i$ is a unit vector in the direction of $\mathbf{R}i = \mathbf{r} - \mathbf{r}i$ (a vector pointing from charges qi to q).[14]

12.4.2   Continuous charge distribution

In this case, the principle of linear superposition is also used. For a continuous charge distribution, an integral over the region containing the charge is equivalent to an infinite summation, treating each infinitesimal element of space as a point charge dq. The distribution of charge is usually linear, surface or volumetric.

For a linear charge distribution (a good approximation for charge in a wire) where $\lambda(\mathbf{r'})$ gives the charge per unit length at position $\mathbf{r'}$, and dl' is an infinitesimal element of length,

$$dq = \lambda(\boldsymbol{r'})dl' .[15]$$

For a surface charge distribution (a good approximation for charge on a plate in a parallel plate capacitor) where $\sigma(\mathbf{r'})$

gives the charge per unit area at position $\mathbf{r'}$, and dA' is an infinitesimal element of area,

$$dq = \sigma(\mathbf{r'})\, dA'.$$

For a volume charge distribution (such as charge within a bulk metal) where $\varrho(\mathbf{r'})$ gives the charge per unit volume at position $\mathbf{r'}$, and dV' is an infinitesimal element of volume,

$$dq = \rho(\mathbf{r'})\, dV'. \quad [14]$$

The force on a small test charge q' at position $\mathbf{r}$ in vacuum is given by the integral over the distribution of charge:

$$\mathbf{F} = \frac{q'}{4\pi\varepsilon_0} \int dq\, \frac{\mathbf{r} - \mathbf{r'}}{|\mathbf{r} - \mathbf{r'}|^3}.$$

12.5 Simple experiment to verify Coulomb's law

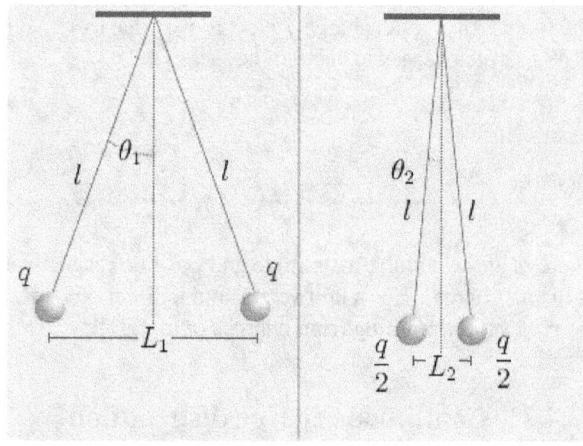

Experiment to verify Coulomb's law.

It is possible to verify Coulomb's law with a simple experiment. Consider two small spheres of mass m and same-sign charge q, hanging from two ropes of negligible mass of length l. The forces acting on each sphere are three: the weight mg, the rope tension T and the electric force F.

In the equilibrium state:

and:

Dividing (1) by (2):

Let L_1 be the distance between the charged spheres; the repulsion force between them F_1, assuming Coulomb's law is correct, is equal to

so:

If we now discharge one of the spheres, and we put it in contact with the charged sphere, each one of them acquires a charge $q/2$. In the equilibrium state, the distance between the charges will be $L_2 < L_1$ and the repulsion force between them will be:

We know that $F_2 = mg \tan\theta_2$. And:

$$\frac{\frac{q^2}{4}}{4\pi\epsilon_0 L_2^2} = mg \tan\theta_2$$

Dividing (4) by (5), we get:

Measuring the angles θ_1 and θ_2 and the distance between the charges L_1 and L_2 is sufficient to verify that the equality is true taking into account the experimental error. In practice, angles can be difficult to measure, so if the length of the ropes is sufficiently great, the angles will be small enough to make the following approximation:

Using this approximation, the relationship (6) becomes the much simpler expression:

In this way, the verification is limited to measuring the distance between the charges and check that the division approximates the theoretical value.

12.6 Electrostatic approximation

In either formulation, Coulomb's law is fully accurate only when the objects are stationary, and remains approximately correct only for slow movement. These conditions are collectively known as the electrostatic approximation. When movement takes place, magnetic fields that alter the force on the two objects are produced. The magnetic interaction between moving charges may be thought of as a manifestation of the force from the electrostatic field but with Einstein's theory of relativity taken into consideration.

12.6.1 Atomic forces

Coulomb's law holds even within atoms, correctly describing the force between the positively charged atomic nucleus and each of the negatively charged electrons. This simple law also correctly accounts for the forces that bind atoms together to form molecules and for the forces that bind atoms and molecules together to form solids and liquids. Generally, as the distance between ions increases, the energy of attraction approaches zero and ionic bonding is less favorable. As the magnitude of opposing charges increases, energy increases and ionic bonding is more favorable.

12.7   See also

- Biot–Savart law

- Darwin Lagrangian

- Electromagnetic force

- Gauss's law

- Method of image charges

- Molecular modelling

- Newton's law of universal gravitation, which uses a similar structure, but for mass instead of charge

- Static forces and virtual-particle exchange

12.8   Notes

[1] Stewart, Joseph (2001). *Intermediate Electromagnetic Theory*. World Scientific. p. 50. ISBN 981-02-4471-1

[2] Simpson, Brian (2003). *Electrical Stimulation and the Relief of Pain*. Elsevier Health Sciences. pp. 6–7. ISBN 0-444-51258-6

[3] Baigrie, Brian (2006). *Electricity and Magnetism: A Historical Perspective*. Greenwood Press. pp. 7–8. ISBN 0-313-33358-0

[4] Chalmers, Gordon (1937). "The Lodestone and the Understanding of Matter in Seventeenth Century England". *Philosophy of Science*. **4** (1): 75–95. doi:10.1086/286445

[5] Socin, Abel (1760). *Acta Helvetica Physico-Mathematico-Anatomico-Botanico-Medica* (in Latin). **4**. Basileae. pp. 224, 225.

[6] Heilbron, J.L. (1979). *Electricity in the 17th and 18th Centuries: A Study of Early Modern Physics*. Los Angeles, California: University of California Press. pp. 460–462 and 464 (including footnote 44). ISBN 0486406881.

[7] Schofield, Robert E. (1997). *The Enlightenment of Joseph Priestley: A Study of his Life and Work from 1733 to 1773*. University Park: Pennsylvania State University Press. pp. 144–56. ISBN 0-271-01662-0.

[8] Priestley, Joseph (1767). *The History and Present State of Electricity, with Original Experiments*. London, England. p. 732.

> May we not infer from this experiment, that the attraction of electricity is subject to the same laws with that of gravitation, and is therefore according to the squares of the distances; since it is easily demonstrated, that were the earth in the form of a shell, a body in the inside of it would not be attracted to one side more than another?

[9] Elliott, Robert S. (1999). *Electromagnetics: History, Theory, and Applications*. ISBN 978-0-7803-5384-8.

[10] Robison, John (1822). Murray, John, ed. *A System of Mechanical Philosophy*. **4**. London, England.
On page 68, the author states that in 1769 he announced his findings regarding the force between spheres of like charge. On page 73, the author states the force between spheres of like charge varies as $x^{-2.06}$:

> The result of the whole was, that the mutual repulsion of two spheres, electrified positively or negatively, was very nearly in the inverse proportion of the squares of the distances of their centres, or rather in a proportion somewhat greater, approaching to $x^{-2.06}$.

When making experiments with charged spheres of opposite charge the results were similar, as stated on page 73:

> When the experiments were repeated with balls having opposite electricities, and which therefore attracted each other, the results were not altogether so regular and a few irregularities amounted to $1/6$ of the whole; but these anomalies were as often on one side of the medium as on the other. This series of experiments gave a result which deviated as little as the former (or rather less) from the inverse duplicate ratio of the distances; but the deviation was in defect as the other was in excess.

Nonetheless, on page 74 the author infers that the actual action is related exactly to the inverse duplicate of the distance:

> We therefore think that it may be concluded, that the action between two spheres is exactly in the inverse duplicate ratio of the distance of their centres, and that this difference between the observed attractions and repulsions is owing to some unperceived cause in the form of the experiment.

On page 75, the authour compares the electric and gravitational forces:

> Therefore we may conclude, that the law of electric attraction and repulsion is similar to that of gravitation, and that each of those forces diminishes in the same proportion that the square of the distance between the particles increases.

[11] Maxwell, James Clerk, ed. (1967) [1879]. "Experiments on Electricity: Experimental determination of the law of electric force.". *The Electrical Researches of the Honourable Henry Cavendish...* (1st ed.). Cambridge, England: Cambridge University Press. pp. 104–113.
On pages 111 and 112 the author states:

> We may therefore conclude that the electric attraction and repulsion must be inversely as some power of the distance between that of the $2 + \frac{1}{50}$ th and that of the $2 - \frac{1}{50}$ th, and there is no reason to think that it differs at all from the inverse duplicate ratio.

[12] Coulomb (1785a) "Premier mémoire sur l'électricité et le magnétisme," *Histoire de l'Académie Royale des Sciences*, pages 569-577 — Coulomb studied the repulsive force between bodies having electrical charges of the same sign:

> *Il résulte donc de ces trois essais, que l'action répulsive que les deux balles électrifées de la même nature d'électricité exercent l'une sur l'autre, suit la raison inverse du carré des distances.* Translation: It follows therefore from these three tests, that the repulsive force that the two balls — [that were] electrified with the same kind of electricity — exert on each other, follows the inverse proportion of the square of the distance.

Coulomb also showed that oppositely charged bodies obey an inverse-square law of attraction.

[13] Coulomb's law, Hyperphysics

[14] Coulomb's law, University of Texas

[15] Charged rods, PhysicsLab.org

12.9   References

- Coulomb, Charles Augustin (1788) [1785]. "Premier mémoire sur l'électricité et le magnétisme". *Histoire de l'Académie Royale des Sciences*. Imprimerie Royale. pp. 569–577.

- Coulomb, Charles Augustin (1788) [1785]. "Second mémoire sur l'électricité et le magnétisme". *Histoire de l'Académie Royale des Sciences*. Imprimerie Royale. pp. 578–611.

- Griffiths, David J. (1998). *Introduction to Electrodynamics* (3rd ed.). Prentice Hall. ISBN 0-13-805326-X.

- Tipler, Paul A.; Mosca, Gene (2008). *Physics for Scientists and Engineers* (6th ed.). New York: W. H. Freeman and Company. ISBN 0-7167-8964-7. LCCN 2007010418.

- Young, Hugh D.; Freedman, Roger A. (2010). *Sears and Zemansky's University Physics : With Modern Physics* (13th ed.). Addison-Wesley (Pearson). ISBN 978-0-321-69686-1.

12.10   External links

- *Coulomb's Law* on Project PHYSNET

- Electricity and the Atom—a chapter from an online textbook

- A maze game for teaching Coulomb's Law—a game created by the Molecular Workbench software

- Electric Charges, Polarization, Electric Force, Coulomb's Law Walter Lewin, *8.02 Electricity and Magnetism, Spring 2002: Lecture 1* (video). MIT OpenCourseWare. License: Creative Commons Attribution-Noncommercial-Share Alike.

Chapter 13

Lorentz force

In physics, particularly in electromagnetism, the **Lorentz force** is the combination of electric and magnetic force on a point charge due to electromagnetic fields. If a particle of charge q moves with velocity $\mathbf{v}$ in the presence of an electric field $\mathbf{E}$ and a magnetic field $\mathbf{B}$, then it will experience a force

$$\mathbf{F} = q\mathbf{E} + q\mathbf{v} \times \mathbf{B}$$

(in SI units). Variations on this basic formula describe the magnetic force on a current-carrying wire (sometimes called *Laplace force*), the electromotive force in a wire loop moving through a magnetic field (an aspect of Faraday's law of induction), and the force on a charged particle which might be travelling near the speed of light (relativistic form of the Lorentz force).

The first derivation of the Lorentz force is commonly attributed to Oliver Heaviside in 1889,[1] although other historians suggest an earlier origin in an 1865 paper by James Clerk Maxwell.[2] Hendrik Lorentz derived it a few years after Heaviside.

13.1 Equation (SI units)

See also: SI units

13.1.1 Charged particle

The force $\mathbf{F}$ acting on a particle of electric charge q with instantaneous velocity $\mathbf{v}$, due to an external electric field $\mathbf{E}$ and magnetic field $\mathbf{B}$, is given by:[3]

where × is the vector cross product. All boldface quantities are vectors. More explicitly stated:

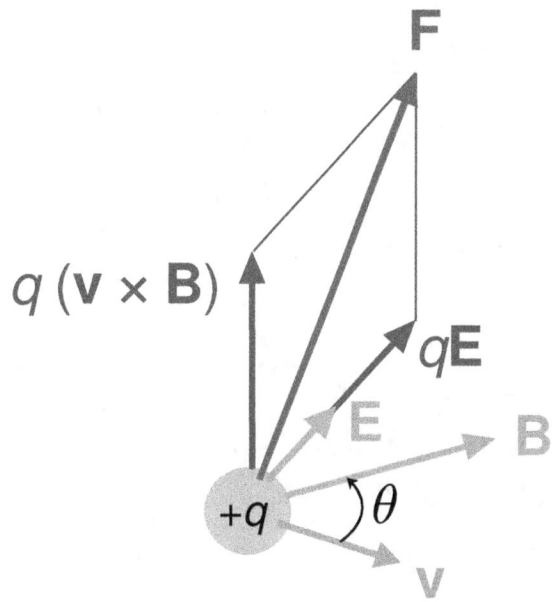

Lorentz force $\mathbf{F}$ on a charged particle (of charge q) in motion (instantaneous velocity v). The E field and B field vary in space and time.

$$\mathbf{F}(\mathbf{r}, \dot{\mathbf{r}}, t, q) = q[\mathbf{E}(\mathbf{r}, t) + \dot{\mathbf{r}} \times \mathbf{B}(\mathbf{r}, t)]$$

in which $\mathbf{r}$ is the position vector of the charged particle, t is time, and the overdot is a time derivative.

A positively charged particle will be accelerated in the *same* linear orientation as the $\mathbf{E}$ field, but will curve perpendicularly to both the instantaneous velocity vector $\mathbf{v}$ and the $\mathbf{B}$ field according to the right-hand rule (in detail, if the fingers of the right hand are extended to point in the direction of $\mathbf{v}$ and are then curled to point in the direction of $\mathbf{B}$, then the extended thumb will point in the direction of $\mathbf{F}$).

The term $q\mathbf{E}$ is called the **electric force**, while the term $q\mathbf{v} \times \mathbf{B}$ is called the **magnetic force**.[4] According to some definitions, the term "Lorentz force" refers specifically to the formula for the magnetic force,[5] with the *total* electromagnetic force (including the electric force) given some other (nonstandard) name. This article will *not* follow this nomenclature: In what follows, the term "Lorentz force" will refer only to the expression for the total force.

The magnetic force component of the Lorentz force manifests itself as the force that acts on a current-carrying wire in a magnetic field. In that context, it is also called the **Laplace force**.

13.1.2 Continuous charge distribution

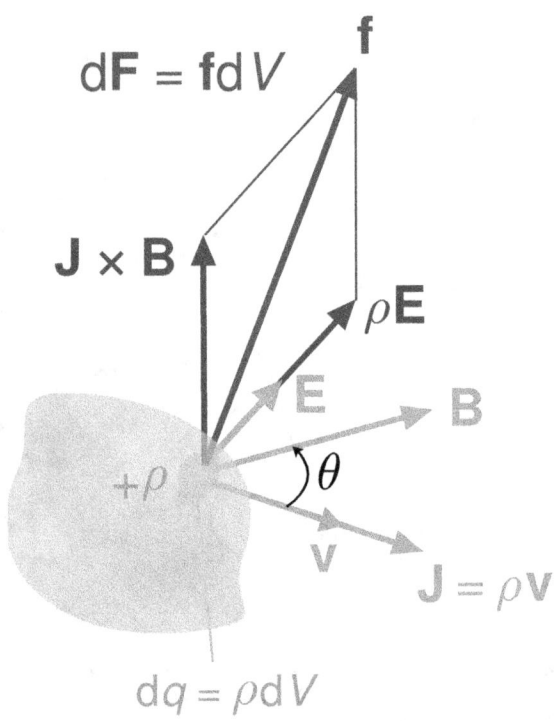

$$dF = fdV$$

Lorentz force (per unit 3-volume) f on a continuous charge distribution (charge density ϱ) in motion. The 3-current density J corresponds to the motion of the charge element dq *in volume element* dV *and varies throughout the continuum.*

For a continuous charge distribution in motion, the Lorentz force equation becomes:

$$d\mathbf{F} = dq\,(\mathbf{E} + \mathbf{v} \times \mathbf{B})$$

where $d\mathbf{F}$ is the force on a small piece of the charge distribution with charge dq. If both sides of this equation are divided by the volume of this small piece of the charge distribution dV, the result is:

$$\mathbf{f} = \rho\,(\mathbf{E} + \mathbf{v} \times \mathbf{B})$$

where $\mathbf{f}$ is the *force density* (force per unit volume) and ϱ is the charge density (charge per unit volume). Next, the current density corresponding to the motion of the charge continuum is

$$\mathbf{J} = \rho\mathbf{v}$$

so the continuous analogue to the equation is[6]

The total force is the volume integral over the charge distribution:

$$\mathbf{F} = \iiint (\rho\mathbf{E} + \mathbf{J} \times \mathbf{B})\,dV.$$

By eliminating ρ and $\mathbf{J}$, using Maxwell's equations, and manipulating using the theorems of vector calculus, this form of the equation can be used to derive the Maxwell stress tensor $\boldsymbol{\sigma}$, in turn this can be combined with the Poynting vector $\mathbf{S}$ to obtain the electromagnetic stress–energy tensor $\mathbf{T}$ used in general relativity.[6]

In terms of $\boldsymbol{\sigma}$ and $\mathbf{S}$, another way to write the Lorentz force (per unit volume) is[6]

$$\mathbf{f} = \nabla \cdot \boldsymbol{\sigma} - \frac{1}{c^2}\frac{\partial \mathbf{S}}{\partial t}$$

where c is the speed of light and $\nabla\cdot$ denotes the divergence of a tensor field. Rather than the amount of charge and its velocity in electric and magnetic fields, this equation relates the energy flux (flow of *energy* per unit time per unit distance) in the fields to the force exerted on a charge distribution. See Covariant formulation of classical electromagnetism for more details.

13.2 History

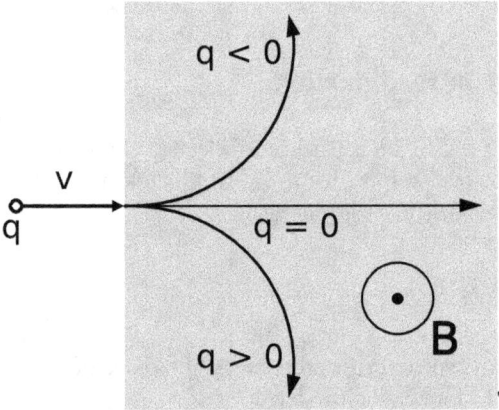

Trajectory

of a particle with a positive or negative charge q under the influence of a magnetic field B, which is directed perpendicularly out of the screen.

Beam of electrons moving in a circle, due to the presence of a magnetic field. Purple light is emitted along the electron path, due to the electrons colliding with gas molecules in the bulb. A Teltron tube is used in this example.
Charged particles experiencing the Lorentz force.

Early attempts to quantitatively describe the electromagnetic force were made in the mid-18th century. It was proposed that the force on magnetic poles, by Johann Tobias Mayer and others in 1760, and electrically charged objects, by Henry Cavendish in 1762, obeyed an inverse-square law. However, in both cases the experimental proof was neither complete nor conclusive. It was not until 1784 when Charles-Augustin de Coulomb, using a torsion balance, was able to definitively show through experiment that this was true.[7] Soon after the discovery in 1820 by H. C. Ørsted that a magnetic needle is acted on by a voltaic current, André-Marie Ampère that same year was able to devise through experimentation the formula for the angular dependence of the force between two current elements.[8][9] In all these descriptions, the force was always given in terms of the properties of the objects involved and the distances between them rather than in terms of electric and magnetic fields.[10]

The modern concept of electric and magnetic fields first arose in the theories of Michael Faraday, particularly his idea of lines of force, later to be given full mathematical description by Lord Kelvin and James Clerk Maxwell.[11] From a modern perspective it is possible to identify in Maxwell's 1865 formulation of his field equations a form of the Lorentz force equation in relation to electric currents,[2] however, in the time of Maxwell it was not evident how his equations related to the forces on moving charged objects. J. J. Thomson was the first to attempt to derive from Maxwell's field equations the electromagnetic forces on a moving charged object in terms of the object's properties and external fields. Interested in determining the electromagnetic behavior of the charged particles in cathode rays, Thomson published a paper in 1881 wherein he gave the

force on the particles due to an external magnetic field as[1]

$$\mathbf{F} = \frac{q}{2}\mathbf{v} \times \mathbf{B}.$$

Thomson derived the correct basic form of the formula, but, because of some miscalculations and an incomplete description of the displacement current, included an incorrect scale-factor of a half in front of the formula. It was Oliver Heaviside, who had invented the modern vector notation and applied them to Maxwell's field equations, that in 1885 and 1889 fixed the mistakes of Thomson's derivation and arrived at the correct form of the magnetic force on a moving charged object.[1][12][13] Finally, in 1892, Hendrik Lorentz derived the modern form of the formula for the electromagnetic force which includes the contributions to the total force from both the electric and the magnetic fields. Lorentz began by abandoning the Maxwellian descriptions of the ether and conduction. Instead, Lorentz made a distinction between matter and the luminiferous aether and sought to apply the Maxwell equations at a microscopic scale. Using Heaviside's version of the Maxwell equations for a stationary ether and applying Lagrangian mechanics (see below), Lorentz arrived at the correct and complete form of the force law that now bears his name.[14][15]

13.3 Trajectories of particles due to the Lorentz force

Main article: Guiding center
In many cases of practical interest, the motion in a magnetic field of an electrically charged particle (such as an electron or ion in a plasma) can be treated as the superposition of a relatively fast circular motion around a point called the **guiding center** and a relatively slow **drift** of this point. The drift speeds may differ for various species depending on their charge states, masses, or temperatures, possibly resulting in electric currents or chemical separation.

13.4 Significance of the Lorentz force

While the modern Maxwell's equations describe how electrically charged particles and currents or moving charged particles give rise to electric and magnetic fields, the Lorentz force law completes that picture by describing the force acting on a moving point charge q in the presence of electromagnetic fields.[3][16] The Lorentz force law describes the effect of $\mathbf{E}$ and $\mathbf{B}$ upon a point charge, but such electromagnetic forces are not the entire picture. Charged

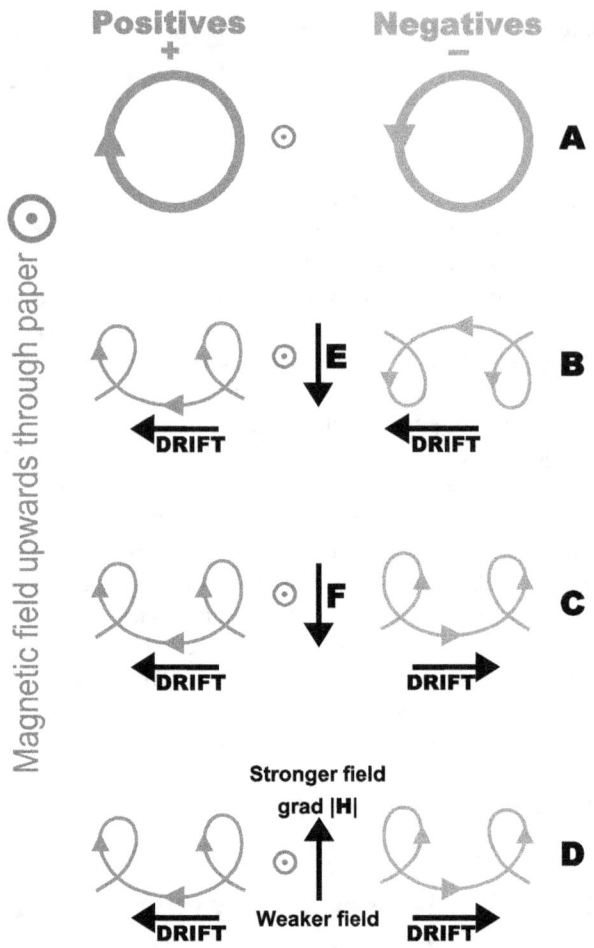

Positives +   **Negatives** −

Magnetic field upwards through paper ⊙

A

B ⊙ E DRIFT ... DRIFT

C ⊙ F DRIFT ... DRIFT

Stronger field
grad |H|

D ⊙ Weaker field DRIFT ... DRIFT

Charged particle drifts *in a homogeneous magnetic field. (A) No disturbing force (B) With an electric field, E (C) With an independent force, F (e.g. gravity) (D) In an inhomogeneous magnetic field, grad H*

Green's function (many-body theory).

13.5 Lorentz force law as the definition of E and B

In many textbook treatments of classical electromagnetism, the Lorentz force Law is used as the *definition* of the electric and magnetic fields **E** and **B**.[17][18][19] To be specific, the Lorentz force is understood to be the following empirical statement:

> *The electromagnetic force F on a test charge at a given point and time is a certain function of its charge* q *and velocity* v, *which can be parameterized by exactly two vectors E and B, in the functional form:*

$$\mathbf{F} = q(\mathbf{E} + \mathbf{v} \times \mathbf{B})$$

This is valid, even for particles approaching the speed of light (that is, magnitude of $\mathbf{v} = |\mathbf{v}| = c$).[20] So the two vector fields **E** and **B** are thereby defined throughout space and time, and these are called the "electric field" and "magnetic field". Note that the fields are defined everywhere in space and time with respect to what force a test charge would receive regardless of whether a charge is present to experience the force.

Note also that as a definition of **E** and **B**, the Lorentz force is only a definition in principle because a real particle (as opposed to the hypothetical "test charge" of infinitesimally-small mass and charge) would generate its own finite **E** and **B** fields, which would alter the electromagnetic force that it experiences. In addition, if the charge experiences acceleration, as if forced into a curved trajectory by some external agency, it emits radiation that causes braking of its motion. See for example Bremsstrahlung and synchrotron light. These effects occur through both a direct effect (called the radiation reaction force) and indirectly (by affecting the motion of nearby charges and currents). Moreover, net force must include gravity, electroweak, and any other forces aside from electromagnetic force.

13.6 Force on a current-carrying wire

When a wire carrying an electric current is placed in a magnetic field, each of the moving charges, which comprise

particles are possibly coupled to other forces, notably gravity and nuclear forces. Thus, Maxwell's equations do not stand separate from other physical laws, but are coupled to them via the charge and current densities. The response of a point charge to the Lorentz law is one aspect; the generation of **E** and **B** by currents and charges is another.

In real materials the Lorentz force is inadequate to describe the collective behavior of charged particles, both in principle and as a matter of computation. The charged particles in a material medium not only respond to the **E** and **B** fields but also generate these fields. Complex transport equations must be solved to determine the time and spatial response of charges, for example, the Boltzmann equation or the Fokker–Planck equation or the Navier–Stokes equations. For example, see magnetohydrodynamics, fluid dynamics, electrohydrodynamics, superconductivity, stellar evolution. An entire physical apparatus for dealing with these matters has developed. See for example, Green–Kubo relations and

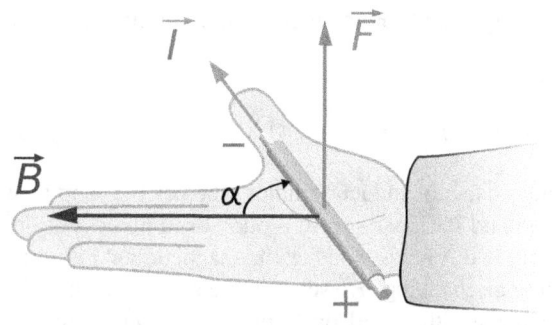

Right-hand rule for a current-carrying wire in a magnetic field B

the current, experiences the Lorentz force, and together they can create a macroscopic force on the wire (sometimes called the **Laplace force**). By combining the Lorentz force law above with the definition of electric current, the following equation results, in the case of a straight, stationary wire:

$$\mathbf{F} = I\boldsymbol{\ell} \times \mathbf{B}$$

where $\boldsymbol{\ell}$ is a vector whose magnitude is the length of wire, and whose direction is along the wire, aligned with the direction of conventional current flow I.

If the wire is not straight but curved, the force on it can be computed by applying this formula to each infinitesimal segment of wire $d\boldsymbol{\ell}$, then adding up all these forces by integration. Formally, the net force on a stationary, rigid wire carrying a steady current I is

$$\mathbf{F} = I \int d\boldsymbol{\ell} \times \mathbf{B}$$

This is the net force. In addition, there will usually be torque, plus other effects if the wire is not perfectly rigid.

One application of this is Ampère's force law, which describes how two current-carrying wires can attract or repel each other, since each experiences a Lorentz force from the other's magnetic field. For more information, see the article: Ampère's force law.

13.7 EMF

The magnetic force ($q\,\mathbf{v} \times \mathbf{B}$) component of the Lorentz force is responsible for *motional* electromotive force (or *motional EMF*), the phenomenon underlying many electrical generators. When a conductor is moved through a magnetic field, the magnetic force tries to push electrons through the wire, and this creates the EMF. The term "motional EMF"

is applied to this phenomenon, since the EMF is due to the *motion* of the wire.

In other electrical generators, the magnets move, while the conductors do not. In this case, the EMF is due to the electric force ($q\mathbf{E}$) term in the Lorentz Force equation. The electric field in question is created by the changing magnetic field, resulting in an *induced* EMF, as described by the Maxwell–Faraday equation (one of the four modern Maxwell's equations).[21]

Both of these EMF's, despite their different origins, can be described by the same equation, namely, the EMF is the rate of change of magnetic flux through the wire. (This is Faraday's law of induction, see below.) Einstein's special theory of relativity was partially motivated by the desire to better understand this link between the two effects.[21] In fact, the electric and magnetic fields are different faces of the same electromagnetic field, and in moving from one inertial frame to another, the solenoidal vector field portion of the E-field can change in whole or in part to a B-field or *vice versa*.[22]

13.8 Lorentz force and Faraday's law of induction

Main article: Faraday's law of induction

Given a loop of wire in a magnetic field, Faraday's law of induction states the induced electromotive force (EMF) in the wire is:

$$\mathcal{E} = -\frac{d\Phi_B}{dt}$$

where

$$\Phi_B = \iint_{\Sigma(t)} d\mathbf{A} \cdot \mathbf{B}(\mathbf{r}, t)$$

is the magnetic flux through the loop, $\mathbf{B}$ is the magnetic field, $\Sigma(t)$ is a surface bounded by the closed contour $\partial\Sigma(t)$, at all at time t, $d\mathbf{A}$ is an infinitesimal vector area element of $\Sigma(t)$ (magnitude is the area of an infinitesimal patch of surface, direction is orthogonal to that surface patch).

The *sign* of the EMF is determined by Lenz's law. Note that this is valid for not only a *stationary* wire — but also for a *moving* wire.

From Faraday's law of induction (that is valid for a moving wire, for instance in a motor) and the Maxwell Equations, the Lorentz Force can be deduced. The reverse is also true, the Lorentz force and the Maxwell Equations can be used to derive the Faraday Law.

Let $\Sigma(t)$ be the moving wire, moving together without rotation and with constant velocity $\mathbf{v}$ and $\Sigma(t)$ be the internal surface of the wire. The EMF around the closed path $\partial\Sigma(t)$ is given by:[23]

$$\mathcal{E} = \oint_{\partial\Sigma(t)} d\boldsymbol{\ell} \cdot \mathbf{F}/q$$

where

$$\mathbf{E} = \mathbf{F}/q$$

is the electric field and $d\boldsymbol{\ell}$ is an infinitesimal vector element of the contour $\partial\Sigma(t)$.

NB: Both $d\boldsymbol{\ell}$ and $d\mathbf{A}$ have a sign ambiguity; to get the correct sign, the right-hand rule is used, as explained in the article Kelvin–Stokes theorem.

The above result can be compared with the version of Faraday's law of induction that appears in the modern Maxwell's equations, called here the *Maxwell–Faraday equation*:

$$\nabla \times \mathbf{E} = -\frac{\partial\mathbf{B}}{\partial t} .$$

The Maxwell–Faraday equation also can be written in an *integral form* using the Kelvin–Stokes theorem.[24]

So we have, the Maxwell Faraday equation:

$$\oint_{\partial\Sigma(t)} d\boldsymbol{\ell} \cdot \mathbf{E}(\mathbf{r},\,t) = -\iint_{\Sigma(t)} d\mathbf{A} \cdot \frac{d\,\mathbf{B}(\mathbf{r},\,t)}{dt}$$

and the Faraday Law,

$$\oint_{\partial\Sigma(t)} d\boldsymbol{\ell} \cdot \mathbf{F}/q(\mathbf{r},\,t) = -\frac{d}{dt}\iint_{\Sigma(t)} d\mathbf{A} \cdot \mathbf{B}(\mathbf{r},\,t).$$

The two are equivalent if the wire is not moving. Using the Leibniz integral rule and that *div* $\mathbf{B} = 0$, results in,

$$\oint_{\partial\Sigma(t)} d\boldsymbol{\ell}\cdot\mathbf{F}/q(\mathbf{r},t) = -\iint_{\Sigma(t)} d\mathbf{A}\cdot\frac{\partial}{\partial t}\mathbf{B}(\mathbf{r},t)+\oint_{\partial\Sigma(t)} \mathbf{v}\times\mathbf{B}\,d\boldsymbol{\ell}$$

and using the Maxwell Faraday equation,

$$\oint_{\partial\Sigma(t)} d\boldsymbol{\ell}\cdot\mathbf{F}/q(\mathbf{r},\,t) = \oint_{\partial\Sigma(t)} d\boldsymbol{\ell}\cdot\mathbf{E}(\mathbf{r},\,t)+\oint_{\partial\Sigma(t)} \mathbf{v}\times\mathbf{B}(\mathbf{r},\,t)\,d\boldsymbol{\ell}$$

since this is valid for any wire position it implies that,

$$\mathbf{F} = q\,\mathbf{E}(\mathbf{r},\,t) + q\,\mathbf{v} \times \mathbf{B}(\mathbf{r},\,t).$$

Faraday's law of induction holds whether the loop of wire is rigid and stationary, or in motion or in process of deformation, and it holds whether the magnetic field is constant in time or changing. However, there are cases where Faraday's law is either inadequate or difficult to use, and application of the underlying Lorentz force law is necessary. See inapplicability of Faraday's law.

If the magnetic field is fixed in time and the conducting loop moves through the field, the magnetic flux ΦB linking the loop can change in several ways. For example, if the **B**-field varies with position, and the loop moves to a location with different **B**-field, ΦB will change. Alternatively, if the loop changes orientation with respect to the **B**-field, the **B** • d**A** differential element will change because of the different angle between **B** and d**A**, also changing ΦB. As a third example, if a portion of the circuit is swept through a uniform, time-independent **B**-field, and another portion of the circuit is held stationary, the flux linking the entire closed circuit can change due to the shift in relative position of the circuit's component parts with time (surface $\partial\Sigma(t)$ time-dependent). In all three cases, Faraday's law of induction then predicts the EMF generated by the change in ΦB.

Note that the Maxwell Faraday's equation implies that the Electric Field **E** is non conservative when the Magnetic Field **B** varies in time, and is not expressible as the gradient of a scalar field, and not subject to the gradient theorem since its rotational is not zero.[23][25]

13.9 Lorentz force in terms of potentials

See also: Mathematical descriptions of the electromagnetic field, Maxwell's equations, and Helmholtz decomposition

The **E** and **B** fields can be replaced by the magnetic vector potential **A** and (scalar) electrostatic potential ϕ by

$$\mathbf{E} = -\nabla\phi - \frac{\partial\mathbf{A}}{\partial t}$$

$$\mathbf{B} = \nabla \times \mathbf{A}$$

where ∇ is the gradient, $\nabla\bullet$ is the divergence, $\nabla\times$ is the curl.

The force becomes

$$F = q \left[-\nabla\phi - \frac{\partial \mathbf{A}}{\partial t} + \mathbf{v} \times (\nabla \times \mathbf{A}) \right]$$

and using an identity for the triple product simplifies to

using the chain rule, the total derivative of $\mathbf{A}$ is:

$$\frac{d\mathbf{A}}{dt} = \frac{\partial \mathbf{A}}{\partial t} + (\mathbf{v} \cdot \nabla)\mathbf{A}$$

so the above expression can be rewritten as;

$$F = q \left[-\nabla(\phi - \mathbf{v} \cdot \mathbf{A}) - \frac{d\mathbf{A}}{dt} \right]$$

which can take the convenient Euler–Lagrange form

13.10 Lorentz force and analytical mechanics

See also: Momentum

The Lagrangian for a charged particle of mass m and charge q in an electromagnetic field equivalently describes the dynamics of the particle in terms of its *energy*, rather than the force exerted on it. The classical expression is given by:[26]

$$L = \frac{m}{2}\mathbf{r} \cdot \mathbf{r} + q\mathbf{A} \cdot \mathbf{r} - q\phi$$

where $\mathbf{A}$ and ϕ are the potential fields as above. Using Lagrange's equations, the equation for the Lorentz force can be obtained.

The potential energy depends on the velocity of the particle, so the force is velocity dependent, so it is not conservative.

The relativistic Lagrangian is

$$L = -mc^2 \sqrt{1 - \left(\frac{\dot{\mathbf{r}}}{c}\right)^2} + e\mathbf{A}(\mathbf{r}) \cdot \dot{\mathbf{r}} - e\phi(\mathbf{r})$$

The action is the relativistic arclength of the path of the particle in space time, minus the potential energy contribution, plus an extra contribution which quantum mechanically is an extra phase a charged particle gets when it is moving along a vector potential.

13.11 Equation (cgs units)

See also: cgs units

The above-mentioned formulae use SI units which are the most common among experimentalists, technicians, and engineers. In cgs-Gaussian units, which are somewhat more common among theoretical physicists, one has instead

$$F = q_{\text{cgs}} \left(\mathbf{E}_{\text{cgs}} + \frac{\mathbf{v}}{c} \times \mathbf{B}_{\text{cgs}} \right).$$

where c is the speed of light. Although this equation looks slightly different, it is completely equivalent, since one has the following relations:

$$q_{\text{cgs}} = \frac{q_{\text{SI}}}{\sqrt{4\pi\epsilon_0}}, \quad \mathbf{E}_{\text{cgs}} = \sqrt{4\pi\epsilon_0}\,\mathbf{E}_{\text{SI}}, \quad \mathbf{B}_{\text{cgs}} = \sqrt{4\pi/\mu_0}\,\mathbf{B}_{\text{SI}}$$

where ϵ_0 is the vacuum permittivity and μ_0 the vacuum permeability. In practice, the subscripts "cgs" and "SI" are always omitted, and the unit system has to be assessed from context.

13.12 Relativistic form of the Lorentz force

13.12.1 Covariant form of the Lorentz force

Field tensor

Main articles: Covariant formulation of classical electromagnetism and Mathematical descriptions of the electromagnetic field

Using the metric signature $(1, -1, -1, -1)$, The Lorentz force for a charge q can be written in[27] covariant form:

where p^α is the four-momentum, defined as

$$p^\alpha = (p_0, p_1, p_2, p_3) = (\gamma mc, p_x, p_y, p_z) \,,$$

τ the proper time of the particle, $F^{\alpha\beta}$ the contravariant electromagnetic tensor

$$F^{\alpha\beta} = \begin{pmatrix} 0 & -E_x/c & -E_y/c & -E_z/c \\ E_x/c & 0 & -B_z & B_y \\ E_y/c & B_z & 0 & -B_x \\ E_z/c & -B_y & B_x & 0 \end{pmatrix}$$

and U is the covariant 4-velocity of the particle, defined as:

$$U_\beta = (U_0, U_1, U_2, U_3) = \gamma \left(c, -v_x, -v_y, -v_z \right) \,,$$

in which

$$\gamma(v) = \frac{1}{\sqrt{1 - \frac{v^2}{c^2}}} = \frac{1}{\sqrt{1 - \frac{v_x^2 + v_y^2 + v_z^2}{c^2}}}$$

is the Lorentz factor.

The fields are transformed to a frame moving with constant relative velocity by:

$$F'^{\mu\nu} = \Lambda^\mu{}_\alpha \Lambda^\nu{}_\beta F^{\alpha\beta} \,,$$

where $\Lambda^\mu{}_\alpha$ is the Lorentz transformation tensor.

Translation to vector notation

The $\alpha = 1$ component (x-component) of the force is

$$\frac{dp^1}{d\tau} = q U_\beta F^{1\beta} = q \left(U_0 F^{10} + U_1 F^{11} + U_2 F^{12} + U_3 F^{13} \right).$$

Substituting the components of the covariant electromagnetic tensor F yields

$$\frac{dp^1}{d\tau} = q \left[U_0 \left(\frac{-E_x}{c} \right) + U_2 (B_z) + U_3 (-B_y) \right].$$

Using the components of covariant four-velocity yields

$$\frac{dp^1}{d\tau} = q\gamma \left[-c \left(\frac{-E_x}{c} \right) + v_y B_z + v_z (-B_y) \right]$$
$$= q\gamma \left(E_x + v_y B_z - v_z B_y \right)$$
$$= q\gamma \left[E_x + (\mathbf{v} \times \mathbf{B})_x \right].$$

The calculation for $\alpha = 2, 3$ (force components in the y and z directions) yields similar results, so collecting the 3 equations into one:

$$\frac{d\mathbf{p}}{d\tau} = q\gamma \left(\mathbf{E} + \mathbf{v} \times \mathbf{B} \right) \,,$$

and since differentials in coordinate time dt and proper time $d\tau$ are related by the Lorentz factor,

$$dt = \gamma(v) d\tau \,,$$

so we arrive at

$$\frac{d\mathbf{p}}{dt} = q \left(\mathbf{E} + \mathbf{v} \times \mathbf{B} \right) \,.$$

This is precisely the Lorentz force law, however, it is important to note that $\mathbf{p}$ is the relativistic expression,

$$\mathbf{p} = \gamma(v) m_0 \mathbf{v} \,.$$

13.12.2 Lorentz force in spacetime algebra (STA)

The electric and magnetic fields are dependent on the velocity of an observer, so the relativistic form of the Lorentz force law can best be exhibited starting from a coordinate-independent expression for the electromagnetic and magnetic fields $\mathcal{F}$, and an arbitrary time-direction, γ_0 . This can be settled through Space-Time Algebra (or the geometric algebra of space-time), a type of Clifford's Algebra defined on a pseudo-euclidian space,[28] as

$$\mathbf{E} = (\mathcal{F} \cdot \gamma_0) \gamma_0$$

and

$$i\mathbf{B} = (\mathcal{F} \wedge \gamma_0) \gamma_0$$

$\mathcal{F}$ is a space-time bivector (an oriented plane segment, just like a vector is an oriented line segment), which has six degrees of freedom corresponding to boosts (rotations in space-time planes) and rotations (rotations in space-space planes). The dot product with the vector γ_0 pulls a vector (in the space algebra) from the translational part, while the wedge-product creates a trivector (in the space algebra) who is dual to a vector which is the usual magnetic field vector.

The relativistic velocity is given by the (time-like) changes in a time-position vector $v = \dot{x}$, where

$$v^2 = 1,$$

(which shows our choice for the metric) and the velocity is

$$\mathbf{v} = cv \wedge \gamma_0/(v \cdot \gamma_0).$$

The proper (invariant is an inadequate term because no transformation has been defined) form of the Lorentz force law is simply

Note that the order is important because between a bivector and a vector the dot product is anti-symmetric. Upon a space time split like one can obtain the velocity, and fields as above yielding the usual expression.

13.13   Applications

The Lorentz force occurs in many devices, including:

- Cyclotrons and other circular path particle accelerators
- Mass spectrometers
- Velocity Filters
- Magnetrons
- Lorentz force velocimetry

In its manifestation as the Laplace force on an electric current in a conductor, this force occurs in many devices including:

- Electric motors
- Railguns
- Linear motors
- Loudspeakers}
- Magnetoplasmadynamic thrusters
- Electrical generators
- Homopolar generators
- Linear alternators

13.14   See also

- Hall effect
- Electromagnetism
- Gravitomagnetism
- Ampère's force law
- Hendrik Lorentz
- Maxwell's equations
- Formulation of Maxwell's equations in special relativity
- Moving magnet and conductor problem
- Abraham–Lorentz force
- Larmor formula
- Cyclotron radiation
- Magnetic potential
- Magnetoresistance
- Scalar potential
- Helmholtz decomposition
- Guiding center
- Field line
- Coulomb's law

13.15   Footnotes

[1] Oliver Heaviside By Paul J. Nahin, p120

[2] Huray, Paul G. (2009). *Maxwell's Equations*. Wiley-IEEE. p. 22. ISBN 0-470-54276-4.

[3] See Jackson page 2. The book lists the four modern Maxwell's equations, and then states, "Also essential for consideration of charged particle motion is the Lorentz force equation, $\mathbf{F} = q(\mathbf{E} + \mathbf{v} \times \mathbf{B})$, which gives the force acting on a point charge q in the presence of electromagnetic fields."

[4] See Griffiths page 204.

[5] For example, see the website of the "Lorentz Institute": \, or Griffiths.

[6] Griffiths, David J. (1999). *Introduction to electrodynamics*. reprint. with corr. (3rd ed.). Upper Saddle River, New Jersey [u.a.]: Prentice Hall. ISBN 978-0-13-805326-0.

[7] Meyer, Herbert W. (1972). *A History of Electricity and Magnetism*. Norwalk, Connecticut: Burndy Library. pp. 30–31. ISBN 0-262-13070-X.

[8] Verschuur, Gerrit L. (1993). *Hidden Attraction : The History And Mystery Of Magnetism*. New York: Oxford University Press. pp. 78–79. ISBN 0-19-506488-7.

[9] Darrigol, Olivier (2000). *Electrodynamics from Ampère to Einstein*. Oxford, [England]: Oxford University Press. pp. 9, 25. ISBN 0-19-850593-0.

[10] Verschuur, Gerrit L. (1993). *Hidden Attraction : The History And Mystery Of Magnetism*. New York: Oxford University Press. p. 76. ISBN 0-19-506488-7.

[11] Darrigol, Olivier (2000). *Electrodynamics from Ampère to Einstein*. Oxford, [England]: Oxford University Press. pp. 126–131, 139–144. ISBN 0-19-850593-0.

[12] Darrigol, Olivier (2000). *Electrodynamics from Ampère to Einstein*. Oxford, [England]: Oxford University Press. pp. 200, 429–430. ISBN 0-19-850593-0.

[13] Heaviside, Oliver. "On the Electromagnetic Effects due to the Motion of Electrification through a Dielectric". *Philosophical Magazine, April 1889, p. 324.*

[14] Darrigol, Olivier (2000). *Electrodynamics from Ampère to Einstein*. Oxford, [England]: Oxford University Press. p. 327. ISBN 0-19-850593-0.

[15] Whittaker, E. T. (1910). *A History of the Theories of Aether and Electricity: From the Age of Descartes to the Close of the Nineteenth Century*. Longmans, Green and Co. pp. 420–423. ISBN 1-143-01208-9.

[16] See Griffiths page 326, which states that Maxwell's equations, "together with the [Lorentz] force law...summarize the entire theoretical content of classical electrodynamics".

[17] See, for example, Jackson pp 777–8.

[18] J.A. Wheeler; C. Misner; K.S. Thorne (1973). *Gravitation*. W.H. Freeman & Co. pp. 72–73. ISBN 0-7167-0344-0.. These authors use the Lorentz force in tensor form as definer of the electromagnetic tensor *F*, in turn the fields **E** and **B**.

[19] I.S. Grant; W.R. Phillips; Manchester Physics (2008). *Electromagnetism* (2nd ed.). John Wiley & Sons. p. 122. ISBN 978-0-471-92712-9.

[20] I.S. Grant; W.R. Phillips; Manchester Physics (2008). *Electromagnetism (2nd Edition)*. John Wiley & Sons. p. 123. ISBN 978-0-471-92712-9.

[21] See Griffiths pages 301–3.

[22] Tai L. Chow (2006). *Electromagnetic theory*. Sudbury MA: Jones and Bartlett. p. 395. ISBN 0-7637-3827-1.

[23] Landau, L. D., Lifshiţs, E. M., & Pitaevskiĭ, L. P. (1984). *Electrodynamics of continuous media; Volume 8* Course of Theoretical Physics (Second ed.). Oxford: Butterworth-Heinemann. p. §63 (§49 pp. 205–207 in 1960 edition). ISBN 0-7506-2634-8.

[24] Roger F Harrington (2003). *Introduction to electromagnetic engineering*. Mineola, New York: Dover Publications. p. 56. ISBN 0-486-43241-6.

[25] M N O Sadiku (2007). *Elements of electromagnetics* (Fourth ed.). NY/Oxford: Oxford University Press. p. 391. ISBN 0-19-530048-3.

[26] Classical Mechanics (2nd Edition), T.W.B. Kibble, European Physics Series, Mc Graw Hill (UK), 1973, ISBN 0-07-084018-0.

[27] Jackson, J.D. Chapter 11

[28] Hestenes, David. "SpaceTime Calculus".

13.16 References

The numbered references refer in part to the list immediately below.

- Feynman, Richard Phillips; Leighton, Robert B.; Sands, Matthew L. (2006). *The Feynman lectures on physics (3 vol.)*. Pearson / Addison-Wesley. ISBN 0-8053-9047-2.: volume 2.

- Griffiths, David J. (1999). *Introduction to electrodynamics* (3rd ed.). Upper Saddle River, [NJ.]: Prentice-Hall. ISBN 0-13-805326-X.

- Jackson, John David (1999). *Classical electrodynamics* (3rd ed.). New York, [NY.]: Wiley. ISBN 0-471-30932-X.

- Serway, Raymond A.; Jewett, John W., Jr. (2004). *Physics for scientists and engineers, with modern physics*. Belmont, [CA.]: Thomson Brooks/Cole. ISBN 0-534-40846-X.

- Srednicki, Mark A. (2007). *Quantum field theory*. Cambridge, [England] ; New York [NY.]: Cambridge University Press. ISBN 978-0-521-86449-7.

13.17 External links

- Interactive Java tutorial on the Lorentz force National High Magnetic Field Laboratory

- Lorentz force (demonstration)

- Faraday's law: Tankersley and Mosca

- Notes from Physics and Astronomy HyperPhysics at Georgia State University; see also home page

- Interactive Java applet on the magnetic deflection of a particle beam in a homogeneous magnetic field by Wolfgang Bauer

Chapter 14

Electromagnetism

"Electromagnetic" redirects here. Electromagnetic may also refer to the use of an electromagnet.

"Electromagnetic Force" redirects here. For a description of the force exerted on particles due to electromagnetic fields, see Lorentz Force.

Electromagnetism is a branch of physics which involves the study of the **electromagnetic force**, a type of physical interaction that occurs between electrically charged particles. The electromagnetic force usually exhibits electromagnetic fields, such as electric fields, magnetic fields, and light. The electromagnetic force is one of the four fundamental interactions (commonly called forces) in nature. The other three fundamental interactions are the strong interaction, the weak interaction, and gravitation.[1]

Lightning is an electrostatic discharge that travels between two charged regions.

The word *electromagnetism* is a compound form of two Greek terms, ἤλεκτρον, *ēlektron*, "amber", and μαγνῆτις λίθος *magnētis lithos*, which means "magnesian stone", a type of iron ore. Electromagnetic phenomena is defined in terms of the electromagnetic force, sometimes called the Lorentz force, which includes both electricity and magnetism as different manifestations of the same phe-

nomenon.

The electromagnetic force plays a major role in determining the internal properties of most objects encountered in daily life. Ordinary matter takes its form as a result of intermolecular forces between individual atoms and molecules in matter, and are a manifestation of the electromagnetic force. Electrons are bound by the electromagnetic force to atomic nuclei, and their orbital shapes and their influence on nearby atoms with their electrons is described by quantum mechanics. The electromagnetic force governs the processes involved in chemistry, which arise from interactions between the electrons of neighboring atoms.

There are numerous mathematical descriptions of the electromagnetic field. In classical electrodynamics, electric fields are described as electric potential and electric current. In Faraday's law, magnetic fields are associated with electromagnetic induction and magnetism, and Maxwell's equations describe how electric and magnetic fields are generated and altered by each other and by charges and currents.

The theoretical implications of electromagnetism, in particular the establishment of the speed of light based on properties of the "medium" of propagation (permeability and permittivity), led to the development of special relativity by Albert Einstein in 1905.

Although electromagnetism is considered one of the four fundamental forces, at high energy the weak force and electromagnetic force are unified as a single electroweak force. In the history of the universe, during the quark epoch the unified force broke into the two separate forces as the universe cooled.

14.1 History of the theory

See also: History of electromagnetic theory

Originally, electricity and magnetism were thought of as

Hans Christian Ørsted.

two separate forces. This view changed, however, with the publication of James Clerk Maxwell's 1873 *A Treatise on Electricity and Magnetism* in which the interactions of positive and negative charges were shown to be mediated by one force. There are four main effects resulting from these interactions, all of which have been clearly demonstrated by experiments:

1. Electric charges attract or repel one another with a force inversely proportional to the square of the distance between them: unlike charges attract, like ones repel.

2. Magnetic poles (or states of polarization at individual points) attract or repel one another in a manner similar to positive and negative charges and always exist as pairs: every north pole is yoked to a south pole.

3. An electric current inside a wire creates a corresponding circumferential magnetic field outside the wire. Its direction (clockwise or counter-clockwise) depends on the direction of the current in the wire.

4. A current is induced in a loop of wire when it is moved toward or away from a magnetic field, or a magnet is moved towards or away from it; the direction of current depends on that of the movement.

André-Marie Ampère

While preparing for an evening lecture on 21 April 1820, Hans Christian Ørsted made a surprising observation. As he was setting up his materials, he noticed a compass needle deflected away from magnetic north when the electric current from the battery he was using was switched on and off. This deflection convinced him that magnetic fields radiate from all sides of a wire carrying an electric current, just as light and heat do, and that it confirmed a direct relationship between electricity and magnetism.

At the time of discovery, Ørsted did not suggest any satisfactory explanation of the phenomenon, nor did he try to represent the phenomenon in a mathematical framework. However, three months later he began more intensive investigations. Soon thereafter he published his findings, proving that an electric current produces a magnetic field as it flows through a wire. The CGS unit of magnetic induction (oersted) is named in honor of his contributions to the field of electromagnetism.

His findings resulted in intensive research throughout the scientific community in electrodynamics. They influenced French physicist André-Marie Ampère's developments of a single mathematical form to represent the magnetic forces between current-carrying conductors. Ørsted's discovery

Michael Faraday

James Clerk Maxwell

published in 1802 in an Italian newspaper, but it was largely overlooked by the contemporary scientific community.[2]

14.2 Fundamental forces

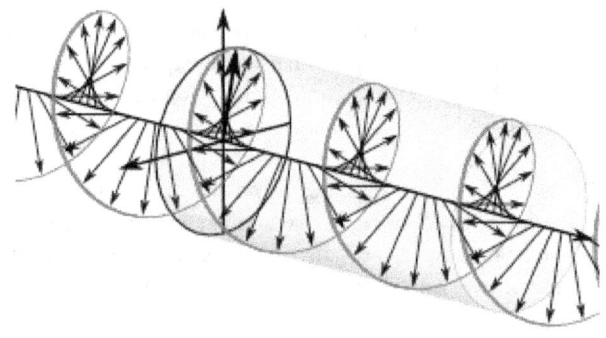

Representation of the electric field vector of a wave of circularly polarized electromagnetic radiation.

also represented a major step toward a unified concept of energy.

This unification, which was observed by Michael Faraday, extended by James Clerk Maxwell, and partially reformulated by Oliver Heaviside and Heinrich Hertz, is one of the key accomplishments of 19th century mathematical physics. It had far-reaching consequences, one of which was the understanding of the nature of light. Unlike what was proposed by electromagnetic theory of that time, light and other electromagnetic waves are at the present seen as taking the form of quantized, self-propagating oscillatory electromagnetic field disturbances called photons. Different frequencies of oscillation give rise to the different forms of electromagnetic radiation, from radio waves at the lowest frequencies, to visible light at intermediate frequencies, to gamma rays at the highest frequencies.

Ørsted was not the only person to examine the relationship between electricity and magnetism. In 1802, Gian Domenico Romagnosi, an Italian legal scholar, deflected a magnetic needle using electrostatic charges. Actually, no galvanic current existed in the setup and hence no electromagnetism was present. An account of the discovery was

The electromagnetic force is one of the four known fundamental forces. The other fundamental forces are:

- the weak nuclear force, which binds to all known particles in the Standard Model, and causes certain forms of radioactive decay. (In particle physics though, the electroweak interaction is the unified description of two of the four known fundamental interactions of nature: electromagnetism and the weak interaction);

- the strong nuclear force, which binds quarks to form nucleons, and binds nucleons to form nuclei

- the gravitational force.

All other forces (e.g., friction, contact forces) are derived from these four fundamental forces (including momentum which is carried by the movement of particles).

The electromagnetic force is the one responsible for practically all the phenomena one encounters in daily life above the nuclear scale, with the exception of gravity. Roughly speaking, all the forces involved in interactions between atoms can be explained by the electromagnetic force acting between the electrically charged atomic nuclei and electrons of the atoms. Electromagnetic forces also explain how these particles carry momentum by their movement. This includes the forces we experience in "pushing" or "pulling" ordinary material objects, which result from the intermolecular forces that act between the individual molecules in our bodies and those in the objects. The electromagnetic force is also involved in all forms of chemical phenomena.

A necessary part of understanding the intra-atomic and intermolecular forces is the effective force generated by the momentum of the electrons' movement, such that as electrons move between interacting atoms they carry momentum with them. As a collection of electrons becomes more confined, their minimum momentum necessarily increases due to the Pauli exclusion principle. The behaviour of matter at the molecular scale including its density is determined by the balance between the electromagnetic force and the force generated by the exchange of momentum carried by the electrons themselves.

14.3   Classical electrodynamics

Main article: Classical electrodynamics

The scientist William Gilbert proposed, in his *De Magnete* (1600), that electricity and magnetism, while both capable of causing attraction and repulsion of objects, were distinct effects. Mariners had noticed that lightning strikes had the ability to disturb a compass needle, but the link between lightning and electricity was not confirmed until Benjamin Franklin's proposed experiments in 1752. One of the first to discover and publish a link between man-made electric

current and magnetism was Romagnosi, who in 1802 noticed that connecting a wire across a voltaic pile deflected a nearby compass needle. However, the effect did not become widely known until 1820, when Ørsted performed a similar experiment.[3] Ørsted's work influenced Ampère to produce a theory of electromagnetism that set the subject on a mathematical foundation.

A theory of electromagnetism, known as classical electromagnetism, was developed by various physicists over the course of the 19th century, culminating in the work of James Clerk Maxwell, who unified the preceding developments into a single theory and discovered the electromagnetic nature of light. In classical electromagnetism, the behavior of the electromagnetic field is described by a set of equations known as Maxwell's equations, and the electromagnetic force is given by the Lorentz force law.

One of the peculiarities of classical electromagnetism is that it is difficult to reconcile with classical mechanics, but it is compatible with special relativity. According to Maxwell's equations, the speed of light in a vacuum is a universal constant that is dependent only on the electrical permittivity and magnetic permeability of free space. This violates Galilean invariance, a long-standing cornerstone of classical mechanics. One way to reconcile the two theories (electromagnetism and classical mechanics) is to assume the existence of a luminiferous aether through which the light propagates. However, subsequent experimental efforts failed to detect the presence of the aether. After important contributions of Hendrik Lorentz and Henri Poincaré, in 1905, Albert Einstein solved the problem with the introduction of special relativity, which replaced classical kinematics with a new theory of kinematics compatible with classical electromagnetism. (For more information, see History of special relativity.)

In addition, relativity theory implies that in moving frames of reference a magnetic field transforms to a field with a nonzero electric component and conversely, a moving electric field transforms to a nonzero magnetic component, thus firmly showing that the phenomena are two sides of the same coin. Hence the term "electromagnetism". (For more information, see Classical electromagnetism and special relativity and Covariant formulation of classical electromagnetism.

14.4   Quantum mechanics

14.4.1   Photoelectric effect

Main article: Photoelectric effect

In a second paper published in 1905, Albert Einstein undermined the very foundations of classical electromagnetism. In his theory of the photoelectric effect, for which he won the Nobel prize in physics, he posited that light could exist in discrete particle-like quantities, which later came to be called photons. Einstein's theory of the photoelectric effect extended the insights that appeared in the solution of the ultraviolet catastrophe presented by Max Planck in 1900 and who coined the term "quanta" . In his work, Planck showed that hot objects emit electromagnetic radiation in discrete packets ("quanta"), which leads to a finite total energy emitted as black body radiation. Both of these results were in direct contradiction with the classical view of light as a continuous wave. Planck's and Einstein's theories were progenitors of quantum mechanics, which, when formulated in 1925, necessitated the invention of a quantum theory of electromagnetism. This theory, completed in the 1940s-1950s, is known as quantum electrodynamics (or "QED"), and, in situations where perturbation theory can be applied, is one of the most accurate theories known to physics.

14.4.2 Quantum electrodynamics

Main article: Quantum electrodynamics

All electromagnetic phenomena can be explained in terms of quantum mechanics, specifically by quantum electrodynamics (which includes classical electrodynamics as a limiting case) and this accounts for almost all physical phenomena observable to the unaided human senses, including electromagnetic radiation (light), all of chemistry, most of mechanics (excepting gravitation), and, of course, magnetism and electricity.

14.4.3 Electroweak interaction

Main article: Electroweak interaction

The **electroweak interaction** is the unified field theory description of two of the four known fundamental interactions of nature: electromagnetism and the weak interaction. Although these two forces appear very different at everyday low energies, the theory models them as two different aspects of the same force. At energies greater than 100 GeV, called the unification energy, the two forces merge into a single **electroweak force**. Thus when the universe was hot enough (approximately 10^{15} K, a temperature that was exceeded until shortly after the Big Bang) the electromagnetic force and weak force were merged into the electroweak force. With the cooling of the universe, during the electroweak epoch, the electroweak force separated from the strong force. Following that it was still too hot for quarks to combine into hadrons and they moved about freely.

14.5 Quantities and units

See also: List of physical quantities and List of electromagnetism equations

Electromagnetic units are part of a system of electrical units based primarily upon the magnetic properties of electric currents, the fundamental SI unit being the ampere. The units are:

- ampere (electric current)

- coulomb (electric charge)

- farad (capacitance)

- henry (inductance)

- ohm (resistance)

- siemens (conductance)

- tesla (magnetic flux density)

- volt (electric potential)

- watt (power)

- weber (magnetic flux)

In the electromagnetic cgs system, electric current is a fundamental quantity defined via Ampère's law and takes the permeability as a dimensionless quantity (relative permeability) whose value in a vacuum is unity. As a consequence, the square of the speed of light appears explicitly in some of the equations interrelating quantities in this system.

Formulas for physical laws of electromagnetism (such as Maxwell's equations) need to be adjusted depending on what system of units one uses. This is because there is no one-to-one correspondence between electromagnetic units in SI and those in CGS, as is the case for mechanical units. Furthermore, within CGS, there are several plausible choices of electromagnetic units, leading to different unit "sub-systems", including Gaussian, "ESU", "EMU", and Heaviside–Lorentz. Among these choices, Gaussian units are the most common today, and in fact the phrase "CGS units" is often used to refer specifically to CGS-Gaussian units.

14.6   See also

- Abraham–Lorentz force
- Aeromagnetic surveys
- Computational electromagnetics
- Double-slit experiment
- Electromagnet
- Electromagnetic induction
- Electromagnetic wave equation
- Electromechanics
- Geophysics
- Magnetostatics
- Magnetoquasistatic field
- Optics
- Relativistic electromagnetism
- Wheeler–Feynman absorber theory

14.7   References

[1] Ravaioli, Fawwaz T. Ulaby, Eric Michielssen, Umberto (2010). *Fundamentals of applied electromagnetics* (6th ed.). Boston: Prentice Hall. p. 13. ISBN 978-0-13-213931-1.

[2] Martins, Roberto de Andrade. "Romagnosi and Volta's Pile: Early Difficulties in the Interpretation of Voltaic Electricity". In Fabio Bevilacqua and Lucio Fregonese (eds). *Nuova Voltiana: Studies on Volta and his Times* (PDF). vol. 3. Università degli Studi di Pavia. pp. 81–102. Retrieved 2010-12-02.

[3] Stern, Dr. David P.; Peredo, Mauricio (2001-11-25). "Magnetic Fields -- History". NASA Goddard Space Flight Center. Retrieved 2009-11-27.

[4] International Union of Pure and Applied Chemistry (1993). *Quantities, Units and Symbols in Physical Chemistry*, 2nd edition, Oxford: Blackwell Science. ISBN 0-632-03583-8. pp. 14–15. Electronic version.

14.8   Further reading

14.8.1   Web sources

- Nave, R. "Electricity and magnetism". *HyperPhysics*. Georgia State University. Retrieved 2013-11-12.

- Khutoryansky, E. "Electromagnetism - Maxwell's Laws". Retrieved 2014-12-28.

14.8.2   Textbooks

- G.A.G. Bennet (1974). *Electricity and Modern Physics* (2nd ed.). Edward Arnold (UK). ISBN 0-7131-2459-8.

- Dibner, Bern (2012). *Oersted and the discovery of electromagnetism*. Literary Licensing, LLC. ISBN 9781258335557.

- Durney, Carl H.; Johnson, Curtis C. (1969). *Introduction to modern electromagnetics*. McGraw-Hill. ISBN 0-07-018388-0.

- Feynman, Richard P. (1970). *The Feynman Lectures on Physics Vol II*. Addison Wesley Longman. ISBN 978-0-201-02115-8.

- Fleisch, Daniel (2008). *A Student's Guide to Maxwell's Equations*. Cambridge, UK: Cambridge University Press. ISBN 978-0-521-70147-1.

- I.S. Grant; W.R. Phillips; Manchester Physics (2008). *Electromagnetism* (2nd ed.). John Wiley & Sons. ISBN 978-0-471-92712-9.

- Griffiths, David J. (1998). *Introduction to Electrodynamics* (3rd ed.). Prentice Hall. ISBN 0-13-805326-X.

- Jackson, John D. (1998). *Classical Electrodynamics* (3rd ed.). Wiley. ISBN 0-471-30932-X.

- Moliton, André (2007). *Basic electromagnetism and materials. 430 pages*. New York City: Springer-Verlag New York, LLC. ISBN 978-0-387-30284-3.

- Purcell, Edward M. (1985). *Electricity and Magnetism Berkeley Physics Course Volume 2 (2nd ed.)*. McGraw-Hill. ISBN 0-07-004908-4.

- Rao, Nannapaneni N. (1994). *Elements of engineering electromagnetics (4th ed.)*. Prentice Hall. ISBN 0-13-948746-8.

- Rothwell, Edward J.; Cloud, Michael J. (2001). *Electromagnetics*. CRC Press. ISBN 0-8493-1397-X.

- Tipler, Paul (1998). *Physics for Scientists and Engineers: Vol. 2: Light, Electricity and Magnetism* (4th ed.). W. H. Freeman. ISBN 1-57259-492-6.

- Wangsness, Roald K.; Cloud, Michael J. (1986). *Electromagnetic Fields (2nd Edition)*. Wiley. ISBN 0-471-81186-6.

14.8.3 General references

- A. Beiser (1987). *Concepts of Modern Physics* (4th ed.). McGraw-Hill (International). ISBN 0-07-100144-1.

- L.H. Greenberg (1978). *Physics with Modern Applications*. Holt-Saunders International W.B. Saunders and Co. ISBN 0-7216-4247-0.

- R.G. Lerner; G.L. Trigg (2005). *Encyclopaedia of Physics* (2nd ed.). VHC Publishers, Hans Warlimont, Springer. pp. 12–13. ISBN 978-0-07-025734-4.

- J.B. Marion; W.F. Hornyak (1984). *Principles of Physics*. Holt-Saunders International Saunders College. ISBN 4-8337-0195-2.

- H.J. Pain (1983). *The Physics of Vibrations and Waves* (3rd ed.). John Wiley & Sons,. ISBN 0-471-90182-2.

- C.B. Parker (1994). *McGraw Hill Encyclopaedia of Physics* (2nd ed.). McGraw Hill. ISBN 0-07-051400-3.

- R. Penrose (2007). *The Road to Reality*. Vintage books. ISBN 0-679-77631-1.

- P.A. Tipler; G. Mosca (2008). *Physics for Scientists and Engineers: With Modern Physics* (6th ed.). W.H. Freeman and Co. ISBN 9-781429-202657.

- P.M. Whelan; M.J. Hodgeson (1978). *Essential Principles of Physics* (2nd ed.). John Murray. ISBN 0-7195-3382-1.

14.9 External links

- Oppelt, Arnulf (2006-11-02). "magnetic field strength". Retrieved 2007-06-04.

- "magnetic field strength converter". Retrieved 2007-06-04.

- Electromagnetic Force - from Eric Weisstein's World of Physics

- Goudarzi, Sara (2006-08-15). "Ties That Bind Atoms Weaker Than Thought". *LiveScience.com*. Retrieved 2013-11-12.

- Quarked Electromagnetic force - A good introduction for kids

- The Deflection of a Magnetic Compass Needle by a Current in a Wire (video) on YouTube

- Electromagnetism abridged

Chapter 15

Gravity

For other uses, see Gravity (disambiguation).
"Gravitation" and "Law of Gravity" redirect here. For other uses, see Gravitation (disambiguation) and Law of Gravity (disambiguation).

Gravity or **gravitation** is a natural phenomenon by which

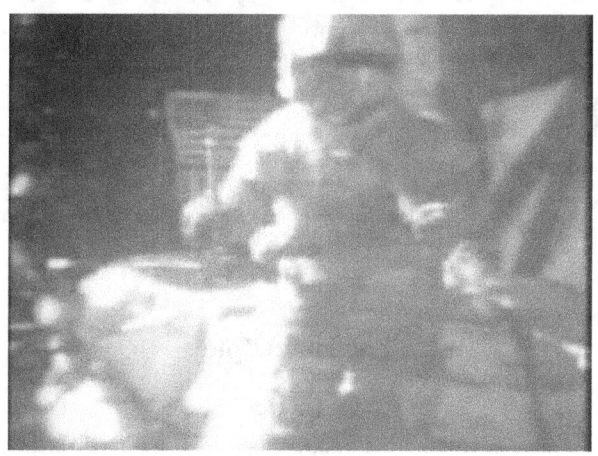

Hammer and feather drop: Apollo 15 astronaut David Scott on the Moon enacting the legend of Galileo's gravity experiment. (1.38 MB, ogg/Theora format).

all things with energy are brought toward (or *gravitate* toward) one another, including stars, planets, galaxies and even light and sub-atomic particles. Gravity is responsible for many of the structures in the Universe, by creating spheres of hydrogen — where hydrogen fuses under pressure to form stars — and grouping them into galaxies. On Earth, gravity gives weight to physical objects and causes the tides. Gravity has an infinite range, although its effects become increasingly weaker on farther objects.

Gravity is most accurately described by the general theory of relativity (proposed by Albert Einstein in 1915) which describes gravity not as a force but as a consequence of the curvature of spacetime caused by the uneven distribution of mass/energy; and resulting in gravitational time dilation, where time lapses more slowly in lower (stronger) gravitational potential. However, for most applications,

gravity is well approximated by Newton's law of universal gravitation, which postulates that gravity causes a force where two bodies of mass are directly drawn (or 'attracted') to each other according to a mathematical relationship, where the attractive force is proportional to the product of their masses and inversely proportional to the square of the distance between them. This is considered to occur over an infinite range, such that all bodies (with mass) in the universe are drawn to each other no matter how far they are apart.

Gravity is the weakest of the four fundamental interactions of nature. The gravitational attraction is approximately 10^{-38} times the strength of the strong force (i.e. gravity is 38 orders of magnitude weaker), 10^{-36} times the strength of the electromagnetic force, and 10^{-29} times the strength of the weak force. As a consequence, gravity has a negligible influence on the behavior of subatomic particles, and plays no role in determining the internal properties of everyday matter (but see quantum gravity). On the other hand, gravity is the dominant interaction at the macroscopic scale, and is the cause of the formation, shape, and trajectory (orbit) of astronomical bodies. It is responsible for various phenomena observed on Earth and throughout the universe; for example, it causes the Earth and the other planets to orbit the Sun, the Moon to orbit the Earth, the formation of tides, and the formation and evolution of galaxies, stars and the Solar System.

In pursuit of a theory of everything, the merging of general relativity and quantum mechanics (or quantum field theory) into a more general theory of quantum gravity has become an area of research.

15.1 History of gravitational theory

Main article: History of gravitational theory

15.1.1 Scientific revolution

Modern work on gravitational theory began with the work of Galileo Galilei in the late 16th and early 17th centuries. In his famous (though possibly apocryphal[1]) experiment dropping balls from the Tower of Pisa, and later with careful measurements of balls rolling down inclines, Galileo showed that gravitational acceleration is the same for all objects. This was a major departure from Aristotle's belief that heavier objects have a higher gravitational acceleration.[2] Galileo postulated air resistance as the reason that objects with less mass may fall slower in an atmosphere. Galileo's work set the stage for the formulation of Newton's theory of gravity.

15.1.2 Newton's theory of gravitation

Main article: Newton's law of universal gravitation

In 1687, English mathematician Sir Isaac Newton published-

Sir Isaac Newton, an English physicist who lived from 1642 to 1727

lished *Principia*, which hypothesizes the inverse-square law of universal gravitation. In his own words, "I deduced that the forces which keep the planets in their orbs must [be] reciprocally as the squares of their distances from the centers about which they revolve: and thereby compared the force requisite to keep the Moon in her Orb with the force of gravity at the surface of the Earth; and found them answer pretty nearly."[3] The equation is the following:

$$F = G\frac{m_1 m_2}{r^2}$$

Where F is the force, m_1 and m_2 are the masses of the objects interacting, r is the distance between the centers of the masses and G is the gravitational constant.

Newton's theory enjoyed its greatest success when it was used to predict the existence of Neptune based on motions of Uranus that could not be accounted for by the actions of the other planets. Calculations by both John Couch Adams and Urbain Le Verrier predicted the general position of the planet, and Le Verrier's calculations are what led Johann Gottfried Galle to the discovery of Neptune.

A discrepancy in Mercury's orbit pointed out flaws in Newton's theory. By the end of the 19th century, it was known that its orbit showed slight perturbations that could not be accounted for entirely under Newton's theory, but all searches for another perturbing body (such as a planet orbiting the Sun even closer than Mercury) had been fruitless. The issue was resolved in 1915 by Albert Einstein's new theory of general relativity, which accounted for the small discrepancy in Mercury's orbit.

Although Newton's theory has been superseded by the Einstein's general relativity, most modern non-relativistic gravitational calculations are still made using Newton's theory because it is simpler to work with and it gives sufficiently accurate results for most applications involving sufficiently small masses, speeds and energies.

15.1.3 Equivalence principle

The equivalence principle, explored by a succession of researchers including Galileo, Loránd Eötvös, and Einstein, expresses the idea that all objects fall in the same way, and that the effects of gravity are indistinguishable from certain aspects of acceleration and deceleration. The simplest way to test the weak equivalence principle is to drop two objects of different masses or compositions in a vacuum and see whether they hit the ground at the same time. Such experiments demonstrate that all objects fall at the same rate when other forces (such as air resistance and electromagnetic effects) are negligible. More sophisticated tests use a torsion balance of a type invented by Eötvös. Satellite experiments, for example STEP, are planned for more accurate experiments in space.[4]

Formulations of the equivalence principle include:

- The weak equivalence principle: *The trajectory of a point mass in a gravitational field depends only on its initial position and velocity, and is independent of its composition.*[5]

- The Einsteinian equivalence principle: *The outcome of any local non-gravitational experiment in a freely falling laboratory is independent of the velocity of the laboratory and its location in spacetime.*[6]

- The strong equivalence principle requiring both of the above.

15.1.4   General relativity

See also: Introduction to general relativity

In general relativity, the effects of gravitation are ascribed

Two-dimensional analogy of spacetime distortion generated by the mass of an object. Matter changes the geometry of spacetime, this (curved) geometry being interpreted as gravity. White lines do not represent the curvature of space but instead represent the coordinate system imposed on the curved spacetime, which would be rectilinear in a flat spacetime.

to spacetime curvature instead of a force. The starting point for general relativity is the equivalence principle, which equates free fall with inertial motion and describes free-falling inertial objects as being accelerated relative to non-inertial observers on the ground.[7][8] In Newtonian physics, however, no such acceleration can occur unless at least one of the objects is being operated on by a force.

Einstein proposed that spacetime is curved by matter, and that free-falling objects are moving along locally straight paths in curved spacetime. These straight paths are called geodesics. Like Newton's first law of motion, Einstein's theory states that if a force is applied on an object, it would deviate from a geodesic. For instance, we are no longer following geodesics while standing because the mechanical resistance of the Earth exerts an upward force on us, and we are non-inertial on the ground as a result. This explains why moving along the geodesics in spacetime is considered inertial.

Einstein discovered the field equations of general relativity, which relate the presence of matter and the curvature of spacetime and are named after him. The Einstein

field equations are a set of 10 simultaneous, non-linear, differential equations. The solutions of the field equations are the components of the metric tensor of spacetime. A metric tensor describes a geometry of spacetime. The geodesic paths for a spacetime are calculated from the metric tensor.

Solutions

Notable solutions of the Einstein field equations include:

- The Schwarzschild solution, which describes spacetime surrounding a spherically symmetric non-rotating uncharged massive object. For compact enough objects, this solution generated a black hole with a central singularity. For radial distances from the center which are much greater than the Schwarzschild radius, the accelerations predicted by the Schwarzschild solution are practically identical to those predicted by Newton's theory of gravity.

- The Reissner-Nordström solution, in which the central object has an electrical charge. For charges with a geometrized length which are less than the geometrized length of the mass of the object, this solution produces black holes with double event horizons.

- The Kerr solution for rotating massive objects. This solution also produces black holes with multiple event horizons.

- The Kerr-Newman solution for charged, rotating massive objects. This solution also produces black holes with multiple event horizons.

- The cosmological Friedmann-Lemaître-Robertson-Walker solution, which predicts the expansion of the universe.

Tests

The tests of general relativity included the following:[9]

- General relativity accounts for the anomalous perihelion precession of Mercury.[10]

- The prediction that time runs slower at lower potentials (gravitational time dilation) has been confirmed by the Pound–Rebka experiment (1959), the Hafele–Keating experiment, and the GPS.

- The prediction of the deflection of light was first confirmed by Arthur Stanley Eddington from his observations during the Solar eclipse of May 29, 1919.[11][12] Eddington measured starlight deflections twice those predicted by Newtonian corpuscular theory, in accordance with the predictions of general relativity. However, his interpretation of the results was later disputed.[13] More recent tests using radio interferometric measurements of quasars passing behind the Sun have more accurately and consistently confirmed the deflection of light to the degree predicted by general relativity.[14] See also gravitational lens.

- The time delay of light passing close to a massive object was first identified by Irwin I. Shapiro in 1964 in interplanetary spacecraft signals.

- Gravitational radiation has been indirectly confirmed through studies of binary pulsars. On 11 February 2016, the LIGO and Virgo collaborations announced the first observation of a gravitational wave.

- Alexander Friedmann in 1922 found that Einstein equations have non-stationary solutions (even in the presence of the cosmological constant). In 1927 Georges Lemaître showed that static solutions of the Einstein equations, which are possible in the presence of the cosmological constant, are unstable, and therefore the static universe envisioned by Einstein could not exist. Later, in 1931, Einstein himself agreed with the results of Friedmann and Lemaître. Thus general relativity predicted that the Universe had to be non-static—it had to either expand or contract. The expansion of the universe discovered by Edwin Hubble in 1929 confirmed this prediction.[15]

- The theory's prediction of frame dragging was consistent with the recent Gravity Probe B results.[16]

- General relativity predicts that light should lose its energy when traveling away from massive bodies through gravitational redshift. This was verified on earth and in the solar system around 1960.

15.1.5   Gravity and quantum mechanics

Main articles: Graviton and Quantum gravity

In the decades after the discovery of general relativity, it was realized that general relativity is incompatible with quantum mechanics.[17] It is possible to describe gravity in the framework of quantum field theory like the other fundamental forces, such that the attractive force of gravity arises due to exchange of virtual gravitons, in the same way as the electromagnetic force arises from exchange of virtual photons.[18][19] This reproduces general relativity in the classical limit. However, this approach fails at short distances of the order of the Planck length,[17] where a more

complete theory of quantum gravity (or a new approach to quantum mechanics) is required.

15.2   Specifics

15.2.1   Earth's gravity

Main article: Earth's gravity

Every planetary body (including the Earth) is surrounded by its own gravitational field, which can be conceptualized with Newtonian physics as exerting an attractive force on all objects. Assuming a spherically symmetrical planet, the strength of this field at any given point above the surface is proportional to the planetary body's mass and inversely proportional to the square of the distance from the center of the body.

  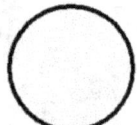

If an object with comparable mass to that of the Earth were to fall towards it, then the corresponding acceleration of the Earth would be observable.

The strength of the gravitational field is numerically equal to the acceleration of objects under its influence. The rate of acceleration of falling objects near the Earth's surface varies very slightly depending on latitude, surface features such as mountains and ridges, and perhaps unusually high or low sub-surface densities.[20] For purposes of weights and measures, a standard gravity value is defined by the International Bureau of Weights and Measures, under the International System of Units (SI).

That value, denoted g, is $g = 9.80665$ m/s^2 (32.1740 ft/s^2).[21][22]

The standard value of 9.80665 m/s^2 is the one originally adopted by the International Committee on Weights and Measures in 1901 for 45° latitude, even though it has been shown to be too high by about five parts in ten thousand.[23] This value has persisted in meteorology and in some standard atmospheres as the value for 45° latitude even though it applies more precisely to latitude of 45°32′33".[24]

Assuming the standardized value for g and ignoring air resistance, this means that an object falling freely near the Earth's surface increases its velocity by 9.80665 m/s (32.1740 ft/s or 22 mph) for each second of its descent. Thus, an object starting from rest will attain a velocity of 9.80665 m/s (32.1740 ft/s) after one second, approximately 19.62 m/s (64.4 ft/s) after two seconds, and so on, adding 9.80665 m/s (32.1740 ft/s) to each resulting velocity. Also, again ignoring air resistance, any and all objects, when dropped from the same height, will hit the ground at the same time.

According to Newton's 3rd Law, the Earth itself experiences a force equal in magnitude and opposite in direction to that which it exerts on a falling object. This means that the Earth also accelerates towards the object until they collide. Because the mass of the Earth is huge, however, the acceleration imparted to the Earth by this opposite force is negligible in comparison to the object's. If the object doesn't bounce after it has collided with the Earth, each of them then exerts a repulsive contact force on the other which effectively balances the attractive force of gravity and prevents further acceleration.

The force of gravity on Earth is the resultant (vector sum) of two forces: (a) The gravitational attraction in accordance with Newton's universal law of gravitation, and (b) the centrifugal force, which results from the choice of an earthbound, rotating frame of reference. At the equator, the force of gravity is the weakest due to the centrifugal force caused by the Earth's rotation. The force of gravity varies with latitude and increases from about 9.780 m/s^2 at the Equator to about 9.832 m/s^2 at the poles.

15.2.2 Equations for a falling body near the surface of the Earth

Main article: Equations for a falling body

Under an assumption of constant gravitational attraction, Newton's law of universal gravitation simplifies to $F = mg$, where m is the mass of the body and g is a constant vector with an average magnitude of 9.81 m/s^2 on Earth. This resulting force is the object's weight. The acceleration due to gravity is equal to this g. An initially stationary object which is allowed to fall freely under gravity drops a distance which is proportional to the square of the elapsed time. The image on the right, spanning half a second, was captured with a stroboscopic flash at 20 flashes per second. During the first $^1/_{20}$ of a second the ball drops one unit of distance (here, a unit is about 12 mm); by $^2/_{20}$ it has dropped at total of 4 units; by $^3/_{20}$, 9 units and so on.

Under the same constant gravity assumptions, the potential energy, Ep, of a body at height h is given by $Ep = mgh$ (or $Ep = Wh$, with W meaning weight). This expression is valid only over small distances h from the surface of the Earth. Similarly, the expression $h = \frac{v^2}{2g}$ for the maximum height reached by a vertically projected body with initial velocity v is useful for small heights and small initial velocities only.

15.2.3 Gravity and astronomy

The application of Newton's law of gravity has enabled the acquisition of much of the detailed information we have about the planets in the Solar System, the mass of the Sun, and details of quasars; even the existence of dark matter is inferred using Newton's law of gravity. Although we have not traveled to all the planets nor to the Sun, we know their masses. These masses are obtained by applying the laws of gravity to the measured characteristics of the orbit. In space an object maintains its orbit because of the force of gravity acting upon it. Planets orbit stars, stars orbit galactic centers, galaxies orbit a center of mass in clusters, and clusters orbit in superclusters. The force of gravity exerted on one object by another is directly proportional to the product of those objects' masses and inversely proportional to the square of the distance between them.

15.2.4 Gravitational radiation

Main article: Gravitational wave

According to general relativity, gravitational radiation is generated in situations where the curvature of spacetime is oscillating, such as is the case with co-orbiting objects. The gravitational radiation emitted by the Solar System is far too small to measure. However, gravitational radiation has been indirectly observed as an energy loss over time in binary pulsar systems such as PSR B1913+16. It is believed that neutron star mergers and black hole formation may create detectable amounts of gravitational radiation. Gravitational radiation observatories such as the Laser Interferometer Gravitational Wave Observatory (LIGO) have been created to study the problem. In February 2016, the Advanced LIGO team announced that they had detected gravitational waves from a black hole collision. On September 14, 2015 LIGO registered gravitational waves for the first time, as a result of the collision of two black holes 1.3 billion light-years from Earth.[26][27] This observation confirms the theoretical predictions of Einstein and others that such waves exist. The event confirms that binary black holes exist. It also opens the way for practical observation and understanding of the nature of gravity and events in the Universe including the Big Bang and what happened after it.[28][29]

15.2.5 Speed of gravity

Main article: Speed of gravity

In December 2012, a research team in China announced that it had produced measurements of the phase lag of Earth tides during full and new moons which seem to prove that the speed of gravity is equal to the speed of light.[30] This means that if the Sun suddenly disappeared, the Earth would keep orbiting it normally for 8 minutes, which is the time light takes to travel that distance. The team's findings were released in the Chinese Science Bulletin in February 2013.[31]

15.3 Anomalies and discrepancies

There are some observations that are not adequately accounted for, which may point to the need for better theories of gravity or perhaps be explained in other ways.

- **Extra-fast stars**: Stars in galaxies follow a distribution of velocities where stars on the outskirts are moving faster than they should according to the observed distributions of normal matter. Galaxies within galaxy clusters show a similar pattern. Dark matter, which would interact through gravitation but not electromagnetically, would account for the discrepancy. Various modifications to Newtonian dynamics have also been proposed.

- **Flyby anomaly**: Various spacecraft have experienced greater acceleration than expected during gravity assist maneuvers.

- **Accelerating expansion**: The metric expansion of space seems to be speeding up. Dark energy has been proposed to explain this. A recent alternative explanation is that the geometry of space is not homogeneous (due to clusters of galaxies) and that when the data are reinterpreted to take this into account, the expansion is not speeding up after all,[32] however this conclusion is disputed.[33]

- **Anomalous increase of the astronomical unit**: Recent measurements indicate that planetary orbits are widening faster than if this were solely through the Sun losing mass by radiating energy.

- **Extra energetic photons**: Photons travelling through galaxy clusters should gain energy and then lose it again on the way out. The accelerating expansion of the universe should stop the photons returning all the energy, but even taking this into account photons from the cosmic microwave background radiation gain twice as much energy as expected. This may indicate that gravity falls off *faster* than inverse-squared at certain distance scales.[34]

- **Extra massive hydrogen clouds**: The spectral lines of the Lyman-alpha forest suggest that hydrogen clouds are more clumped together at certain scales than expected and, like dark flow, may indicate that gravity falls off *slower* than inverse-squared at certain distance scales.[34]

- **Power**: Proposed extra dimensions could explain why the gravity force is so weak.[35]

15.4 Alternative theories

Main article: Alternatives to general relativity

15.4.1 Historical alternative theories

- Aristotelian theory of gravity

- Le Sage's theory of gravitation (1784) also called LeSage gravity, proposed by Georges-Louis Le Sage, based on a fluid-based explanation where a light gas fills the entire universe.

- Ritz's theory of gravitation, *Ann. Chem. Phys.* 13, 145, (1908) pp. 267–271, Weber-Gauss electrodynamics applied to gravitation. Classical advancement of perihelia.

- Nordström's theory of gravitation (1912, 1913), an early competitor of general relativity.

- Kaluza Klein theory (1921)

- Whitehead's theory of gravitation (1922), another early competitor of general relativity.

15.4.2 Modern alternative theories

- Brans–Dicke theory of gravity (1961) [36]

- Induced gravity (1967), a proposal by Andrei Sakharov according to which general relativity might arise from quantum field theories of matter

- $f(R)$ gravity (1970)

- Horndeski theory (1974) [37]

- Supergravity (1976)

- String theory

- In the modified Newtonian dynamics (MOND) (1981), Mordehai Milgrom proposes a modification of Newton's Second Law of motion for small accelerations [38]

- The self-creation cosmology theory of gravity (1982) by G.A. Barber in which the Brans-Dicke theory is modified to allow mass creation

- Loop quantum gravity (1988) by Carlo Rovelli, Lee Smolin, and Abhay Ashtekar

- Nonsymmetric gravitational theory (NGT) (1994) by John Moffat

- Conformal gravity[39]

- Tensor–vector–scalar gravity (TeVeS) (2004), a relativistic modification of MOND by Jacob Bekenstein

- Gravity as an entropic force, gravity arising as an emergent phenomenon from the thermodynamic concept of entropy.

- In the superfluid vacuum theory the gravity and curved space-time arise as a collective excitation mode of non-relativistic background superfluid.

- Chameleon theory (2004) by Justin Khoury and Amanda Weltman.

- Pressuron theory (2013) by Olivier Minazzoli and Aurélien Hees.

15.5   See also

- Angular momentum

- Anti-gravity, the idea of neutralizing or repelling gravity

- Artificial gravity

- Birkeland current

- Cosmic gravitational wave background

- Einstein–Infeld–Hoffmann equations

- Escape velocity, the minimum velocity needed to escape from a gravity well

- g-force, a measure of acceleration

- Gauge gravitation theory

- Gauss's law for gravity

- Gravitational binding energy

- Gravitational wave

- Gravitational wave background

- Gravity assist

- Gravity gradiometry

- Gravity Recovery and Climate Experiment

- Gravity Research Foundation

- Jovian–Plutonian gravitational effect

- Kepler's third law of planetary motion

- Lagrangian point

- Micro-g environment, also called microgravity

- Mixmaster dynamics

- n-body problem

- Newton's laws of motion

- Pioneer anomaly

- Scalar theories of gravitation

- Speed of gravity

- Standard gravitational parameter

- Standard gravity

- Weightlessness

15.6   Footnotes

[1] Ball, Phil (June 2005). "Tall Tales". *Nature News*. doi:10.1038/news050613-10.

[2] Galileo (1638), *Two New Sciences*, First Day Salviati speaks: "If this were what Aristotle meant you would burden him with another error which would amount to a falsehood; because, since there is no such sheer height available on earth, it is clear that Aristotle could not have made the experiment; yet he wishes to give us the impression of his having performed it when he speaks of such an effect as one which we see."

[3] - Chandrasekhar, Subrahmanyan (2003). *Newton's Principia for the common reader*. Oxford: Oxford University Press. (pp.1–2). The quotation comes from a memorandum thought to have been written about 1714. As early as 1645 Ismaël Bullialdus had argued that any force exerted by the Sun on distant objects would have to follow an inverse-square law.

However, he also dismissed the idea that any such force did exist. See, for example,

Linton, Christopher M. (2004). *From Eudoxus to Einstein—A History of Mathematical Astronomy*. Cambridge: Cambridge University Press. p. 225. ISBN 978-0-521-82750-8.

[4] M.C.W.Sandford (2008). "STEP: Satellite Test of the Equivalence Principle". Rutherford Appleton Laboratory. Retrieved 2011-10-14.

[5] Paul S Wesson (2006). *Five-dimensional Physics*. World Scientific. p. 82. ISBN 981-256-661-9.

[6] Haugen, Mark P.; C. Lämmerzahl (2001). *Principles of Equivalence: Their Role in Gravitation Physics and Experiments that Test Them*. Springer. arXiv:gr-qc/0103067∂. ISBN 978-3-540-41236-6.

[7] "Gravity and Warped Spacetime". black-holes.org. Retrieved 2010-10-16.

[8] Dmitri Pogosyan. "Lecture 20: Black Holes—The Einstein Equivalence Principle". University of Alberta. Retrieved 2011-10-14.

[9] Pauli, Wolfgang Ernst (1958). "Part IV. General Theory of Relativity". *Theory of Relativity*. Courier Dover Publications. ISBN 978-0-486-64152-2.

[10] Max Born (1924), *Einstein's Theory of Relativity* (The 1962 Dover edition, page 348 lists a table documenting the observed and calculated values for the precession of the perihelion of Mercury, Venus, and Earth.)

[11] Dyson, F.W.; Eddington, A.S.; Davidson, C.R. (1920). "A Determination of the Deflection of Light by the Sun's Gravitational Field, from Observations Made at the Total Eclipse of May 29, 1919". *Phil. Trans. Roy. Soc. A*. **220** (571–581): 291–333. Bibcode:1920RSPTA.220..291D. doi:10.1098/rsta.1920.0009.. Quote, p. 332: "Thus the results of the expeditions to Sobral and Principe can leave little doubt that a deflection of light takes place in the neighbourhood of the sun and that it is of the amount demanded by Einstein's generalised theory of relativity, as attributable to the sun's gravitational field."

[12] Weinberg, Steven (1972). *Gravitation and cosmology*. John Wiley & Sons.. Quote, p. 192: "About a dozen stars in all were studied, and yielded values 1.98 ± 0.11" and 1.61 ± 0.31", in substantial agreement with Einstein's prediction θ☉ = 1.75"."

[13] Earman, John; Glymour, Clark (1980). "Relativity and Eclipses: The British eclipse expeditions of 1919 and their predecessors". *Historical Studies in the Physical Sciences*. **11**: 49–85. doi:10.2307/27757471.

[14] Weinberg, Steven (1972). *Gravitation and cosmology*. John Wiley & Sons. p. 194.

[15] See W.Pauli, 1958, pp.219–220

[16] "NASA's Gravity Probe B Confirms Two Einstein Space-Time Theories". Nasa.gov. Retrieved 2013-07-23.

[17] Randall, Lisa (2005). *Warped Passages: Unraveling the Universe's Hidden Dimensions*. Ecco. ISBN 0-06-053108-8.

[18] Feynman, R. P.; Morinigo, F. B.; Wagner, W. G.; Hatfield, B. (1995). *Feynman lectures on gravitation*. Addison-Wesley. ISBN 0-201-62734-5.

[19] Zee, A. (2003). *Quantum Field Theory in a Nutshell*. Princeton University Press. ISBN 0-691-01019-6.

[20] "Astronomy Picture of the Day".

[21] Bureau International des Poids et Mesures (2006). "The International System of Units (SI)" (PDF) (8th ed.): 131. Retrieved 2009-11-25. Unit names are normally printed in Roman (upright) type ... Symbols for quantities are generally single letters set in an italic font, although they may be qualified by further information in subscripts or superscripts or in brackets.

[22] "SI Unit rules and style conventions". National Institute For Standards and Technology (USA). September 2004. Retrieved 2009-11-25. Variables and quantity symbols are in italic type. Unit symbols are in Roman type.

[23] List, R. J. editor, 1968, Acceleration of Gravity, *Smithsonian Meteorological Tables*, Sixth Ed. Smithsonian Institution, Washington, D.C., p. 68.

[24] U.S. Standard Atmosphere, 1976, U.S. Government Printing Office, Washington, D.C., 1976. (Linked file is very large.)

[25] "Milky Way Emerges as Sun Sets over Paranal". *www.eso.org*. European Southern Obseevatory. Retrieved 29 April 2015.

[26] Clark, Stuart (2016-02-11). "Gravitational waves: scientists announce 'we did it!' – live". *the Guardian*. Retrieved 2016-02-11.

[27] Castelvecchi, Davide; Witze, Witze (February 11, 2016). "Einstein's gravitational waves found at last". *Nature News*. doi:10.1038/nature.2016.19361. Retrieved 2016-02-11.

[28] "Scientists announce finding Gravitational Waves confirming Einstein's theory". WorldBreakingNews.

[29] "WHAT ARE GRAVITATIONAL WAVES AND WHY DO THEY MATTER?". popsci.com. Retrieved 12 February 2016.

[30] Chinese scientists find evidence for speed of gravity, astrowatch.com, 12/28/12.

[31] TANG, Ke Yun; HUA ChangCai; WEN Wu; CHI ShunLiang; YOU QingYu; YU Dan (February 2013). "Observational evidences for the speed of the gravity based on the Earth tide" (PDF). *Chinese Science Bulletin*. **58** (4-5): 474–477. doi:10.1007/s11434-012-5603-3. Retrieved 12 June 2013.

[32] Dark energy may just be a cosmic illusion, *New Scientist*, issue 2646, 7 March 2008.

[33] Swiss-cheese model of the cosmos is full of holes, *New Scientist*, issue 2678, 18 October 2008.

[34] Chown, Marcus (16 March 2009). "Gravity may venture where matter fears to tread". *New Scientist*. Retrieved 4 August 2013.

[35] CERN (20 January 2012). "Extra dimensions, gravitons, and tiny black holes".

[36] Brans, C.H. (Mar 2014). "Jordan-Brans-Dicke Theory". *Scholarpedia*. **9**: 31358. Bibcode:2014Schpj...931358B. doi:10.4249/scholarpedia.31358.

[37] Horndeski, G.W. (Sep 1974). "Second-Order Scalar-Tensor Field Equations in a Four-Dimensional Space". *International Journal of Theoretical Physics*. **88** (10): 363–384. Bibcode:1974IJTP...10..363H. doi:10.1007/BF01807638.

[38] Milgrom, M. (Jun 2014). "The MOND paradigm of modified dynamics". *Scholarpedia*. **9**: 31410. Bibcode:2014SchpJ...931410M. doi:10.4249/scholarpedia.31410.

[39] Einstein gravity from conformal gravity

15.7 References

- Halliday, David; Robert Resnick; Kenneth S. Krane (2001). *Physics v. 1*. New York: John Wiley & Sons. ISBN 0-471-32057-9.

- Serway, Raymond A.; Jewett, John W. (2004). *Physics for Scientists and Engineers* (6th ed.). Brooks/Cole. ISBN 0-534-40842-7.

- Tipler, Paul (2004). *Physics for Scientists and Engineers: Mechanics, Oscillations and Waves, Thermodynamics* (5th ed.). W. H. Freeman. ISBN 0-7167-0809-4.

15.8 Further reading

- Thorne, Kip S.; Misner, Charles W.; Wheeler, John Archibald (1973). *Gravitation*. W.H. Freeman. ISBN 0-7167-0344-0.

15.9 External links

- Hazewinkel, Michiel, ed. (2001), "Gravitation", *Encyclopedia of Mathematics*, Springer, ISBN 978-1-55608-010-4

- Hazewinkel, Michiel, ed. (2001), "Gravitation, theory of", *Encyclopedia of Mathematics*, Springer, ISBN 978-1-55608-010-4

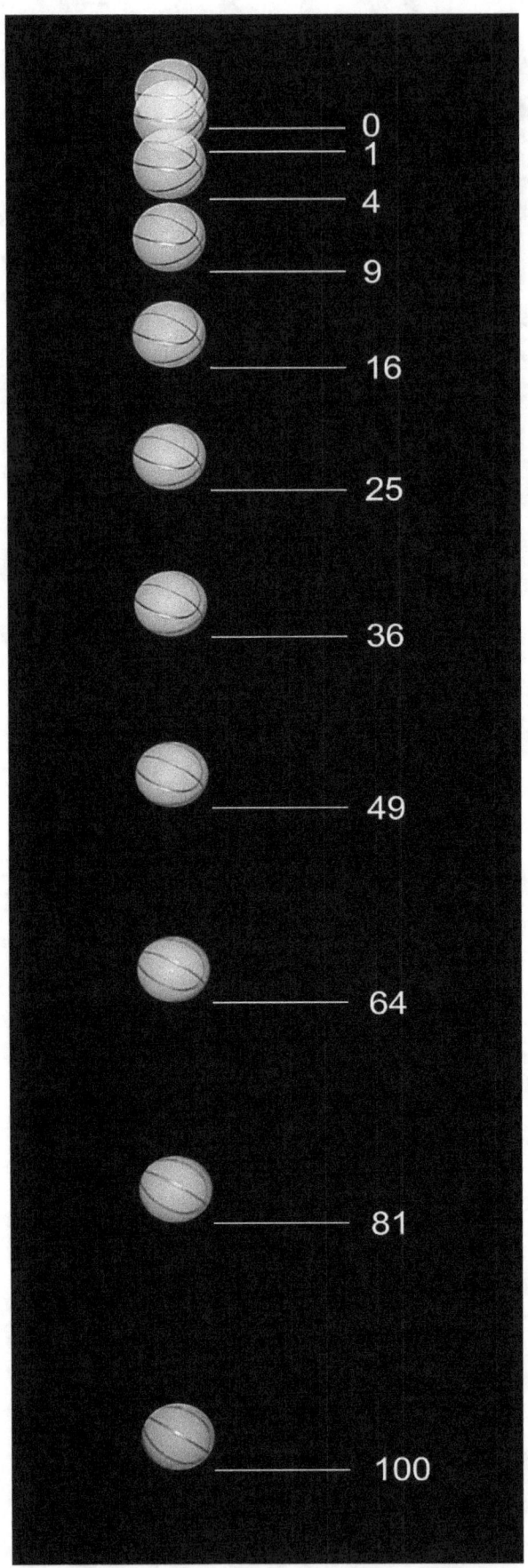

Gravity acts on stars that form our Milky Way.[25]

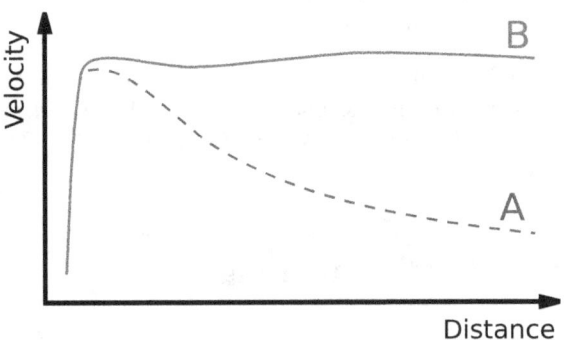

*Rotation curve of a typical spiral galaxy: predicted (**A**) and observed (**B**). The discrepancy between the curves is attributed to dark matter.*

Chapter 16

Newton's law of universal gravitation

Newton's law of universal gravitation states that a particle attracts every other particle in the universe using a force that is directly proportional to the product of their masses and inversely proportional to the square of the distance between them.[note 1] This is a general physical law derived from empirical observations by what Isaac Newton called induction.[1] It is a part of classical mechanics and was formulated in Newton's work *Philosophiæ Naturalis Principia Mathematica* ("the *Principia*"), first published on 5 July 1687. (When Newton's book was presented in 1686 to the Royal Society, Robert Hooke made a claim that Newton had obtained the inverse square law from him; see the History section below.)

In modern language, the law states: Every point mass attracts every single other point mass by a force pointing along the line intersecting both points. The force is proportional to the product of the two masses and inversely proportional to the square of the distance between them.[2] The first test of Newton's theory of gravitation between masses in the laboratory was the Cavendish experiment conducted by the British scientist Henry Cavendish in 1798.[3] It took place 111 years after the publication of Newton's *Principia* and approximately 71 years after his death.

Newton's law of gravitation resembles Coulomb's law of electrical forces, which is used to calculate the magnitude of the electrical force arising between two charged bodies. Both are inverse-square laws, where force is inversely proportional to the square of the distance between the bodies. Coulomb's law has the product of two charges in place of the product of the masses, and the electrostatic constant in place of the gravitational constant.

Newton's law has since been superseded by Einstein's theory of general relativity, but it continues to be used as an excellent approximation of the effects of gravity in most applications. Relativity is required only when there is a need for extreme precision, or when dealing with very strong gravitational fields, such as those found near extremely massive and dense objects, or at very close distances (such as Mercury's orbit around the sun).

16.1 History

16.1.1 Early history

A recent assessment (by Ofer Gal) about the early history of the inverse square law is "by the late 1660s", the assumption of an "inverse proportion between gravity and the square of distance was rather common and had been advanced by a number of different people for different reasons". The same author does credit Hooke with a significant and even seminal contribution, but he treats Hooke's claim of priority on the inverse square point as uninteresting since several individuals besides Newton and Hooke had at least suggested it, and he points instead to the idea of "compounding the celestial motions" and the conversion of Newton's thinking away from "centrifugal" and towards "centripetal" force as Hooke's significant contributions.

16.1.2 Plagiarism dispute

In 1686, when the first book of Newton's *Principia* was presented to the Royal Society, Robert Hooke accused Newton of plagiarism by claiming that he had taken from him the "notion" of "the rule of the decrease of Gravity, being reciprocally as the squares of the distances from the Center". At the same time (according to Edmond Halley's contemporary report) Hooke agreed that "the Demonstration of the Curves generated thereby" was wholly Newton's.[4]

In this way, the question arose as to what, if anything, Newton owed to Hooke. This is a subject extensively discussed since that time and on which some points, outlined below, continue to excite controversy.

16.1.3 Hooke's work and claims

Robert Hooke published his ideas about the "System of the World" in the 1660s, when he read to the Royal Society on March 21, 1666, a paper "On gravity", "concerning the inflection of a direct motion into a curve by a supervening attractive principle", and he published them again in somewhat developed form in 1674, as an addition to "An Attempt to Prove the Motion of the Earth from Observations".[5] Hooke announced in 1674 that he planned to "explain a System of the World differing in many particulars from any yet known", based on three "Suppositions": that "all Celestial Bodies whatsoever, have an attraction or gravitating power towards their own Centers" [and] "they do also attract all the other Celestial Bodies that are within the sphere of their activity";[6] that "all bodies whatsoever that are put into a direct and simple motion, will so continue to move forward in a straight line, till they are by some other effectual powers deflected and bent..."; and that "these attractive powers are so much the more powerful in operating, by how much the nearer the body wrought upon is to their own Centers". Thus Hooke clearly postulated mutual attractions between the Sun and planets, in a way that increased with nearness to the attracting body, together with a principle of linear inertia.

Hooke's statements up to 1674 made no mention, however, that an inverse square law applies or might apply to these attractions. Hooke's gravitation was also not yet universal, though it approached universality more closely than previous hypotheses.[7] He also did not provide accompanying evidence or mathematical demonstration. On the latter two aspects, Hooke himself stated in 1674: "Now what these several degrees [of attraction] are I have not yet experimentally verified"; and as to his whole proposal: "This I only hint at present", "having my self many other things in hand which I would first compleat, and therefore cannot so well attend it" (i.e. "prosecuting this Inquiry").[5] It was later on, in writing on 6 January 1679|80[8] to Newton, that Hooke communicated his "supposition ... that the Attraction always is in a duplicate proportion to the Distance from the Center Reciprocall, and Consequently that the Velocity will be in a subduplicate proportion to the Attraction and Consequently as Kepler Supposes Reciprocall to the Distance."[9] (The inference about the velocity was incorrect.[10])

Hooke's correspondence with Newton during 1679–1680 not only mentioned this inverse square supposition for the decline of attraction with increasing distance, but also, in Hooke's opening letter to Newton, of 24 November 1679, an approach of "compounding the celestial motions of the planets of a direct motion by the tangent & an attractive motion towards the central body".[11]

16.1.4   Newton's work and claims

Newton, faced in May 1686 with Hooke's claim on the inverse square law, denied that Hooke was to be credited as author of the idea. Among the reasons, Newton recalled

that the idea had been discussed with Sir Christopher Wren previous to Hooke's 1679 letter.[12] Newton also pointed out and acknowledged prior work of others,[13] including Bullialdus,[14] (who suggested, but without demonstration, that there was an attractive force from the Sun in the inverse square proportion to the distance), and Borelli[15] (who suggested, also without demonstration, that there was a centrifugal tendency in counterbalance with a gravitational attraction towards the Sun so as to make the planets move in ellipses). D T Whiteside has described the contribution to Newton's thinking that came from Borelli's book, a copy of which was in Newton's library at his death.[16]

Newton further defended his work by saying that had he first heard of the inverse square proportion from Hooke, he would still have some rights to it in view of his demonstrations of its accuracy. Hooke, without evidence in favor of the supposition, could only guess that the inverse square law was approximately valid at great distances from the center. According to Newton, while the 'Principia' was still at pre-publication stage, there were so many a-priori reasons to doubt the accuracy of the inverse-square law (especially close to an attracting sphere) that "without my (Newton's) Demonstrations, to which Mr Hooke is yet a stranger, it cannot believed by a judicious Philosopher to be any where accurate."[17]

This remark refers among other things to Newton's finding, supported by mathematical demonstration, that if the inverse square law applies to tiny particles, then even a large spherically symmetrical mass also attracts masses external to its surface, even close up, exactly as if all its own mass were concentrated at its center. Thus Newton gave a justification, otherwise lacking, for applying the inverse square law to large spherical planetary masses as if they were tiny particles.[18] In addition, Newton had formulated in Propositions 43-45 of Book 1,[19] and associated sections of Book 3, a sensitive test of the accuracy of the inverse square law, in which he showed that only where the law of force is accurately as the inverse square of the distance will the directions of orientation of the planets' orbital ellipses stay constant as they are observed to do apart from small effects attributable to inter-planetary perturbations.

In regard to evidence that still survives of the earlier history, manuscripts written by Newton in the 1660s show that Newton himself had, by 1669, arrived at proofs that in a circular case of planetary motion, "endeavour to recede" (what was later called centrifugal force) had an inverse-square relation with distance from the center.[20] After his 1679-1680 correspondence with Hooke, Newton adopted the language of inward or centripetal force. According to Newton scholar J. Bruce Brackenridge, although much has been made of the change in language and difference of point of view, as between centrifugal or centripetal forces, the actual computations and proofs remained the same either way. They also involved the combination of tangential and radial displacements, which Newton was making in the 1660s. The lesson offered by Hooke to Newton here, although significant, was one of perspective and did not change the analysis.[21] This background shows there was basis for Newton to deny deriving the inverse square law from Hooke.

16.1.5 Newton's acknowledgment

On the other hand, Newton did accept and acknowledge, in all editions of the 'Principia', that Hooke (but not exclusively Hooke) had separately appreciated the inverse square law in the solar system. Newton acknowledged Wren, Hooke and Halley in this connection in the Scholium to Proposition 4 in Book 1.[22] Newton also acknowledged to Halley that his correspondence with Hooke in 1679-80 had reawakened his dormant interest in astronomical matters, but that did not mean, according to Newton, that Hooke had told Newton anything new or original: "yet am I not beholden to him for any light into that business but only for the diversion he gave me from my other studies to think on these things & for his dogmaticalness in writing as if he had found the motion in the Ellipsis, which inclined me to try it ..."[13]

16.1.6 Modern priority controversy

Since the time of Newton and Hooke, scholarly discussion has also touched on the question of whether Hooke's 1679 mention of 'compounding the motions' provided Newton with something new and valuable, even though that was not a claim actually voiced by Hooke at the time. As described above, Newton's manuscripts of the 1660s do show him actually combining tangential motion with the effects of radially directed force or endeavour, for example in his derivation of the inverse square relation for the circular case. They also show Newton clearly expressing the concept of linear inertia—for which he was indebted to Descartes' work, published in 1644 (as Hooke probably was).[23] These matters do not appear to have been learned by Newton from Hooke.

Nevertheless, a number of authors have had more to say about what Newton gained from Hooke and some aspects remain controversial.[24] The fact that most of Hooke's private papers had been destroyed or have disappeared does not help to establish the truth.

Newton's role in relation to the inverse square law was not as it has sometimes been represented. He did not claim to think it up as a bare idea. What Newton did was to show how the inverse-square law of attraction had many necessary mathematical connections with observable features of the motions of bodies in the solar system; and that they were related in such a way that the observational evidence and the

mathematical demonstrations, taken together, gave reason to believe that the inverse square law was not just approximately true but exactly true (to the accuracy achievable in Newton's time and for about two centuries afterwards – and with some loose ends of points that could not yet be certainly examined, where the implications of the theory had not yet been adequately identified or calculated).[25][26]

About thirty years after Newton's death in 1727, Alexis Clairaut, a mathematical astronomer eminent in his own right in the field of gravitational studies, wrote after reviewing what Hooke published, that "One must not think that this idea ... of Hooke diminishes Newton's glory"; and that "the example of Hooke" serves "to show what a distance there is between a truth that is glimpsed and a truth that is demonstrated".[27][28]

16.2   Modern form

In modern language, the law states the following:

Assuming SI units, F is measured in newtons (N), m_1 and m_2 in kilograms (kg), r in meters (m), and the constant G is approximately equal to 6.674×10^{-11} N m^2 kg^{-2}.[29] The value of the constant G was first accurately determined from the results of the Cavendish experiment conducted by the British scientist Henry Cavendish in 1798, although Cavendish did not himself calculate a numerical value for G.[3] This experiment was also the first test of Newton's theory of gravitation between masses in the laboratory. It took place 111 years after the publication of Newton's *Principia* and 71 years after Newton's death, so none of Newton's calculations could use the value of G; instead he could only calculate a force relative to another force.

16.3   Bodies with spatial extent

If the bodies in question have spatial extent (rather than being theoretical point masses), then the gravitational force between them is calculated by summing the contributions of the notional point masses which constitute the bodies. In the limit, as the component point masses become "infinitely small", this entails integrating the force (in vector form, see below) over the extents of the two bodies.

In this way, it can be shown that an object with a spherically-symmetric distribution of mass exerts the same gravitational attraction on external bodies as if all the object's mass were concentrated at a point at its centre.[2] (This is not generally true for non-spherically-symmetrical bodies.)

For points *inside* a spherically-symmetric distribution of matter, Newton's Shell theorem can be used to find the grav-

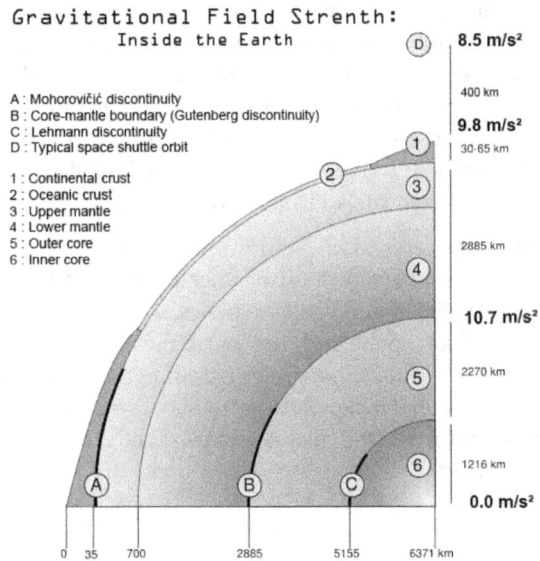

Gravitational field strength within the Earth

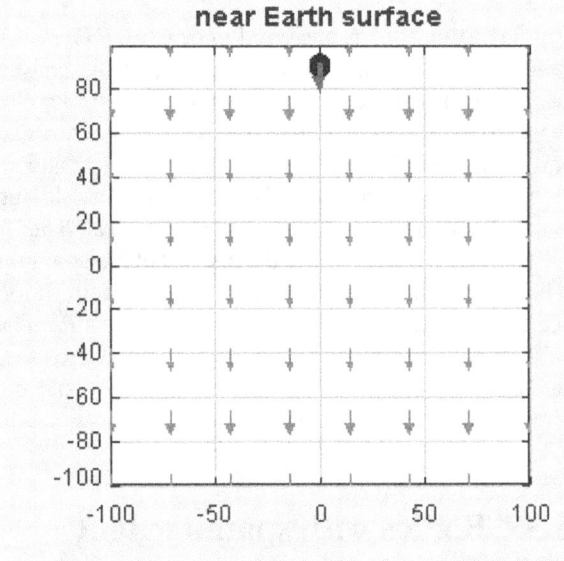

Gravity field near the surface of the Earth – an object is shown accelerating toward the surface

itational force. The theorem tells us how different parts of the mass distribution affect the gravitational force measured at a point located a distance r_0 from the center of the mass distribution:[30]

- The portion of the mass that is located at radii $r < r_0$ causes the same force at r_0 as if all of the mass enclosed within a sphere of radius r_0 was concentrated at the center of the mass distribution (as noted above).

- The portion of the mass that is located at radii $r > r_0$

exerts *no net* gravitational force at the distance r_0 from the center. That is, the individual gravitational forces exerted by the elements of the sphere out there, on the point at r_0, cancel each other out.

As a consequence, for example, within a shell of uniform thickness and density there is *no net* gravitational acceleration anywhere within the hollow sphere.

Furthermore, inside a uniform sphere the gravity increases linearly with the distance from the center; the increase due to the additional mass is 1.5 times the decrease due to the larger distance from the center. Thus, if a spherically symmetric body has a uniform core and a uniform mantle with a density that is less than 2/3 of that of the core, then the gravity initially decreases outwardly beyond the boundary, and if the sphere is large enough, further outward the gravity increases again, and eventually it exceeds the gravity at the core/mantle boundary. The gravity of the Earth may be highest at the core/mantle boundary.

16.4 Vector form

calculating

Field lines drawn for a point mass using 24 field lines

Newton's law of universal gravitation can be written as a vector equation to account for the direction of the gravitational force as well as its magnitude. In this formula, quantities in bold represent vectors.

$$\mathbf{F}_{12} = -G \frac{m_1 m_2}{|\mathbf{r}_{12}|^2} \hat{\mathbf{r}}_{12}$$

where

$\mathbf{F}_{12}$ is the force applied on object 2 due to object 1,

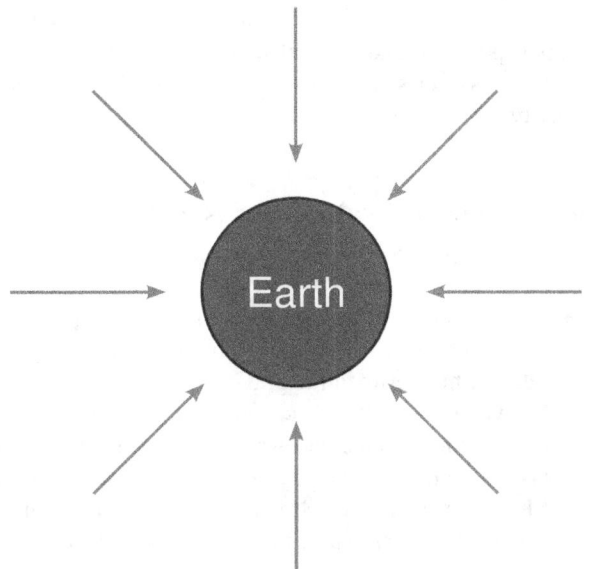

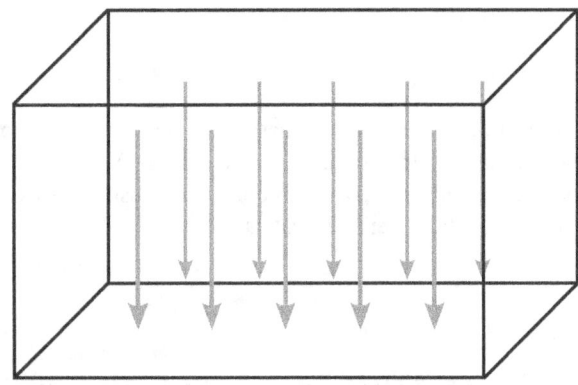

Gravity in a room: the curvature of the Earth is negligible at this scale, and the force lines can be approximated as being parallel and pointing straight down to the center of the Earth

Gravity field surrounding Earth from a macroscopic perspective.

It can be seen that the vector form of the equation is the same as the scalar form given earlier, except that $\mathbf{F}$ is now a vector quantity, and the right hand side is multiplied by the appropriate unit vector. Also, it can be seen that $\mathbf{F}_{12} = -\mathbf{F}_{21}$.

16.5   Gravitational field

Main article: Gravitational field

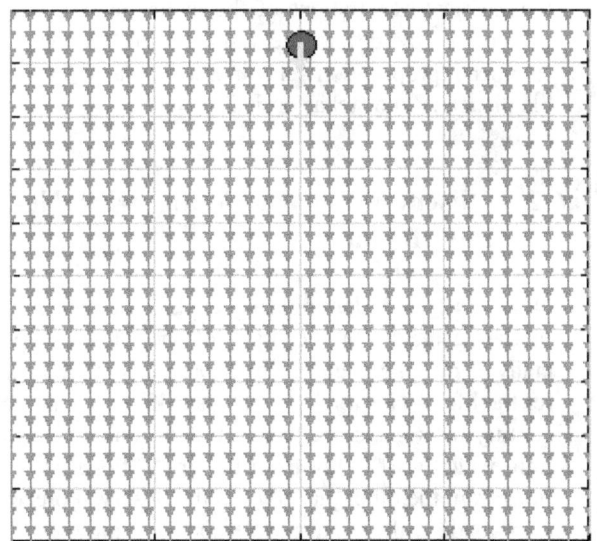

Gravity field lines representation is arbitrary as illustrated here represented in 30x30 grid to 0x0 grid and almost being parallel and pointing straight down to the center of the Earth

The **gravitational field** is a vector field that describes the gravitational force which would be applied on an object in any given point in space, per unit mass. It is actually equal to the gravitational acceleration at that point.

It is a generalisation of the vector form, which becomes particularly useful if more than 2 objects are involved (such as a rocket between the Earth and the Moon). For 2 objects (e.g. object 2 is a rocket, object 1 the Earth), we simply write $\mathbf{r}$ instead of $\mathbf{r}_{12}$ and m instead of m_2 and define the gravitational field $\mathbf{g}(\mathbf{r})$ as:

$$\mathbf{g}(\mathbf{r}) = -G\frac{m_1}{|\mathbf{r}|^2}\hat{\mathbf{r}}$$

so that we can write:

G is the gravitational constant,

m_1 and m_2 are respectively the masses of objects 1 and 2,

$|\mathbf{r}_{12}| = |\mathbf{r}_2 - \mathbf{r}_1|$ is the distance between objects 1 and 2, and

$\hat{\mathbf{r}}_{12} \overset{\text{def}}{=} \frac{\mathbf{r}_2 - \mathbf{r}_1}{|\mathbf{r}_2 - \mathbf{r}_1|}$ is the unit vector from object 1 to 2.

$$\mathbf{F}(\mathbf{r}) = m\mathbf{g}(\mathbf{r}).$$

This formulation is dependent on the objects causing the field. The field has units of acceleration; in SI, this is m/s^2.

Gravitational fields are also conservative; that is, the work done by gravity from one position to another is path-independent. This has the consequence that there exists a gravitational potential field $V(\mathbf{r})$ such that

$$\mathbf{g}(\mathbf{r}) = -\nabla V(\mathbf{r}).$$

If m_1 is a point mass or the mass of a sphere with homogeneous mass distribution, the force field $\mathbf{g}(\mathbf{r})$ outside the sphere is isotropic, i.e., depends only on the distance r from the center of the sphere. In that case

$$V(r) = -G\frac{m_1}{r}.$$

the gravitational field is on, inside and outside of symmetric masses.

As per Gauss Law, field in a symmetric body can be found by the mathematical equation:

where ∂V is a closed surface and M_{enc} is the mass enclosed by the surface.

Hence, for a hollow sphere of radius R and total mass M,

$$|\mathbf{g}(\mathbf{r})| = \begin{cases} 0, & \text{if } r < R \\ \dfrac{GM}{r^2}, & \text{if } r \geq R \end{cases}$$

For a uniform solid sphere of radius R and total mass M,

$$|\mathbf{g}(\mathbf{r})| = \begin{cases} \dfrac{GMr}{R^3}, & \text{if } r < R \\ \dfrac{GM}{r^2}, & \text{if } r \geq R \end{cases}$$

16.6 Problematic aspects

Newton's description of gravity is sufficiently accurate for many practical purposes and is therefore widely used. Deviations from it are small when the dimensionless quantities φ/c^2 and $(v/c)^2$ are both much less than one, where φ is the gravitational potential, v is the velocity of the objects being studied, and c is the speed of light.[31] For example, Newtonian gravity provides an accurate description of the Earth/Sun system, since

$$\frac{\Phi}{c^2} = \frac{GM_{\text{sun}}}{r_{\text{orbit}}c^2} \sim 10^{-8}, \quad \left(\frac{v_{\text{Earth}}}{c}\right)^2 = \left(\frac{2\pi r_{\text{orbit}}}{(1 \text{ yr})c}\right)^2 \sim 10^{-8}$$

where r_{orbit} is the radius of the Earth's orbit around the Sun.

In situations where either dimensionless parameter is large, then general relativity must be used to describe the system.

General relativity reduces to Newtonian gravity in the limit of small potential and low velocities, so Newton's law of gravitation is often said to be the low-gravity limit of general relativity.

16.6.1 Theoretical concerns with Newton's expression

- There is no immediate prospect of identifying the mediator of gravity. Attempts by physicists to identify the relationship between the gravitational force and other known fundamental forces are not yet resolved, although considerable headway has been made over the last 50 years (See: Theory of everything and Standard Model). Newton himself felt that the concept of an inexplicable *action at a distance* was unsatisfactory (see "Newton's reservations" below), but that there was nothing more that he could do at the time.

- Newton's theory of gravitation requires that the gravitational force be transmitted instantaneously. Given the classical assumptions of the nature of space and time before the development of General Relativity, a significant propagation delay in gravity leads to unstable planetary and stellar orbits.

16.6.2 Observations conflicting with Newton's formula

- Newton's Theory does not fully explain the precession of the perihelion of the orbits of the planets, especially of planet Mercury, which was detected long after the life of Newton.[32] There is a 43 arcsecond per century discrepancy between the Newtonian calculation, which arises only from the gravitational attractions from the other planets, and the observed precession, made with advanced telescopes during the 19th Century.

- The predicted angular deflection of light rays by gravity that is calculated by using Newton's Theory is only one-half of the deflection that is actually observed by astronomers. Calculations using General Relativity are in much closer agreement with the astronomical observations.

- In spiral galaxies, the orbiting of stars around their centers seems to strongly disobey Newton's law of universal gravitation. Astrophysicists, however, explain this spectacular phenomenon in the framework of Newton's laws, with the presence of large amounts of Dark matter.

The observed fact that the *gravitational mass* and the *inertial mass* is not the same for all objects is unexplained within Newton's Theories. The equations for Newtonian gravitation and Newtonian inertia use the same "mass", while in those of General Relativity they diverge. The Equivalence Principle is only valid in regimes where the Newtonian model is valid. Examples of this include: the experiments of Galileo Galilei that, decades before Newton, established that objects that have the same air or fluid resistance are accelerated by the force of the Earth's gravity equally, regardless of their different *inertial* masses. In these cases, the forces and energies that are required to accelerate various masses is completely dependent upon their different *inertial* masses, as can be seen from Newton's Second Law of Motion, F = ma. The modern and modified statement of Newtons Second Law gives this vector equation:

$$\vec{F} = \frac{d\vec{p}}{dt},$$

where $\vec{p}$ is the momentum of the system, and $\vec{F}$ is the net (vector sum) force

$$\vec{F} = \frac{d(m\vec{v})}{dt}$$

Substituting the inertial mass dilation expression from Special relativity yields the following expression.

$$F = \frac{d}{dt}\left(\frac{m}{\sqrt{1 - \frac{v^2}{c^2}}} \cdot v\right)$$

When the ratio of velocity to the speed of light is low, then the mass term is constant. When it is near to the speed of light, the denominator approaches zero and the momentum explodes because effective inertial mass explodes.

16.6.3   Newton's reservations

While Newton was able to formulate his law of gravity in his monumental work, he was deeply uncomfortable with the notion of "action at a distance" that his equations implied. In 1692, in his third letter to Bentley, he wrote: *"That one body may act upon another at a distance through a vacuum without the mediation of anything else, by and through which their action and force may be conveyed from one another, is to me so great an absurdity that, I believe, no man who has in philosophic matters a competent faculty of thinking could ever fall into it."*

He never, in his words, "assigned the cause of this power". In all other cases, he used the phenomenon of motion to explain the origin of various forces acting on bodies, but in the case of gravity, he was unable to experimentally identify the motion that produces the force of gravity (although he invented two mechanical hypotheses in 1675 and 1717). Moreover, he refused to even offer a hypothesis as to the cause of this force on grounds that to do so was contrary to sound science. He lamented that "philosophers have hitherto attempted the search of nature in vain" for the source of the gravitational force, as he was convinced "by many reasons" that there were "causes hitherto unknown" that were fundamental to all the "phenomena of nature". These fundamental phenomena are still under investigation and, though hypotheses abound, the definitive answer has yet to be found. And in Newton's 1713 *General Scholium* in the second edition of *Principia*: *"I have not yet been able to discover the cause of these properties of gravity from phenomena and I feign no hypotheses.... It is enough that gravity does really exist and acts according to the laws I have explained, and that it abundantly serves to account for all the motions of celestial bodies."*[33]

16.6.4   Einstein's solution

These objections were explained by Einstein's theory of general relativity, in which gravitation is an attribute of curved spacetime instead of being due to a force propagated between bodies. In Einstein's theory, energy and momentum distort spacetime in their vicinity, and other particles move in trajectories determined by the geometry of spacetime. This allowed a description of the motions of light and mass that was consistent with all available observations. In general relativity, the gravitational force is a fictitious force due to the curvature of spacetime, because the gravitational acceleration of a body in free fall is due to its world line being a geodesic of spacetime.

16.7   Extensions

Newton was the first to consider in his Principia an extended expression of his law of gravity including an inverse-cube term of the form

$$F = G\frac{m_1 m_2}{r^2} + B\frac{m_1 m_2}{r^3}$$

attempting to explain the Moon's apsidal motion. Other extensions were proposed by Laplace (around 1790) and Decombes (1913):[34]

$$F(r) = k\frac{m_1 m_2}{r^2}\exp(-\alpha r)$$

$$F(r) = k \frac{m_1 m_2}{r^2} \left(1 + \frac{\alpha}{r^3} \right)$$

In recent years, quests for non-inverse square terms in the law of gravity have been carried out by neutron interferometry.[35]

16.8 Solutions of Newton's law of universal gravitation

Main article: n-body problem

The *n*-body problem is an ancient, classical problem[36] of predicting the individual motions of a group of celestial objects interacting with each other gravitationally. Solving this problem — from the time of the Greeks and on — has been motivated by the desire to understand the motions of the Sun, planets and the visible stars. In the 20th century, understanding the dynamics of globular cluster star systems became an important *n*-body problem too.[37] The *n*-body problem in general relativity is considerably more difficult to solve.

The classical physical problem can be informally stated as: *given the quasi-steady orbital properties (instantaneous position, velocity and time)*[38] *of a group of celestial bodies, predict their interactive forces; and consequently, predict their true orbital motions for all future times.*[39]

The two-body problem has been completely solved, as has the *Restricted 3-Body Problem.*[40]

16.9 See also

- Bentley's paradox
- Gauss's law for gravity
- Jordan and Einstein frames
- Kepler orbit
- Newton's cannonball
- Newton's laws of motion
- Static forces and virtual-particle exchange

16.10 Notes

[1] It was shown separately that large, spherically symmetrical masses attract and are attracted as if all their mass were concentrated at their centers.

16.11 References

[1] Isaac Newton: "In [experimental] philosophy particular propositions are inferred from the phenomena and afterwards rendered general by induction": "Principia", Book 3, General Scholium, at p.392 in Volume 2 of Andrew Motte's English translation published 1729.

[2] - Proposition 75, Theorem 35: p.956 - I.Bernard Cohen and Anne Whitman, translators: Isaac Newton, *The Principia*: Mathematical Principles of Natural Philosophy. Preceded by *A Guide to Newton's Principia*, by I.Bernard Cohen. University of California Press 1999 ISBN 0-520-08816-6 ISBN 0-520-08817-4

[3] The Michell-Cavendish Experiment, Laurent Hodges

[4] H W Turnbull (ed.), Correspondence of Isaac Newton, Vol 2 (1676-1687), (Cambridge University Press, 1960), giving the Halley-Newton correspondence of May to July 1686 about Hooke's claims at pp.431-448, see particularly page 431.

[5] Hooke's 1674 statement in "An Attempt to Prove the Motion of the Earth from Observations" is available in online facsimile here.

[6] Purrington, Robert D. (2009). *The First Professional Scientist: Robert Hooke and the Royal Society of London.* Springer. p. 168. ISBN 3-0346-0036-4. Extract of page 168

[7] See page 239 in Curtis Wilson (1989), "The Newtonian achievement in astronomy", ch.13 (pages 233-274) in "Planetary astronomy from the Renaissance to the rise of astrophysics: 2A: Tycho Brahe to Newton", CUP 1989.

[8] Calendar (New Style) Act 1750

[9] Page 309 in H W Turnbull (ed.), Correspondence of Isaac Newton, Vol 2 (1676-1687), (Cambridge University Press, 1960), document #239.

[10] See Curtis Wilson (1989) at page 244.

[11] Page 297 in H W Turnbull (ed.), Correspondence of Isaac Newton, Vol 2 (1676-1687), (Cambridge University Press, 1960), document #235, 24 November 1679.

[12] Page 433 in H W Turnbull (ed.), Correspondence of Isaac Newton, Vol 2 (1676-1687), (Cambridge University Press, 1960), document #286, 27 May 1686.

[13] Pages 435-440 in H W Turnbull (ed.), Correspondence of Isaac Newton, Vol 2 (1676-1687), (Cambridge University Press, 1960), document #288, 20 June 1686.

[14] Bullialdus (Ismael Bouillau) (1645), "Astronomia philolaica", Paris, 1645.

[15] Borelli, G. A., "Theoricae Mediceorum Planetarum ex causis physicis deductae", Florence, 1666.

[16] D T Whiteside, "Before the Principia: the maturing of Newton's thoughts on dynamical astronomy, 1664-1684", Journal for the History of Astronomy, i (1970), pages 5-19; especially at page 13.

[17] Page 436, Correspondence, Vol.2, already cited.

[18] Propositions 70 to 75 in Book 1, for example in the 1729 English translation of the *Principia*, start at page 263.

[19] Propositions 43 to 45 in Book 1, in the 1729 English translation of the *Principia*, start at page 177.

[20] D T Whiteside, "The pre-history of the 'Principia' from 1664 to 1686", Notes and Records of the Royal Society of London, 45 (1991), pages 11-61; especially at 13-20.

[21] See J. Bruce Brackenridge, "The key to Newton's dynamics: the Kepler problem and the Principia", (University of California Press, 1995), especially at pages 20-21.

[22] See for example the 1729 English translation of the *Principia*, at page 66.

[23] See page 10 in D T Whiteside, "Before the Principia: the maturing of Newton's thoughts on dynamical astronomy, 1664-1684", Journal for the History of Astronomy, i (1970), pages 5-19.

[24] Discussion points can be seen for example in the following papers: N Guicciardini, "Reconsidering the Hooke-Newton debate on Gravitation: Recent Results", in Early Science and Medicine, 10 (2005), 511-517; Ofer Gal, "The Invention of Celestial Mechanics", in Early Science and Medicine, 10 (2005), 529-534; M Nauenberg, "Hooke's and Newton's Contributions to the Early Development of Orbital mechanics and Universal Gravitation", in Early Science and Medicine, 10 (2005), 518-528.

[25] See for example the results of Propositions 43-45 and 70-75 in Book 1, cited above.

[26] See also G E Smith, in Stanford Encyclopedia of Philosophy, "Newton's Philosophiae Naturalis Principia Mathematica".

[27] The second extract is quoted and translated in W.W. Rouse Ball, "An Essay on Newton's 'Principia'" (London and New York: Macmillan, 1893), at page 69.

[28] The original statements by Clairaut (in French) are found (with orthography here as in the original) in "Explication abregée du systême du monde, et explication des principaux phénomenes astronomiques tirée des Principes de M. Newton" (1759), at Introduction (section IX), page 6: "Il ne faut pas croire que cette idée ... de Hook diminue la gloire de M. Newton", [and] "L'exemple de Hook" [serve] "à faire voir quelle distance il y a entre une vérité entrevue & une vérité démontrée".

[29] Mohr, Peter J.; Taylor, Barry N.; Newell, David B. (2008). "CODATA Recommended Values of the Fundamental Physical Constants: 2006". *Rev. Mod. Phys.* **80** (2): 633–730. arXiv:0801.0028. Bibcode:2008RvMP...80..633M. doi:10.1103/RevModPhys.80.633. Direct link to value..

[30] Equilibrium State

[31] Misner, Charles W.; Thorne, Kip S.; Wheeler, John Archibald (1973). *Gravitation*. New York: W. H.Freeman and Company. ISBN 0-7167-0344-0 Page 1049.

[32] - Max Born (1924), *Einstein's Theory of Relativity* (The 1962 Dover edition, page 348 lists a table documenting the observed and calculated values for the precession of the perihelion of Mercury, Venus, and the Earth.)

[33] - *The Construction of Modern Science: Mechanisms and Mechanics*, by Richard S. Westfall. Cambridge University Press. 1978

[34] http://physicsessays.org/doi/abs/10.4006/1.3038751? journalCode=phes

[35] http://journals.aps.org/prc/abstract/10.1103/PhysRevC.75. 015501

[36] Leimanis and Minorsky: Our interest is with Leimanis, who first discusses some history about the *n*-body problem, especially Ms. Kovalevskaya's ~1868-1888, twenty-year complex-variables approach, failure; **Section 1: The Dynamics of Rigid Bodies and Mathematical Exterior Ballistics** (Chapter 1, *the motion of a rigid body about a fixed point* (**Euler** and **Poisson** *equations*); Chapter 2, *Mathematical Exterior Ballistics*), good precursor background to the *n*-body problem; **Section 2: Celestial Mechanics** (Chapter 1, *The Uniformization of the Three-body Problem* (Restricted Three-body Problem); Chapter 2, *Capture in the Three-Body Problem*; Chapter 3, *Generalized n-body Problem*).

[37] See References sited for Heggie and Hut. This Wikipedia page has made their approach obsolete.

[38] *Quasi-steady* loads refers to the instantaneous inertial loads generated by instantaneous angular velocities and accelerations, as well as translational accelerations (9 variables). It is as though one took a photograph, which also recorded the instantaneous position and properties of motion. In contrast, a *steady-state* condition refers to a system's state being invariant to time; otherwise, the first derivatives and all higher derivatives are zero.

[39] R. M. Rosenberg states the *n*-body problem similarly (see References): *Each particle in a system of a finite number of particles is subjected to a Newtonian gravitational attraction from all the other particles, and to no other forces. If the initial state of the system is given, how will the particles move?* Rosenberg failed to realize, like everyone else, that it is necessary to determine the forces *first* before the motions can be determined.

[40] A general, classical solution in terms of first integrals is known to be impossible. An exact theoretical solution for arbitrary *n* can be approximated via Taylor series, but in practice such an infinite series must be truncated, giving at best only an approximate solution; and an approach now obsolete. In addition, the *n*-body problem may be solved

using numerical integration, but these, too, are approximate solutions; and again obsolete. See Sverre J. Aarseth's book **Gravitational N-body Simulations** listed in the References.

16.12 External links

- Feather & Hammer Drop on Moon on YouTube

- Newton's Law of Universal Gravitation Javascript calculator

Chapter 17

Weight

This page is about the physical concept. In law, commerce, and in colloquial usage *weight* may also refer to mass. For other uses see weight (disambiguation).

In science and engineering, the **weight** of an object is usually taken to be the force on the object due to gravity.[1][2] Weight is a vector whose magnitude (a scalar quantity), often denoted by an italic letter W, is the product of the mass m of the object and the magnitude of the local gravitational acceleration g;[3] thus: $W = mg$. The unit of measurement for weight is that of force, which in the International System of Units (SI) is the newton. For example, an object with a mass of one kilogram has a weight of about 9.8 newtons on the surface of the Earth, and about one-sixth as much on the Moon. In this sense of weight, a body can be weightless only if it is far away (in principle infinitely far away) from any other mass. Although weight and mass are scientifically distinct quantities, the terms are often confused with each other in everyday use (i.e. comparing and converting force weight in pounds to mass in kilograms and vice-versa).[4]

There is also a rival tradition within Newtonian physics and engineering which sees weight as that which is measured when one uses scales. There the weight is a measure of the magnitude of the reaction force exerted on a body. Typically, in measuring an object's weight, the object is placed on scales at rest with respect to the earth, but the definition can be extended to other states of motion. Thus, in a state of free fall, the weight would be zero. In this second sense of weight, terrestrial objects can be weightless. Ignoring air resistance, the famous apple falling from the tree, on its way to meet the ground near Isaac Newton, is weightless.

Further complications in elucidating the various concepts of weight have to do with the theory of relativity according to which gravity is modelled as a consequence of the curvature of spacetime. In the teaching community, a considerable debate has existed for over half a century on how to define weight for their students. The current situation is that a multiple set of concepts co-exist and find use in their various contexts.[2]

17.1 History

Ancient Greek official bronze weights dating from around the 6th century BC, exhibited in the Ancient Agora Museum in Athens, housed in the Stoa of Attalus.

Discussion of the concepts of heaviness (weight) and lightness (levity) date back to the ancient Greek philosophers. These were typically viewed as inherent properties of objects. Plato described weight as the natural tendency of objects to seek their kin. To Aristotle weight and levity represented the tendency to restore the natural order of the basic elements: air, earth, fire and water. He ascribed absolute weight to earth and absolute levity to fire. Archimedes saw weight as a quality opposed to buoyancy, with the conflict between the two determining if an object sinks or floats. The first operational definition of weight was given by Euclid, who defined weight as: "weight is the heaviness or lightness of one thing, compared to another, as measured by a balance."[2] Operational balances (rather than definitions) had, however, been around much longer.[6]

According to Aristotle, weight was the direct cause of the falling motion of an object, the speed of the falling object was supposed to be directly proportionate to the weight of the object. As medieval scholars discovered that in practice the speed of a falling object increased with time, this prompted a change to the concept of weight to maintain

Weighing grain, from the Babur-namah[5]

this cause effect relationship. Weight was split into a "still weight" or *pondus*, which remained constant, and the actual gravity or *gravitas*, which changed as the object fell. The concept of *gravitas* was eventually replaced by Jean Buridan's impetus, a precursor to momentum.[2]

The rise of the Copernican view of the world led to the resurgence of the Platonic idea that like objects attract but in the context of heavenly bodies. In the 17th century, Galileo made significant advances in the concept of weight. He proposed a way to measure the difference between the weight of a moving object and an object at rest. Ultimately, he concluded weight was proportionate to the amount of matter of an object, and not the speed of motion as supposed by the Aristotelean view of physics.[2]

17.1.1 Newton

The introduction of Newton's laws of motion and the development of Newton's law of universal gravitation led to

considerable further development of the concept of weight. Weight became fundamentally separate from mass. Mass was identified as a fundamental property of objects connected to their inertia, while weight became identified with the force of gravity on an object and therefore dependent on the context of the object. In particular, Newton considered weight to be relative to another object causing the gravitational pull, e.g. the weight of the Earth towards the Sun.[2]

Newton considered time and space to be absolute. This allowed him to consider concepts as true position and true velocity. Newton also recognized that weight as measured by the action of weighing was affected by environmental factors such as buoyancy. He considered this a false weight induced by imperfect measurement conditions, for which he introduced the term *apparent weight* as compared to the *true weight* defined by gravity.[2]

Although Newtonian physics made a clear distinction between weight and mass, the term weight continued to be commonly used when people meant mass. This led the 3rd General Conference on Weights and Measures (CGPM) of 1901 to officially declare "The word *weight* denotes a quantity of the same nature as a *force*: the weight of a body is the product of its mass and the acceleration due to gravity", thus distinguishing it from mass for official usage.

17.1.2 Relativity

In the 20th century, the Newtonian concepts of absolute time and space were challenged by relativity. Einstein's principle of equivalence put all observers, moving or accelerating, on the same footing. This led to an ambiguity as to what exactly is meant by the force of gravity and weight. A scale in an accelerating elevator cannot be distinguished from a scale in a gravitational field. Gravitational force and weight thereby became essentially frame-dependent quantities. This prompted the abandonment of the concept as superfluous in the fundamental sciences such as physics and chemistry. Nonetheless, the concept remained important in the teaching of physics. The ambiguities introduced by relativity led, starting in the 1960s, to considerable debate in the teaching community as how to define weight for their students, choosing between a nominal definition of weight as the force due to gravity or an operational definition defined by the act of weighing.[2]

17.2 Definitions

Several definitions exist for *weight*, not all of which are equivalent.[3][7][8][9]

This top-fuel dragster can accelerate from zero to 160 kilometres per hour (99 mph) in 0.86 seconds. This is a horizontal accelera- tion of 5.3 g. Combined with the vertical g-force in the stationary case the Pythagorean theorem yields a g-force of 5.4 g. It is this g- force that causes the driver's weight if one uses the operational def- inition. If one uses the gravitational definition, the driver's weight is unchanged by the motion of the car.

17.2.1   Gravitational definition

The most common definition of weight found in introduc- tory physics textbooks defines weight as the force exerted on a body by gravity.[1][9] This is often expressed in the for- mula $W = mg$, where W is the weight, m the mass of the object, and g gravitational acceleration.

In 1901, the 3rd General Conference on Weights and Mea- sures (CGPM) established this as their official definition of *weight*:

> "The word *weight* denotes a quantity of the same nature[Note 1] as a *force*: the weight of a body is the product of its mass and the acceleration due to gravity."
> — Resolution 2 of the 3rd General Conference on Weights and Measures[11][12]

This resolution defines weight as a vector, since force is a vector quantity. However, some textbooks also take weight to be a scalar by defining:

> "The weight W of a body is equal to the magnitude Fg of the gravitational force on the body."[13]

The gravitational acceleration varies from place to place. Sometimes, it is simply taken to have a standard value of 9.80665 m/s², which gives the standard weight.[11]

The force whose magnitude is equal to mg newtons is also known as the **m kilogram weight** (which term is abbrevi- ated to **kg-wt**)[14]

Measuring weight versus mass

Left: A spring scale measures weight, by seeing how much the object pushes on a spring (inside the device). On the Moon, an object would give a lower reading. Right: A balance scale indirectly measures mass, by comparing an object to references. On the Moon, an object would give the same reading, because the object and references would *both* become lighter.

17.2.2   Operational definition

In the operational definition, the weight of an object is the force measured by the operation of weighing it, which is **the force it exerts on its support**.[7] Since, W=downward force on the body by the centre of earth, and there is no acceleration in the body.So,there exists opposite and equal force by the support on the body.Also it is equal to the force exerted by the body on its support because action and re- action have same numerical value and opposite direction. This can make a considerable difference, depending on the details; for example, an object in free fall exerts little if any force on its support, a situation that is commonly referred to as weightlessness. However, being in free fall does not affect the weight according to the gravitational definition. Therefore, the operational definition is sometimes refined by requiring that the object be at rest. However, this raises the issue of defining "at rest" (usually being at rest with re- spect to the Earth is implied by using standard gravity). In the operational definition, the weight of an object at rest on the surface of the Earth is lessened by the effect of the centrifugal force from the Earth's rotation.

The operational definition, as usually given, does not ex- plicitly exclude the effects of buoyancy, which reduces the measured weight of an object when it is immersed in a fluid such as air or water. As a result, a floating balloon or an object floating in water might be said to have zero weight.

17.2.3 ISO definition

In the ISO International standard ISO 80000-4(2006),[15] describing the basic physical quantities and units in mechanics as a part of the International standard ISO/IEC 80000, the definition of *weight* is given as:

Definition

$F_g = mg$,

where m is mass and g is local acceleration of free fall.

Remarks

- It should be noted that, when the reference frame is Earth, this quantity comprises not only the local gravitational force, but also the local centrifugal force due to the rotation of the Earth, a force which varies with latitude.
- The effect of atmospheric buoyancy is excluded in the weight.
- In common parlance, the name "weight" continues to be used where "mass" is meant, but this practice is deprecated.

— ISO 80000-4 (2006)

The definition is dependent on the chosen frame of reference. When the chosen frame is co-moving with the object in question then this definition precisely agrees with the operational definition.[8] If the specified frame is the surface of the Earth, the weight according to the ISO and gravitational definitions differ only by the centrifugal effects due to the rotation of the Earth.

17.2.4 Apparent weight

Main article: Apparent weight

In many real world situations the act of weighing may produce a result that differs from the ideal value provided by the definition used. This is usually referred to as the apparent weight of the object. A common example of this is the effect of buoyancy, when an object is immersed in a fluid the displacement of the fluid will cause an upward force on the object, making it appear lighter when weighed on a scale.[16] The apparent weight may be similarly affected by levitation and mechanical suspension. When the gravitational definition of weight is used, the operational weight measured by an accelerating scale is often also referred to as the apparent weight.[17]

17.3 Mass

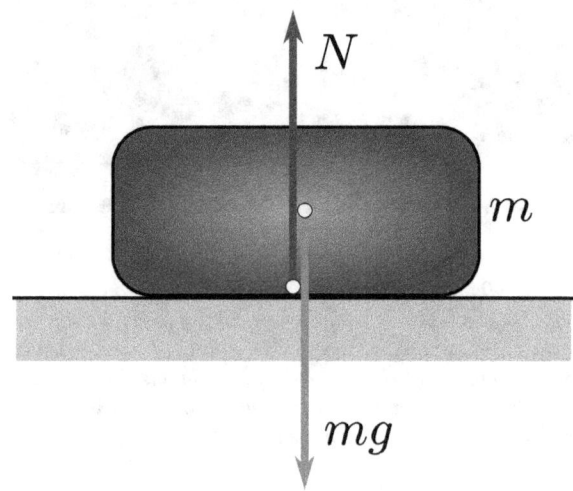

A force diagram showing the forces acting on an object at rest on a surface. Notice that the amount of force that the table is pushing upward on the object (the N vector) is equal to the downward force of the object's weight (shown here as mg, as weight is equal to the object's mass multiplied with the acceleration due to gravity): because these forces are equal, the object is in a state of equilibrium (all the forces acting on it balance to zero).

Main article: Mass versus weight

In modern scientific usage, weight and mass are fundamentally different quantities: mass is an "extrinsic" (extensive) property of matter, whereas weight is a *force* that results from the action of gravity on matter: it measures how strongly the force of gravity pulls on that matter. However, in most practical everyday situations the word "weight" is used when, strictly, "mass" is meant.[4][18] For example, most people would say that an object "weighs one kilogram", even though the kilogram is a unit of mass.

The distinction between mass and weight is unimportant for many practical purposes because the strength of gravity does not vary too much on the surface of the Earth. In a uniform gravitational field, the gravitational force exerted on an object (its weight) is directly proportional to its mass. For example, object A weighs 10 times as much as object B, so therefore the mass of object A is 10 times greater than that of object B. This means that an object's mass can be measured indirectly by its weight, and so, for everyday purposes, weighing (using a weighing scale) is an entirely acceptable way of measuring mass. Similarly, a balance measures mass indirectly by comparing the weight of the measured item to that of an object(s) of known mass. Since the measured item and the comparison mass are in virtually the same location, so experiencing the same gravitational field, the effect of varying gravity does not affect the comparison or the resulting measurement.

The Earth's gravitational field is not uniform but can vary by as much as 0.5%[19] at different locations on Earth (see Earth's gravity). These variations alter the relationship between weight and mass, and must be taken into account in high precision weight measurements that are intended to indirectly measure mass. Spring scales, which measure local weight, must be calibrated at the location at which the objects will be used to show this standard weight, to be legal for commerce.

This table shows the variation of acceleration due to gravity (and hence the variation of weight) at various locations on the Earth's surface.[20]

The historic use of "weight" for "mass" also persists in some scientific terminology – for example, the chemical terms "atomic weight", "molecular weight", and "formula weight", can still be found rather than the preferred "atomic mass" etc.

In a different gravitational field, for example, on the surface of the Moon, an object can have a significantly different weight than on Earth. The gravity on the surface of the Moon is only about one-sixth as strong as on the surface of the Earth. A one-kilogram mass is still a one-kilogram mass (as mass is an extrinsic property of the object) but the downward force due to gravity, and therefore its weight, is only one-sixth of what the object would have on Earth. So a man of mass 180 pounds weighs only about 30 pounds-force when visiting the Moon.

17.3.1   SI units

In most modern scientific work, physical quantities are measured in SI units. The SI unit of weight is the same as that of force: the newton (N) – a derived unit which can also be expressed in SI base units as kg·m/s^2 (kilograms times meters per second squared).[18]

In commercial and everyday use, the term "weight" is usually used to mean mass, and the verb "to weigh" means "to determine the mass of" or "to have a mass of". Used in this sense, the proper SI unit is the kilogram (kg).[18]

17.3.2   Pound and other non-SI units

In United States customary units, the pound can be either a unit of force or a unit of mass.[21] Related units used in some distinct, separate subsystems of units include the poundal and the slug. The poundal is defined as the force necessary to accelerate an object of one-pound *mass* at 1 ft/s^2, and is equivalent to about 1/32.2 of a pound-*force*. The slug

is defined as the amount of mass that accelerates at 1 ft/s^2 when one pound-force is exerted on it, and is equivalent to about 32.2 pounds (mass).

The kilogram-force is a non-SI unit of force, defined as the force exerted by a one kilogram mass in standard Earth gravity (equal to 9.80665 newtons exactly). The dyne is the cgs unit of force and is not a part of SI, while weights measured in the cgs unit of mass, the gram, remain a part of SI.

17.4   Sensation

See also: Apparent weight

The sensation of weight is caused by the force exerted by fluids in the vestibular system, a three-dimensional set of tubes in the inner ear. It is actually the sensation of g-force, regardless of whether this is due to being stationary in the presence of gravity, or, if the person is in motion, the result of any other forces acting on the body such as in the case of acceleration or deceleration of a lift, or centrifugal forces when turning sharply.

17.5   Measuring

Main article: Weighing scale
"Weigh" redirects here. For other uses, see Weigh (disambiguation).

Weight is commonly measured using one of two methods.

A weighbridge, used for weighing trucks

A spring scale or hydraulic or pneumatic scale measures local weight, the local force of gravity on the object (strictly *apparent* weight force). Since the local force of gravity can vary by up to 0.5% at different locations, spring scales will measure slightly different weights for the same object (the same mass) at different locations. To standardize weights, scales are always calibrated to read the weight an object would have at a nominal standard gravity of 9.80665 m/s^2 (approx. 32.174 ft/s^2). However, this calibration is done at the factory. When the scale is moved to another location on Earth, the force of gravity will be different, causing a slight error. So to be highly accurate, and legal for commerce, spring scales must be re-calibrated at the location at which they will be used.

A *balance* on the other hand, compares the weight of an unknown object in one scale pan to the weight of standard masses in the other, using a lever mechanism – a lever-balance. The standard masses are often referred to, non-technically, as *"weights"*. Since any variations in gravity will act equally on the unknown and the known weights, a lever-balance will indicate the same value at any location on Earth. Therefore, balance *"weights"* are usually calibrated and marked in mass units, so the lever-balance measures mass by comparing the Earth's attraction on the unknown object and standard masses in the scale pans. In the absence of a gravitational field, away from planetary bodies (e.g. space), a lever-balance would not work, but on the Moon, for example, it would give the same reading as on Earth. Some balances can be marked in weight units, but since the weights are calibrated at the factory for standard gravity, the balance will measure standard weight, i.e. what the object would weigh at standard gravity, not the actual local force of gravity on the object.

If the actual force of gravity on the object is needed, this can be calculated by multiplying the mass measured by the balance by the acceleration due to gravity – either standard gravity (for everyday work) or the precise local gravity (for precision work). Tables of the gravitational acceleration at different locations can be found on the web.

Gross weight is a term that is generally found in commerce or trade applications, and refers to the total weight of a product and its packaging. Conversely, **net weight** refers to the weight of the product alone, discounting the weight of its container or packaging; and **tare weight** is the weight of the packaging alone.

17.6 Relative weights on the Earth and other celestial bodies

Main articles: Earth's gravity and Surface gravity

The table below shows comparative gravitational accelerations at the surface of the Sun, the Earth's moon, each of the planets in the solar system. The "surface" is taken

to mean the cloud tops of the gas giants (Jupiter, Saturn, Uranus and Neptune). For the Sun, the surface is taken to mean the photosphere. The values in the table have not been de-rated for the centrifugal effect of planet rotation (and cloud-top wind speeds for the gas giants) and therefore, generally speaking, are similar to the actual gravity that would be experienced near the poles.

17.7 See also

- Body weight
- weight, the English unit

17.8 Notes

[1] The phrase "quantity of the same nature" is a literal translation of the French phrase *grandeur de la même nature*. Although this is an authorized translation, VIM 3 of the International Bureau of Weights and Measures recommends translating *grandeurs de même nature* as *quantities of the same kind*.[10]

17.9 References

[1] Richard C. Morrison (1999). "Weight and gravity - the need for consistent definitions". *The Physics Teacher*. **37**: 51. Bibcode:1999PhTea..37...51M. doi:10.1119/1.880152.

[2] Igal Galili (2001). "Weight versus gravitational force: historical and educational perspectives". *International Journal of Science Education*. **23**: 1073. Bibcode:2001IJSEd..23.1073G. doi:10.1080/09500690110038585.

[3] Gat, Uri (1988). "The weight of mass and the mess of weight". In Richard Alan Strehlow. *Standardization of Technical Terminology: Principles and Practice* – second volume. ASTM International. pp. 45–48. ISBN 978-0-8031-1183-7.

[4] The National Standard of Canada, CAN/CSA-Z234.1-89 Canadian Metric Practice Guide, January 1989:

- 5.7.3 Considerable confusion exists in the use of the term "weight." In commercial and everyday use, the term "weight" nearly always means mass. In science and technology "weight" has primarily meant a force due to gravity. In scientific and technical work, the term "weight" should be replaced by the term "mass" or "force," depending on the application.

- 5.7.4 The use of the verb "to weigh" meaning "to determine the mass of," e.g., "I weighed this object and determined its mass to be 5 kg," is correct.

[5] Sur Das (1590s). "Weighing Grain". *Baburnama*.

[6] http://www.averyweigh-tronix.com/museum accessed 29 March 2013.

[7] Allen L. King (1963). "Weight and weightlessness". *American Journal of Physics*. **30**: 387. Bibcode:1962AmJPh..30..387K. doi:10.1119/1.1942032.

[8] A. P. French (1995). "On weightlessness". *American Journal of Physics*. **63**: 105–106. Bibcode:1995AmJPh..63..105F. doi:10.1119/1.17990.

[9] Galili, I.; Lehavi, Y. (2003). "The importance of weightlessness and tides in teaching gravitation" (PDF). *American Journal of Physics*. **71** (11): 1127–1135. Bibcode:2003AmJPh..71.1127G. doi:10.1119/1.1607336.

[10] Working Group 2 of the Joint Committee for Guides in Metrology (JCGM/WG 2) (2008). *International vocabulary of metrology — Basic and general concepts and associated terms (VIM) — Vocabulaire international de métrologie — Concepts fondamentaux et généraux et termes associés (VIM)* (PDF) (JCGM 200:2008) (in English and French) (3rd ed.). BIPM. Note 3 to Section 1.2.

[11] "Resolution of the 3rd meeting of the CGPM (1901)". BIPM.

[12] Barry N. Taylor; Ambler Thompson, eds. (2008). *The International System of Units (SI)* (PDF). NIST Special Publication 330 (2008 ed.). NIST. p. 52.

[13] Halliday, David; Resnick, Robert; Walker, Jearl (2007). *Fundamentals of Physics*. **1** (8th ed.). Wiley. p. 95. ISBN 978-0-470-04473-5.

[14] Chester, W. Mechanics. George Allen & Unwin. London. 1979. ISBN 0-04-510059-4. Section 3.2 at page 83.

[15] ISO 80000-4:2006, Quantities and units - Part 4: Mechanics

[16] Bell, F. (1998). *Principles of mechanics and biomechanics*. Stanley Thornes Ltd. pp. 174–176. ISBN 978-0-7487-3332-3.

[17] Galili, Igal (1993). "Weight and gravity: teachers' ambiguity and students' confusion about the concepts". *International Journal of Science Education*. **15** (2): 149–162. Bibcode:1993IJSEd..15..149G. doi:10.1080/0950069930150204.

[18] A. Thompson & B. N. Taylor (March 3, 2010) [July 2, 2009]. "The NIST Guide for the use of the International System of Units, Section 8: Comments on Some Quantities and Their Units". *Special Publication 811*. NIST. Retrieved 2010-05-22.

[19] Hodgeman, Charles, ed. (1961). *Handbook of Chemistry and Physics* (44th ed.). Cleveland, USA: Chemical Rubber Publishing Co. pp. 3480–3485.

[20] Clark, John B (1964). *Physical and Mathematical Tables*. Oliver and Boyd.

[21] "Common Conversion Factors, Approximate Conversions from U.S. Customary Measures to Metric". National Institute of Standards and Technology. Retrieved 2013-09-03.

[22] This value excludes the adjustment for centrifugal force due to Earth's rotation and is therefore greater than the 9.80665 m/s^2 value of standard gravity.

Chapter 18

Nuclear force

This article is about the force that holds nucleons together in a nucleus. For the force that holds quarks together in a nucleon, see Strong interaction.

Not to be confused with weak nuclear force.

The **nuclear force** (or **nucleon–nucleon interaction** or

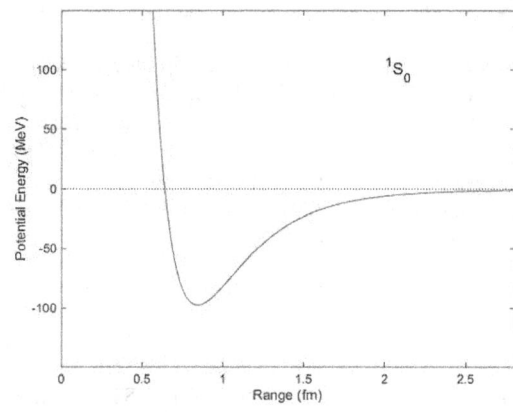

Corresponding potential energy (in units of MeV) of two nucleons as a function of distance as computed from the Reid potential. The potential well is a minimum at a distance of about 0.8 fm. With this potential nucleons can become bound with a negative "binding energy."

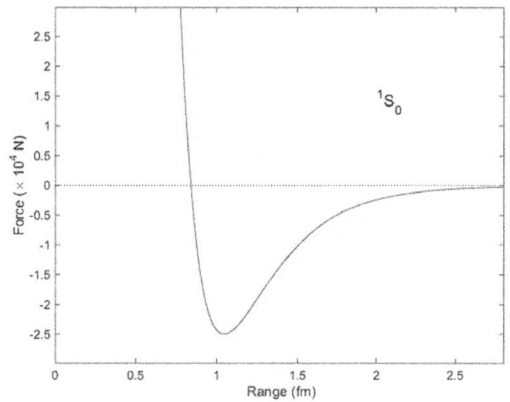

Force (in units of 10,000 N) between two nucleons as a function of distance as computed from the Reid potential (1968).[1] The spins of the neutron and proton are aligned, and they are in the S angular momentum state. The attractive (negative) force has a maximum at a distance of about 1 fm with a force of about 25,000 N. Particles much closer than a distance of 0.8 fm experience a large repulsive (positive) force. Particles separated by a distance greater than 1 fm are still attracted (Yukawa potential), but the force falls as an exponential function of distance.

residual strong force) is the force between protons and neutrons, subatomic particles that are collectively called nucleons. The nuclear force is responsible for binding protons and neutrons into atomic nuclei. Neutrons and protons are affected by the nuclear force almost identically. Since protons have charge $+1\ e$, they experience a strong electric field repulsion (following Coulomb's law) that tends to push them apart, but at short range the attractive nuclear force overcomes the repulsive electromagnetic force. The mass of a nucleus is less than the sum total of the individual masses of the protons and neutrons which form it. The

difference in mass between bound and unbound nucleons is known as the mass defect. Energy is released when some large nuclei break apart, and it is this energy that is used in nuclear power and nuclear weapons.[2][3]

The nuclear force is powerfully attractive between nucleons at distances of about 1 femtometer (fm, or 1.0×10^{-15} metres) between their centers, but rapidly decreases to insignificance at distances beyond about 2.5 fm. At distances less than 0.7 fm, the nuclear force becomes repulsive. This repulsive component is responsible for the physical size of nuclei, since the nucleons can come no closer than the force allows. By comparison, the size of an atom, measured in angstroms (Å, or 1.0×10^{-10} m), is five orders of magnitude larger. The nuclear force is not simple, however, since it depends on the nucleon spins, has a tensor component, and may depend on the relative momentum of the nucleons.[4]

A quantitative description of the nuclear force relies on partially empirical equations that model the internucleon potential energies, or potentials. (Generally, forces within a system of particles can be more simply modeled by describing the system's potential energy; the negative gradient of a potential is equal to the vector force.) The constants for the equations are phenomenological, that is, determined by fitting the equations to experimental data. The internucleon potentials attempt to describe the properties of nucleon–nucleon interaction. Once determined, any given potential can be used in, e.g., the Schrödinger equation to determine the quantum mechanical properties of the nucleon system.

18.1   Description

The nuclear force is only felt between hadrons (particles composed of quarks). At small separations between nucleons (less than ~ 0.7 fm between their centers, depending upon spin alignment) the force becomes repulsive, which keeps the nucleons at a certain average separation, even if they are of different types. This repulsion arises from the Pauli exclusion force for identical nucleons (such as two neutrons or two protons). A Pauli exclusion force also occurs between quarks of the same type within nucleons, when the nucleons are different (a proton and a neutron, for example).

18.1.1   Field Strength

At distances larger than 0.7 fm the force becomes attractive between spin-aligned nucleons, becoming maximal at a center–center distance of about 0.9 fm. Beyond this distance the force drops exponentially, until beyond about 2.0 fm separation, the force is negligible. Nucleons have a radius of about 0.8 fm.[5]

At short distances (less than 1.7 fm or so), the nuclear force is stronger than the Coulomb force between protons; it thus overcomes the repulsion of protons inside the nucleus. However, the Coulomb force between protons has a much larger range due to its falloff as the inverse square of charge separation, and Coulomb repulsion thus becomes the only significant force between protons when their separation exceeds about 2 to 2.5 fm.

The nuclear force has a spin-dependent component. The force is stronger for particles with their spins aligned than for those with their spins anti-aligned. So, for two particles that are the same (such as two neutrons or two protons) the force is not enough to bind the particles, since the spin vectors of two particles of the same type must point in opposite directions when the particles are near each other and are (save for spin) in the same quantum state. This requirement for fermions stems from the Pauli exclusion principle. For fermion particles of different types (such as a proton and neutron), particles may be close to each other and have

aligned spins without violating the Pauli exclusion principle, and the nuclear force may bind them (in this case, into a deuteron), since the nuclear force is much stronger for spin-aligned particles. But if the particles' spins are anti-aligned the nuclear force is too weak to bind them, even if they are of different types.

The nuclear force also has a tensor component which depends on the interaction between the nucleon spins and the angular momentum of the nucleons, leading to deformation from a simple spherical shape.

18.1.2   Nuclear Binding

To disassemble a nucleus into unbound protons and neutrons requires work against the nuclear force. Conversely, energy is released when a nucleus is created from free nucleons or other nuclei: the nuclear binding energy. Because of mass–energy equivalence (i.e. Einstein's famous formula $E = mc^2$), releasing this energy causes the mass of the nucleus to be lower than the total mass of the individual nucleons, leading to the so-called "mass defect".[6]

The nuclear force is nearly independent of whether the nucleons are neutrons or protons. This property is called *charge independence*. The force depends on whether the spins of the nucleons are parallel or antiparallel, and it has a non-central or *tensor* component. This part of the force does not conserve orbital angular momentum, which is a constant of motion under central forces.

The symmetry resulting in the strong force, proposed by Werner Heisenberg, is that protons and neutrons are identical in every respect, other than their charge. This is not completely true, because neutrons are a tiny bit heavier, but it is an approximate symmetry. Protons and neutrons are therefore viewed as the same particle, but with different isospin quantum number. The strong force is invariant under SU(2) transformations, just as particles with intrinsic spin are. Isospin and intrinsic spin are related under this SU(2) symmetry group. There are only strong attractions when the total isospin is 0, as is confirmed by experiment.[7]

The information on nuclear force are obtained by scattering experiments and the study of light nuclei binding energy.

The nuclear force occurs by the exchange of virtual light mesons, such as the virtual pions, as well as two types of virtual mesons with spin (vector mesons), the rho mesons and the omega mesons. The vector mesons account for the spin-dependence of the nuclear force in this "virtual meson" picture.

The nuclear force is separate from what historically was known as the weak nuclear force. The weak interaction is one of the four fundamental interactions, and it refers to such processes as beta decay. The weak force plays no role

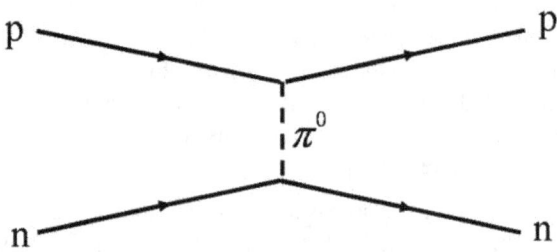

A Feynman diagram of a strong proton–neutron interaction mediated by a neutral pion. Time proceeds from left to right.

in the interaction of nucleons, though it is responsible for the decay of neutrons to protons and vice versa.

18.2 History

The discovery of the neutron in 1932 revealed that atomic nuclei were made of protons and neutrons, held together by an attractive force. By 1935 the nuclear force was conceived to be transmitted by particles called mesons. This theoretical development included a description of the Yukawa potential, an early example of a nuclear potential. Mesons, predicted by theory, were discovered experimentally in 1947. By the 1970s, the quark model had been developed, which showed that the mesons and nucleons were composed of quarks and gluons. By this new model, the nuclear force, resulting from the exchange of mesons between neighboring nucleons, is a residual effect of the strong force.

The nuclear force has been at the heart of nuclear physics ever since the field was born in 1932 with the discovery of the neutron by James Chadwick. The traditional goal of nuclear physics is to understand the properties of atomic nuclei in terms of the 'bare' interaction between pairs of nucleons, or nucleon–nucleon forces (NN forces).

Within months after the discovery of the neutron, Werner Heisenberg[8][9][10] and Dmitri Ivanenko[11] had proposed proton–neutron models for the nucleus.[12] Heisenberg approached the description of protons and neutrons in the nucleus through quantum mechanics, an approach that was not at all obvious at the time. Heisenberg's theory for protons and neutrons in the nucleus was a "major step toward understanding the nucleus as a quantum mechanical system."[13] Heisenberg introduced the first theory of nuclear exchange forces that bind the nucleons. He considered protons and neutrons to be different quantum states of the same particle, i.e., nucleons distinguished by the value of their nuclear isospin quantum numbers.

One of the earliest models for the nucleus was the liquid drop model developed in the 1930s. One property of nu-

clei is that the average binding energy per nucleon is approximately the same for all stable nuclei, which is similar to a liquid drop. The liquid drop model treated the nucleus as a drop of incompressible nuclear fluid, with nucleons behaving like molecules in a liquid. The model was first proposed by George Gamow and then developed by Niels Bohr, Werner Heisenberg and Carl Friedrich von Weizsäcker. This crude model did not explain all the properties of the nucleus, but it did explain the spherical shape of most nuclei. The model also gave good predictions for the nuclear binding energy of nuclei.

In 1934, Hideki Yukawa made the earliest attempt to explain the nature of the nuclear force. According to his theory, massive bosons (mesons) mediate the interaction between two nucleons. Although, in light of quantum chromodynamics (QCD), meson theory is no longer perceived as fundamental, the meson-exchange concept (where hadrons are treated as elementary particles) continues to represent the best working model for a quantitative *NN* potential. The Yukawa potential (also called a screened Coulomb potential) is a potential of the form

$$V_{\text{Yukawa}}(r) = -g^2 \frac{e^{-\mu r}}{r},$$

where g is a magnitude scaling constant, i.e., the amplitude of potential, μ is the Yukawa particle mass, r is the radial distance to the particle. The potential is monotone increasing, implying that the force is always attractive. The constants are determined empirically. The Yukawa potential depends only on the distance between particles, r, hence it models a central force.

Throughout the 1930s a group at Columbia University led by I. I. Rabi developed magnetic resonance techniques to determine the magnetic moments of nuclei. These measurements led to the discovery in 1939 that the deuteron also possessed an electric quadrupole moment.[14][15] This electrical property of the deuteron had been interfering with the measurements by the Rabi group. The deuteron, composed of a proton and a neutron, is one of the simplest nuclear systems. The discovery meant that the physical shape of the deuteron was not symmetric, which provided valuable insight into the nature of the nuclear force binding nucleons. In particular, the result showed that the nuclear force was not a central force, but had a tensor character.[1] Hans Bethe identified the discovery of the deuteron's quadrupole moment as one of the important events during the formative years of nuclear physics.[14]

Historically, the task of describing the nuclear force phenomenologically was formidable. The first semi-empirical quantitative models came in the mid-1950s,[1] such as the Woods–Saxon potential (1954). There was substantial progress in experiment and theory related to the nuclear force in the 1960s and 1970s. One influential model was the Reid potential (1968).[1] In recent years, experimenters have concentrated on the subtleties of the nuclear force, such as its charge dependence, the precise value of the πNN coupling constant, improved phase shift analysis, high-precision NN data, high-precision NN potentials, NN scattering at intermediate and high energies, and attempts to derive the nuclear force from QCD.

18.3 The nuclear force as a residual of the strong force

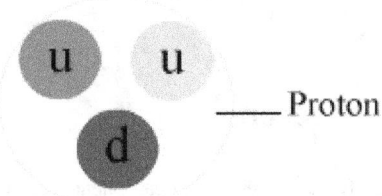

_____ Proton

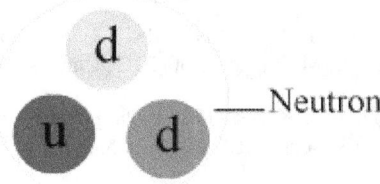

_____ Neutron

An animation of the interaction. The colored double circles are gluons. Anticolors are shown as per this diagram (larger version).

The nuclear force is a residual effect of the more fundamental strong force, or strong interaction. The strong interaction is the attractive force that binds the elementary particles called quarks together to form the nucleons themselves. This more powerful force is mediated by particles called gluons. Gluons hold quarks together with a force like that of electric charge, but of far greater strength. Quarks, gluons and their dynamics are mostly confined within nucleons, but residual influences extend slightly beyond nucleon boundaries to give rise to the nuclear force.

The nuclear forces arising between nucleons are analogous to the forces in chemistry between neutral atoms or molecules called London forces. Such forces between atoms are much weaker than the attractive electrical forces that hold the atoms themselves together (i.e., that bind elec-

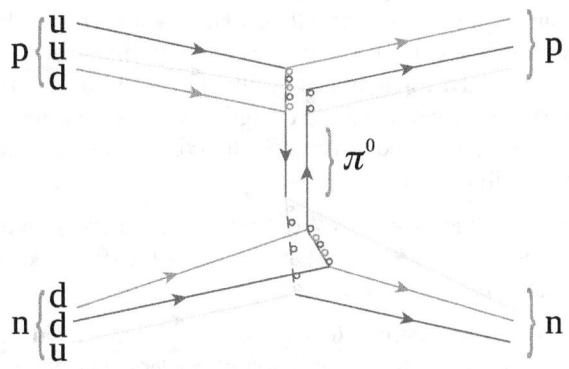

The same diagram as that above with the individual quark constituents shown, to illustrate how the fundamental *strong interaction gives rise to the* **nuclear force**. *Straight lines are quarks, while multi-colored loops are gluons (the carriers of the fundamental force). Other gluons, which bind together the proton, neutron, and pion "in-flight," are not shown.*

trons to the nucleus), and their range between atoms is shorter, because they arise from small separation of charges inside the neutral atom. Similarly, even though nucleons are made of quarks in combinations which cancel most gluon forces (they are "color neutral"), some combinations of quarks and gluons nevertheless leak away from nucleons, in the form of short-range nuclear force fields that extend from one nucleon to another nearby nucleon. These nuclear forces are very weak compared to direct gluon forces ("color forces" or strong forces) inside nucleons, and the nuclear forces extend only over a few nuclear diameters, falling exponentially with distance. Nevertheless, they are strong enough to bind neutrons and protons over short distances, and overcome the electrical repulsion between protons in the nucleus.

Sometimes, the nuclear force is called the **residual strong force**, in contrast to the strong interactions which arise from QCD. This phrasing arose during the 1970s when QCD was being established. Before that time, the *strong nuclear force* referred to the inter-nucleon potential. After the verification of the quark model, *strong interaction* has come to mean QCD.

18.4 Nucleon–nucleon potentials

Two-nucleon systems such as the deuteron, the nucleus of a deuterium atom, as well as proton–proton or neutron–proton scattering are ideal for studying the NN force. Such systems can be described by attributing a *potential* (such as the Yukawa potential) to the nucleons and using the potentials in a Schrödinger equation. The form of the potential is derived phenomenologically (by measurement), although for the long-range interaction, meson-exchange theories help to construct the potential. The parameters of the potential are determined by fitting to experimental data such as the deuteron binding energy or *NN* elastic scattering cross sections (or, equivalently in this context, so-called *NN* phase shifts).

The most widely used *NN* potentials are the Paris potential, the Argonne AV18 potential ,[16] the CD-Bonn potential and the Nijmegen potentials.

A more recent approach is to develop effective field theories for a consistent description of nucleon–nucleon and three-nucleon forces. Quantum hadrodynamics is an effective field theory of the nuclear force, comparable to QCD for color interactions and QED for electromagnetic interactions. Additionally, chiral symmetry breaking can be analyzed in terms of an effective field theory (called chiral perturbation theory) which allows perturbative calculations of the interactions between nucleons with pions as exchange particles.

18.4.1 From nucleons to nuclei

The ultimate goal of nuclear physics would be to describe all nuclear interactions from the basic interactions between nucleons. This is called the *microscopic* or *ab initio* approach of nuclear physics. There are two major obstacles to overcome before this dream can become reality:

- Calculations in many-body systems are difficult and require advanced computation techniques.

- There is evidence that three-nucleon forces (and possibly higher multi-particle interactions) play a significant role. This means that three-nucleon potentials must be included into the model.

This is an active area of research with ongoing advances in computational techniques leading to better first-principles calculations of the nuclear shell structure. Two- and three-nucleon potentials have been implemented for nuclides up to $A = 12$.

18.4.2 Nuclear potentials

A successful way of describing nuclear interactions is to construct one potential for the whole nucleus instead of considering all its nucleon components. This is called the *macroscopic* approach. For example, scattering of neutrons from nuclei can be described by considering a plane wave in the potential of the nucleus, which comprises a real part and an imaginary part. This model is often called the op-

tical model since it resembles the case of light scattered by an opaque glass sphere.

Nuclear potentials can be *local* or *global*: local potentials are limited to a narrow energy range and/or a narrow nuclear mass range, while global potentials, which have more parameters and are usually less accurate, are functions of the energy and the nuclear mass and can therefore be used in a wider range of applications.

18.5 See also

- Strong interaction

- Standard Model

18.6 References

[1] Reid, R.V. (1968). "Local phenomenological nucleon–nucleon potentials". *Annals of Physics*. **50**: 411–448. Bibcode:1968AnPhy..50..411R. doi:10.1016/0003-4916(68)90126-7.

[2] Binding Energy, Mass Defect, Furry Elephant physics educational site, retr 2012 7 1

[3] Chapter 4 NUCLEAR PROCESSES, THE STRONG FORCE, M. Ragheb 1/30/2013, University of Illinois

[4] Kenneth S. Krane (1988). *Introductory Nuclear Physics*. Wiley & Sons. ISBN 0-471-80553-X.

[5] Povh, B.; Rith, K.; Scholz, C.; Zetsche, F. (2002). *Particles and Nuclei: An Introduction to the Physical Concepts*. Berlin: Springer-Verlag. p. 73. ISBN 978-3-540-43823-6.

[6] Stern, Dr. Swapnil Nikam (February 11, 2009). "Nuclear Binding Energy". *"From Stargazers to Starships"*. NASA website. Retrieved 2010-12-30.

[7] Griffiths, David, Introduction to Elementary Particles

[8] Heisenberg, W. (1932). "Über den Bau der Atomkerne. I". *Z. Phys.* **77**: 1–11. doi:10.1007/BF01342433.

[9] Heisenberg, W. (1932). "Über den Bau der Atomkerne. II". *Z. Phys.* **78** (3–4): 156–164. doi:10.1007/BF01337585.

[10] Heisenberg, W. (1933). "Über den Bau der Atomkerne. III". *Z. Phys.* **80** (9–10): 587–596. doi:10.1007/BF01335696.

[11] Iwanenko, D.D., The neutron hypothesis, Nature **129** (1932) 798.

[12] Miller A. I. *Early Quantum Electrodynamics: A Sourcebook*, Cambridge University Press, Cambridge, 1995, ISBN 0521568919, pp. 84–88.

[13] Brown, L.M.; Rechenberg, H. (1996). *The Origin of the Concept of Nuclear Forces*. Bristol and Philadelphia: Institute of Physics Publishing. ISBN 0750303735.

[14] John S. Rigden (1987). *Rabi, Scientist and Citizen*. New York: Basic Books, Inc. pp. 99–114. ISBN 9780674004351. Retrieved May 9, 2015.

[15] Kellogg, J.M.; Rabi, I.I.; Ramsey, N.F.; Zacharias, J.R. (1939). "An electrical quadrupole moment of the deuteron". *Physical Review*. **55**: 318–319. Bibcode:1939PhRv...55..318K. doi:10.1103/physrev.55.318. Retrieved May 9, 2015.

[16] Wiringa, R. B.; Stoks, V. G. J.; Schiavilla, R. (1995). "Accurate nucleon–nucleon potential with charge-independence breaking". *Physical Review C*. **51**: 38. arXiv:nucl-th/9408016. Bibcode:1995PhRvC..51...38W. doi:10.1103/PhysRevC.51.38.

18.7   Bibliography

- Gerald Edward Brown and A. D. Jackson, *The Nucleon–Nucleon Interaction*, (1976) North-Holland Publishing, Amsterdam ISBN 0-7204-0335-9

- R. Machleidt and I. Slaus, "The nucleon–nucleon interaction", *J. Phys.* G **27** (2001) R69 *(topical review)*.

- E.A. Nersesov, *Fundamentals of atomic and nuclear physics*, (1990), Mir Publishers, Moscow, ISBN 5-06-001249-2

- P. Navrátil and W.E. Ormand, "Ab initio shell model with a genuine three-nucleon force for the p-shell nuclei", Phys. Rev. C **68**, 034305 (2003).

18.8   Further reading

- Nuclear Forces Ruprecht Machleidt, Scholarpedia, 9(1):30710. doi:10.4249/scholarpedia.30710

Chapter 19

Strong interaction

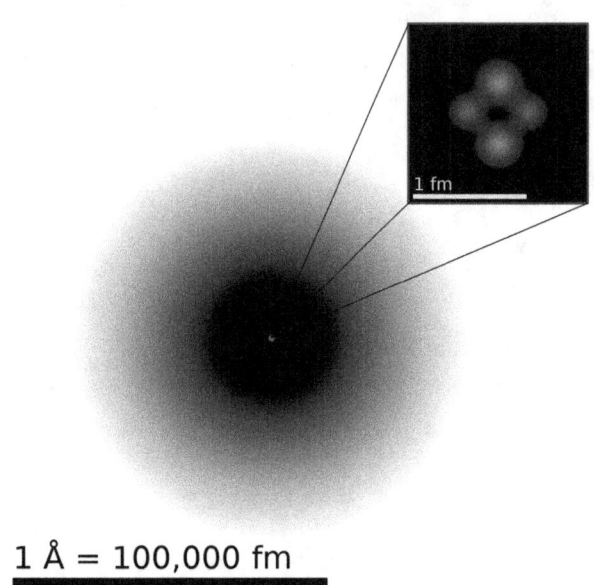

1 Å = 100,000 fm

The nucleus of a helium atom. The two protons have the same charge, but still stay together due to the residual nuclear force

In particle physics, the **strong interaction** is the mechanism responsible for the **strong nuclear force** (also called the **strong force**, **nuclear strong force**), one of the four known fundamental interactions, the others are electromagnetism, the weak interaction and gravitation. At the range of a femtometer, it is the strongest force, is approximately 137 times stronger than electromagnetism, a million times stronger than weak interaction and 10^{38} times stronger than gravitation.[1] The strong nuclear force ensures the stability of ordinary matter, confining quarks into hadron particles, such as the proton and neutron, and the further binding of neutrons and protons into atomic nuclei. Most of the mass-energy of a common proton or neutron is in the form of the strong force field energy; the individual quarks provide only about 1% of the mass-energy of a proton.

The strong interaction is observable at two ranges: on a larger scale (about 1 to 3 femtometers (fm)), it is the force

that binds protons and neutrons (nucleons) together to form the nucleus of an atom. On the smaller scale (less than about 0.8 fm, the radius of a nucleon), it is the force (carried by gluons) that holds quarks together to form protons, neutrons, and other hadron particles. In the latter context, it is often known as the **color force**. The strong force inherently has such a high strength that hadrons bound by the strong force can produce new massive particles. Thus, if hadrons are struck by high-energy particles, they give rise to new hadrons instead of emitting freely moving radiation (gluons). This property of the strong force is called color confinement, and it prevents the free "emission" of the strong force: instead, in practice, jets of massive particles are observed.

In the context of binding protons and neutrons together to form atomic nuclei, the strong interaction is called the nuclear force (or *residual strong force*). In this case, it is the residuum of the strong interaction between the quarks that make up the protons and neutrons. As such, the residual strong interaction obeys a quite different distance-dependent behavior between nucleons, from when it is acting to bind quarks within nucleons. The binding energy that is partly released on the breakup of a nucleus is related to the residual strong force and is harnessed in nuclear power and fission-type nuclear weapons.[2][3]

The strong interaction is hypothesized to be mediated by massless particles called gluons, that are exchanged between quarks, antiquarks, and other gluons. Gluons, in turn, are thought to interact with quarks and gluons as all carry a type of charge called color charge. Color charge is analogous to electromagnetic charge, but it comes in three types rather than one (+/- red, +/- green, +/- blue) that results in a different type of force, with different rules of behavior. These rules are detailed in the theory of quantum chromodynamics (QCD), which is the theory of quark-gluon interactions.

After the Big Bang, during the electroweak epoch, the electroweak force separated from the strong force. A Grand Unified Theory is hypothesized to exist to describe this, no such theory has been successfully formulated yet, and the unification remains an unsolved problem in physics.

19.1   History

Before the 1970s, physicists were uncertain as to how the atomic nucleus was bound together. It was known that the nucleus was composed of protons and neutrons and that protons possessed positive electric charge, while neutrons were electrically neutral. By the understanding of physics at that time, positive charges would repel one another and the positively charged protons should cause the nucleus to fly apart. However, this was never observed. New physics was needed to explain this phenomenon.

A stronger attractive force was postulated to explain how the atomic nucleus was bound despite the protons' mutual electromagnetic repulsion. This hypothesized force was called the *strong force*, which was believed to be a fundamental force that acted on the protons and neutrons that make up the nucleus.

It was later discovered that protons and neutrons were not fundamental particles, but were made up of constituent particles called quarks. The strong attraction between nucleons was the side-effect of a more fundamental force that bound the quarks together into protons and neutrons. The theory of quantum chromodynamics explains that quarks carry what is called a color charge, although it has no relation to visible color.[4] Quarks with unlike color charge attract one another as a result of the **strong interaction**, and the particle that mediated this was called the gluon.

19.2   Details

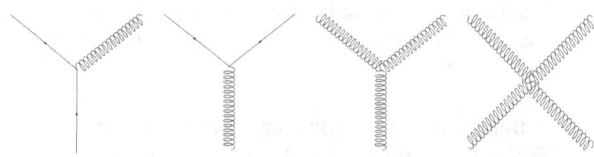

The fundamental couplings of the strong interaction, from left to right: gluon radiation, gluon splitting and gluon self-coupling.

The word *strong* is used since the strong interaction is the "strongest" of the four fundamental forces; its strength is around 137 times that of the electromagnetic force, some 10^6 times as great as that of the weak force, and about 10^{38} times that of gravitation, at a distance of 10^{-15} meter (femtometer) or less.

19.2.1   Behaviour of the strong force

The strong force is described by quantum chromodynamics (QCD), a part of the standard model of particle physics. Mathematically, QCD is a non-Abelian gauge theory based on a local (gauge) symmetry group called SU(3).

Quarks and gluons are the only fundamental particles that carry non-vanishing color charge, and hence participate in strong interactions. The strong force itself acts directly only on elementary quark and gluon particles.

All quarks and gluons in QCD interact with each other through the strong force. The strength of interaction is parametrized by the strong coupling constant. This strength is modified by the gauge color charge of the particle, a group theoretical property.

The strong force acts between quarks. Unlike all other forces (electromagnetic, weak, and gravitational), the strong force does not diminish in strength with increasing distance. After a limiting distance (about the size of a hadron) has been reached, it remains at a strength of about 10,000 newtons, no matter how much farther the distance between the quarks.[5] In QCD, this phenomenon is called color confinement; as a result only hadrons, not individual free quarks, can be observed. The explanation is that the amount of work done against a force of 10,000 newtons is enough to create particle-antiparticle pairs within a very short distance of that interaction. In simple terms, the very energy added to the system required to pull two quarks apart would create a pair of new quarks that will pair up with the original ones. The failure of all experiments that have searched for free quarks is considered to be evidence of this phenomenon.

The elementary quark and gluon particles involved in a high energy collision are not directly observable. They instead emerge as jets of newly created hadrons, whenever sufficient energy is deposited into a quark-quark bond, as when a quark in one proton is struck by a very fast quark of another impacting proton during a particle accelerator experiment. However, quark–gluon plasmas have been observed.

Every quark in the universe does not attract every other quark in the above distance independent manner, since color-confinement implies that the strong force acts without distance-diminishment only between pairs of single quarks, and that in collections of bound quarks (i.e., hadrons), the net color-charge of the quarks cancels out, resulting in a limit of the action of the forces. Collections of quarks (hadrons) therefore appear nearly without color-charge, and the strong force is therefore nearly absent between those hadrons (i.e., between baryons or mesons). However, the cancellation is not quite perfect. A small residual force remains (described below) known as the **residual strong force**. This residual force *does* diminish rapidly with distance, and is thus very short-range (effectively a few femtometers). It manifests as a force between the "colorless" hadrons, and is sometimes known as the **strong nuclear force** or simply nuclear force.

19.2.2 Residual strong force

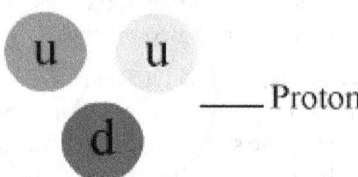

_____ Proton

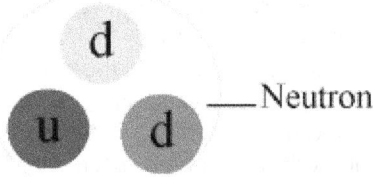

_____ Neutron

An animation of the nuclear force (or residual strong force) inter-
action between a proton and a neutron. The small colored double
circles are gluons, which can be seen binding the proton and neutron
together. These gluons also hold the quark-antiquark combination
called the pion together, and thus help transmit a residual part of the
strong force even between colorless hadrons. Anticolors are shown
as per this diagram. For a larger version, click here

The residual effect of the strong force is called the nuclear force. The nuclear force acts between hadrons, such as mesons or the baryons (such as nucleons) in atomic nuclei. This "residual strong force", acting indirectly, transmits gluons that form part of the virtual pi and rho mesons, which, in turn, transmit the nuclear force between nucleons.

The residual strong force is thus a minor residuum of the strong force that binds quarks together into protons and neutrons. This same force is much weaker *between* neutrons and protons, because it is mostly neutralized *within* them, in the same way that electromagnetic forces between neutral atoms (van der Waals forces) are much weaker than the electromagnetic forces that hold the atoms internally together.[6]

Unlike the strong force itself, the nuclear force, or residual strong force, *does* diminish in strength, and it in fact diminishes rapidly with distance. The decrease is approximately as a negative exponential power of distance, though there is no simple expression known for this; see Yukawa potential. The rapid decrease with distance of the attractive residual force and the less-rapid decrease of the repulsive electromagnetic force acting between protons within a nucleus, causes the instability of larger atomic nuclei, such as all those with atomic numbers larger than 82 (the element lead).

19.3   See also

- Nuclear binding energy

- Color charge

- Coupling constant

- Nuclear physics

- QCD matter

- Quantum field theory and Gauge theory

- Standard model of particle physics and Standard Model (mathematical formulation)

- Weak interaction, electromagnetism and gravity

- Intermolecular force

- Vortex

- Yukawa interaction

19.4   References

[1] Relative strength of interaction varies with distance. See for instance Matt Strassler's essay, "The strength of the known forces".

[2] on Binding energy: see Binding Energy, Mass Defect, Furry Elephant physics educational site, retr 2012 7 1

[3] on Binding energy: see Chapter 4 NUCLEAR PROCESSES, THE STRONG FORCE, M. Ragheb 1/27/2012, University of Illinois

[4] Feynman, R. P. (1985). *QED: The Strange Theory of Light and Matter*. Princeton University Press. p. 136. ISBN 0-691-08388-6. The idiot physicists, unable to come up with any wonderful Greek words anymore, call this type of polarization by the unfortunate name of 'color,' which has nothing to do with color in the normal sense.

[5] Fritzsch, op. cite, p. 164. The author states that the force between differently colored quarks remains constant at any distance after they travel only a tiny distance from each other, and is equal to that need to raise one ton, which is 1000 kg x 9.8 m/s^2 = ~10,000 N.

[6] Fritzsch, H. (1983). *Quarks: The Stuff of Matter*. Basic Books. pp. 167–168. ISBN 978-0-465-06781-7.

19.5   Further reading

- Christman, J. R. (2001). "MISN-0-280: *The Strong Interaction*" (PDF). *Project PHYSNET*. External link in |work= (help)

- Griffiths, David (1987). *Introduction to Elementary Particles*. John Wiley & Sons. ISBN 0-471-60386-4.

- Halzen, F.; Martin, A. D. (1984). *Quarks and Leptons: An Introductory Course in Modern Particle Physics*. John Wiley & Sons. ISBN 0-471-88741-2.

- Kane, G. L. (1987). *Modern Elementary Particle Physics*. Perseus Books. ISBN 0-201-11749-5.

- Morris, R. (2003). *The Last Sorcerers: The Path from Alchemy to the Periodic Table*. Joseph Henry Press. ISBN 0-309-50593-3.

19.6   External links

- Strong force at *Encyclopædia Britannica*

Chapter 20

Weak interaction

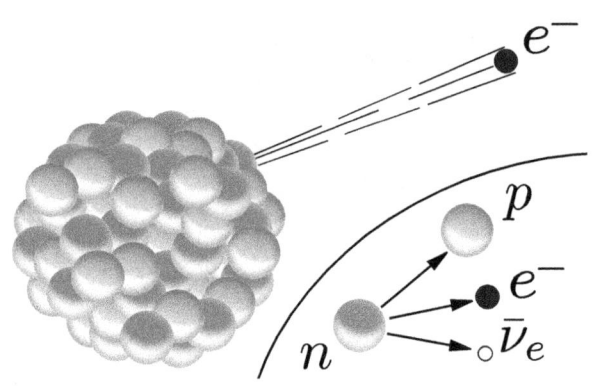

The radioactive beta decay is possibly due to the weak interaction, which transforms a neutron into: a proton, an electron, and an electron antineutrino.

In particle physics, the **weak interaction**, the **weak force** or **weak nuclear force**, is one of the four known fundamental interactions of nature, alongside the strong interaction, electromagnetism, and gravitation. The weak interaction is responsible for radioactive decay, which plays an essential role in nuclear fission. The theory of the weak interaction is sometimes called **quantum flavordynamics** (**QFD**), in analogy with the terms QCD and QED, but the term is rarely used because the weak force is best understood in terms of electro-weak theory (EWT).[1]

In the Standard Model of particle physics, the weak interaction is caused by the emission or absorption of the force carriers, the W and Z bosons. All known fermions interact through the weak interaction. Fermions are particles that have half-integer spin. Spin is one of the fundamental properties of particles. A fermion can be an elementary particle, such as the electron, or it can be a composite particle, such as the proton. The masses of W+, W−, and Z bosons are each far greater than that of interacting protons or neutrons, which is consistent with the short range of the weak force. The force is termed *weak* because its field strength over a given distance is typically several orders of magnitude less than that of the strong nuclear force and electromagnetic force.

During the quark epoch of the early universe, the electroweak force separated into the electromagnetic and weak forces. Important examples of the weak interaction include beta decay, and the fusion of hydrogen into deuterium that powers the Sun's thermonuclear process. Most fermions will decay by a weak interaction over time. Such decay makes radiocarbon dating possible, as carbon-14 decays through the weak interaction to nitrogen-14. It can also create radioluminescence, commonly used in tritium illumination, and in the related field of betavoltaics.[2]

Quarks, which make up composite particles like neutrons and protons, come in six "flavours" – up, down, strange, charm, top and bottom – which give those composite particles their properties. The weak interaction is unique in that it allows for quarks to swap their flavour for another. The swapping of those properties is mediated by the force carrier bosons. For example, during beta minus decay, a down quark within a neutron is changed into an up quark, converting the neutron to a proton and resulting in the emission of an electron and an electron antineutrino. Also, the weak interaction is the only fundamental interaction that breaks parity-symmetry, and similarly, the only one to break charge parity symmetry.

20.1   History

In 1933, Enrico Fermi proposed the first theory of the weak interaction, known as Fermi's interaction. He suggested that beta decay could be explained by a four-fermion interaction, involving a contact force with no range.[3][4]

However, it is better described as a non-contact force field having a finite range, albeit very short. In 1968, Sheldon Glashow, Abdus Salam and Steven Weinberg unified the electromagnetic force and the weak interaction by showing them to be two aspects of a single force, now termed the electro-weak force.

The existence of the W and Z bosons was not directly confirmed until 1983.

20.2 Properties

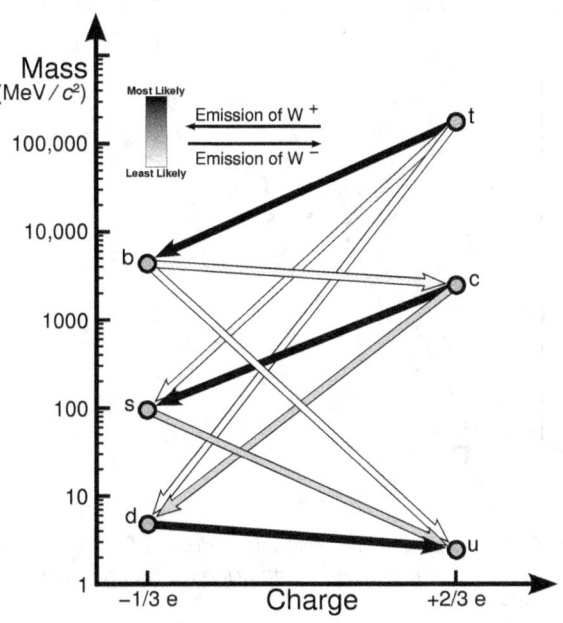

A diagram depicting the various decay routes due to the weak interaction and some indication of their likelihood. The intensity of the lines are given by the CKM parameters.

The weak interaction is unique in a number of respects:

- It is the only interaction capable of changing the flavor of quarks (i.e., of changing one type of quark into another).

- It is the only interaction that violates **P** or parity-symmetry. It is also the only one that violates **CP** symmetry.

- It is propagated by force carrier particles that have significant masses, an unusual feature which is explained in the Standard Model by the Higgs mechanism.

Due to their large mass (approximately 90 GeV/c^2[5]) these carrier particles, termed the W and Z bosons, are short-lived with a lifetime of under 10^{-24} seconds.[6] The weak interaction has a coupling constant (an indicator of interaction strength) of between 10^{-7} and 10^{-6}, compared to the strong interaction's coupling constant of 1 and the electromagnetic coupling constant of about 10^{-2};[7] consequently the weak interaction is weak in terms of strength.[8] The weak interaction has a very short range (around 10^{-17} to 10^{-16} m[8]).[7] At distances around 10^{-18} meters, the weak interaction has a strength of a similar magnitude to the electromagnetic force, but this starts to decrease exponentially with increasing distance. At distances of around 3×10^{-17}

m, the weak interaction is 10,000 times weaker than the electromagnetic.[9]

The weak interaction affects all the fermions of the Standard Model, as well as the Higgs boson; neutrinos interact through gravity and the weak interaction only, and neutrinos were the original reason for the name *weak force*.[8] The weak interaction does not produce bound states (nor does it involve binding energy) – something that gravity does on an astronomical scale, that the electromagnetic force does at the atomic level, and that the strong nuclear force does inside nuclei.[10]

Its most noticeable effect is due to its first unique feature: flavor changing. A neutron, for example, is heavier than a proton (its sister nucleon), but it cannot decay into a proton without changing the flavor (type) of one of its two *down* quarks to an *up* quark. Neither the strong interaction nor electromagnetism permit flavour changing, so this proceeds by **weak decay**; without weak decay, quark properties such as strangeness and charm (associated with the quarks of the same name) would also be conserved across all interactions. All mesons are unstable because of weak decay.[11] In the process known as beta decay, a *down* quark in the neutron can change into an *up* quark by emitting a virtual W− boson which is then converted into an electron and an electron antineutrino.[12] Another example is the electron capture, a common variant of radioactive decay, wherein a proton and an electron within an atom interact, and are changed to a neutron (an up quark is changed to a down quark) and an electron neutrino is emitted.

Due to the large mass of a boson, weak decay occurs more slowly. Hence, weak decay is much less likely to occur before either strong or electromagnetic decay, as they proceed more rapidly. For example, a neutral pion (which decays electromagnetically) has a life of about 10^{-16} seconds, while a charged pion (which decays through the weak interaction) lives about 10^{-8} seconds, a hundred million times longer.[13] In contrast, a free neutron (which also decays through the weak interaction) lives about 15 minutes.[12]

20.2.1 Weak isospin and weak hypercharge

Main article: Weak isospin

All particles have a property called weak isospin (T_3), which serves as a quantum number and governs how that particle behaves in the weak interaction. Weak isospin plays the same role in the weak interaction as does electric charge in electromagnetism, and color charge in the strong interaction. All fermions have a weak isospin value of either $+\frac{1}{2}$ or $-\frac{1}{2}$. For example, the up quark has a T_3 of $+\frac{1}{2}$ and the down quark $-\frac{1}{2}$. A quark never decays through the weak interaction into a quark of the same T_3: quarks with a T_3 of $+\frac{1}{2}$ decay into quarks with a T_3 of $-\frac{1}{2}$ and vice versa.

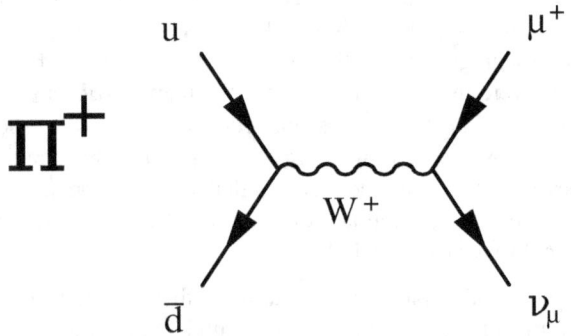

π+ decay through the weak interaction

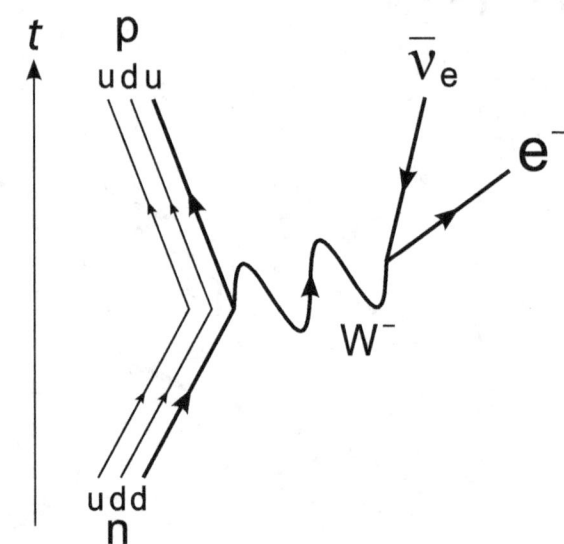

The Feynman diagram for beta-minus decay of a neutron into a proton, electron and electron anti-neutrino, via an intermediate heavy W− boson

In any given interaction, weak isospin is conserved: the sum of the weak isospin numbers of the particles entering the interaction equals the sum of the weak isospin numbers of the particles exiting that interaction. For example, a (left-handed) π+, with a weak isospin of 1 normally decays into a ν

μ (+1/2) and a μ+ (as a right-handed antiparticle, +1/2).[13]

Following the development of the electroweak theory, another property, weak hypercharge, was developed. It is dependent on a particle's electrical charge and weak isospin, and is defined as:

$$Y_W = 2(Q - T_3)$$

where YW is the weak hypercharge of a given type of particle, Q is its electrical charge (in elementary charge units) and T_3 is its weak isospin. Whereas some particles have a weak isospin of zero, all particles, except gluons, have non-zero weak hypercharge. Weak hypercharge is the generator of the U(1) component of the electroweak gauge group.

20.3 Interaction types

There are two types of weak interaction (called *vertices*). The first type is called the "charged-current interaction" because it is mediated by particles that carry an electric charge (the W+ or W− bosons), and is responsible for the beta decay phenomenon. The second type is called the "neutral-current interaction" because it is mediated by a neutral particle, the Z boson.

20.3.1 Charged-current interaction

In one type of charged current interaction, a charged lepton (such as an electron or a muon, having a charge of −1) can

absorb a W+ boson (a particle with a charge of +1) and be thereby converted into a corresponding neutrino (with a charge of 0), where the type ("flavour") of neutrino (electron, muon or tau) is the same as the type of lepton in the interaction, for example:

$$\mu^- + W^+ \to \nu_\mu$$

Similarly, a down-type quark (d with a charge of $-\frac{1}{3}$) can be converted into an up-type quark (u, with a charge of $+\frac{2}{3}$), by emitting a W− boson or by absorbing a W+ boson. More precisely, the down-type quark becomes a quantum superposition of up-type quarks: that is to say, it has a possibility of becoming any one of the three up-type quarks, with the probabilities given in the CKM matrix tables. Conversely, an up-type quark can emit a W+ boson, or absorb a W− boson, and thereby be converted into a down-type quark, for example:

$$d \to u + W^-$$
$$d + W^+ \to u$$
$$c \to s + W^+$$
$$c + W^- \to s$$

The W boson is unstable so will rapidly decay, with a very short lifetime. For example:

$$W^- \rightarrow e^- + \bar{\nu}_e$$
$$W^+ \rightarrow e^+ + \nu_e$$

Decay of the W boson to other products can happen, with varying probabilities.[15]

In the so-called beta decay of a neutron (see picture, above), a down quark within the neutron emits a virtual W− boson and is thereby converted into an up quark, converting the neutron into a proton. Because of the energy involved in the process (i.e., the mass difference between the down quark and the up quark), the W− boson can only be converted into an electron and an electron-antineutrino.[16] At the quark level, the process can be represented as:

$$d \rightarrow u + e^- + \bar{\nu}_e$$

20.3.2   Neutral-current interaction

In neutral current interactions, a quark or a lepton (e.g., an electron or a muon) emits or absorbs a neutral Z boson. For example:

$$e^- \rightarrow e^- + Z^0$$

Like the W boson, the Z boson also decays rapidly,[15] for example:

$$Z^0 \rightarrow b + \bar{b}$$

20.4   Electroweak theory

Main article: Electroweak interaction

The Standard Model of particle physics describes the electromagnetic interaction and the weak interaction as two different aspects of a single electroweak interaction, the theory of which was developed around 1968 by Sheldon Glashow, Abdus Salam and Steven Weinberg. They were awarded the 1979 Nobel Prize in Physics for their work.[17] The Higgs mechanism provides an explanation for the presence of three massive gauge bosons (the three carriers of the weak interaction) and the massless photon of the electromagnetic interaction.[18]

According to the electroweak theory, at very high energies, the universe has four massless gauge boson fields similar to

the photon and a complex scalar Higgs field doublet. However, at low energies, gauge symmetry is spontaneously broken down to the U(1) symmetry of electromagnetism (one of the Higgs fields acquires a vacuum expectation value). This symmetry breaking would produce three massless bosons, but they become integrated by three photon-like fields (through the Higgs mechanism) giving them mass. These three fields become the W+, W− and Z bosons of the weak interaction, while the fourth gauge field, which remains massless, is the photon of electromagnetism.[18]

This theory has made a number of predictions, including a prediction of the masses of the Z and W bosons before their discovery. On 4 July 2012, the CMS and the ATLAS experimental teams at the Large Hadron Collider independently announced that they had confirmed the formal discovery of a previously unknown boson of mass between 125–127 GeV/c^2, whose behaviour so far was "consistent with" a Higgs boson, while adding a cautious note that further data and analysis were needed before positively identifying the new boson as being a Higgs boson of some type. By 14 March 2013, the Higgs boson was tentatively confirmed to exist .[19]

20.5   Violation of symmetry

Left- and right-handed particles: p is the particle's momentum and S is its spin. Note the lack of reflective symmetry between the states.

The laws of nature were long thought to remain the same under mirror reflection. The results of an experiment viewed via a mirror were expected to be identical to the results of a mirror-reflected copy of the experimental apparatus. This so-called law of parity conservation was known to be respected by classical gravitation, electromagnetism and the strong interaction; it was assumed to be a universal law.[20] However, in the mid-1950s Chen Ning Yang and Tsung-Dao Lee suggested that the weak interaction might violate this law. Chien Shiung Wu and collaborators in 1957 discovered that the weak interaction violates parity, earning Yang and Lee the 1957 Nobel Prize in Physics.[21]

Although the weak interaction was once described by Fermi's theory, the discovery of parity violation and renormalization theory suggested that a new approach was needed. In 1957, Robert Marshak and George Sudarshan and, somewhat later, Richard Feynman and Murray Gell-

Mann proposed a **V−A** (vector minus axial vector or left-handed) Lagrangian for weak interactions. In this theory, the weak interaction acts only on left-handed particles (and right-handed antiparticles). Since the mirror reflection of a left-handed particle is right-handed, this explains the maximal violation of parity. Interestingly, the **V−A** theory was developed before the discovery of the Z boson, so it did not include the right-handed fields that enter in the neutral current interaction.

However, this theory allowed a compound symmetry **CP** to be conserved. **CP** combines parity **P** (switching left to right) with charge conjugation **C** (switching particles with antiparticles). Physicists were again surprised when in 1964, James Cronin and Val Fitch provided clear evidence in kaon decays that CP symmetry could be broken too, winning them the 1980 Nobel Prize in Physics.[22] In 1973, Makoto Kobayashi and Toshihide Maskawa showed that CP violation in the weak interaction required more than two generations of particles,[23] effectively predicting the existence of a then unknown third generation. This discovery earned them half of the 2008 Nobel Prize in Physics.[24] Unlike parity violation, CP violation occurs in only a small number of instances, but remains widely held as an answer to the difference between the amount of matter and antimatter in the universe; it thus forms one of Andrei Sakharov's three conditions for baryogenesis.[25]

20.6   See also

- Weakless Universe – the postulate that weak interactions are not anthropically necessary

- Gravity

- Nuclear force

- Electromagnetism

20.7   References

20.7.1   Citations

[1] Griffiths, David (2009). *Introduction to Elementary Particles*. pp. 59–60. ISBN 978-3-527-40601-2.

[2] "The Nobel Prize in Physics 1979: Press Release". *NobelPrize.org*. Nobel Media. Retrieved 22 March 2011.

[3] Fermi, Enrico (1934). "Versuch einer Theorie der β-Strahlen. I". *Zeitschrift für Physik A*. **88** (3–4): 161–177. Bibcode:1934ZPhy...88..161F. doi:10.1007/BF01351864.

[4] Wilson, Fred L. (December 1968). "Fermi's Theory of Beta Decay". *American Journal of Physics*. **36** (12): 1150–1160. Bibcode:1968AmJPh..36.1150W. doi:10.1119/1.1974382.

[5] W.-M. Yao *et al.* (Particle Data Group) (2006). "Review of Particle Physics: Quarks" (PDF). *Journal of Physics G*. **33**: 1–1232. arXiv:astro-ph/0601168ⓐ. Bibcode:2006JPhG...33....1Y. doi:10.1088/0954-3899/33/1/001.

[6] Peter Watkins (1986). *Story of the W and Z*. Cambridge: Cambridge University Press. p. 70. ISBN 978-0-521-31875-4.

[7] "Coupling Constants for the Fundamental Forces". *HyperPhysics*. Georgia State University. Retrieved 2 March 2011.

[8] J. Christman (2001). "The Weak Interaction" (PDF). *Physnet*. Michigan State University.

[9] "Electroweak". *The Particle Adventure*. Particle Data Group. Retrieved 3 March 2011.

[10] Walter Greiner; Berndt Müller (2009). *Gauge Theory of Weak Interactions*. Springer. p. 2. ISBN 978-3-540-87842-1.

[11] Cottingham & Greenwood (1986, 2001), p.29

[12] Cottingham & Greenwood (1986, 2001), p.28

[13] Cottingham & Greenwood (1986, 2001), p.30

[14] Baez, John C.; Huerta, John (2009). "The Algebra of Grand Unified Theories". *Bull.Am.Math.Soc*. **0904**: 483–552. arXiv:0904.1556ⓐ. Bibcode:2009arXiv0904.1556B. doi:10.1090/s0273-0979-10-01294-2. Retrieved 15 October 2013.

[15] K. Nakamura *et al.* (Particle Data Group) (2010). "Gauge and Higgs Bosons" (PDF). *Journal of Physics G*. **37**. Bibcode:2010JPhG...37g5021N. doi:10.1088/0954-3899/37/7a/075021.

[16] K. Nakamura *et al.* (Particle Data Group) (2010). "n" (PDF). *Journal of Physics G*. **37**: 7. Bibcode:2010JPhG...37g5021N. doi:10.1088/0954-3899/37/7a/075021.

[17] "The Nobel Prize in Physics 1979". *NobelPrize.org*. Nobel Media. Retrieved 26 February 2011.

[18] C. Amsler *et al.* (Particle Data Group) (2008). "Review of Particle Physics – Higgs Bosons: Theory and Searches" (PDF). *Physics Letters B*. **667**: 1–6. Bibcode:2008PhLB..667....1P. doi:10.1016/j.physletb.2008.07.018.

[19] "New results indicate that new particle is a Higgs boson | CERN". Home.web.cern.ch. Retrieved 20 September 2013.

[20] Charles W. Carey (2006). "Lee, Tsung-Dao". *American scientists*. Facts on File Inc. p. 225. ISBN 9781438108070.

[21] "The Nobel Prize in Physics 1957". *NobelPrize.org*. Nobel Media. Retrieved 26 February 2011.

[22] "The Nobel Prize in Physics 1980". *NobelPrize.org*. Nobel Media. Retrieved 26 February 2011.

[23] M. Kobayashi; T. Maskawa (1973). "CP-Violation in the Renormalizable Theory of Weak Interaction". *Progress of Theoretical Physics*. **49** (2): 652–657. Bibcode:1973PThPh..49..652K. doi:10.1143/PTP.49.652.

[24] "The Nobel Prize in Physics 1980". *NobelPrize.org*. Nobel Media. Retrieved 17 March 2011.

[25] Paul Langacker (2001) [1989]. "Cp Violation and Cosmology". In Cecilia Jarlskog. *CP violation*. London, River Edge: World Scientific Publishing Co. p. 552. ISBN 9789971505615.

20.7.2   General readers

- R. Oerter (2006). *The Theory of Almost Everything: The Standard Model, the Unsung Triumph of Modern Physics*. Plume. ISBN 978-0-13-236678-6.

- B.A. Schumm (2004). *Deep Down Things: The Breathtaking Beauty of Particle Physics*. Johns Hopkins University Press. ISBN 0-8018-7971-X.

20.7.3   Texts

- D.A. Bromley (2000). *Gauge Theory of Weak Interactions*. Springer. ISBN 3-540-67672-4.

- G.D. Coughlan; J.E. Dodd; B.M. Gripaios (2006). *The Ideas of Particle Physics: An Introduction for Scientists* (3rd ed.). Cambridge University Press. ISBN 978-0-521-67775-2.

- W. N. Cottingham; D. A. Greenwood (2001) [1986]. *An introduction to nuclear physics* (2nd ed.). Cambridge University Press. p. 30. ISBN 978-0-521-65733-4.

- D.J. Griffiths (1987). *Introduction to Elementary Particles*. John Wiley & Sons. ISBN 0-471-60386-4.

- G.L. Kane (1987). *Modern Elementary Particle Physics*. Perseus Books. ISBN 0-201-11749-5.

- D.H. Perkins (2000). *Introduction to High Energy Physics*. Cambridge University Press. ISBN 0-521-62196-8.

Chapter 21

Mechanics of planar particle motion

For general derivations and discussion of fictitious forces, see Fictitious force.
See also: Classical mechanics and Analytical mechanics

This article describes a **particle in planar motion**[1] when observed from non-inertial reference frames.[2] [3][4] The most famous examples of planar motion are related to the motion of two spheres that are gravitationally attracted to one another, and the generalization of this problem to planetary motion.[5] See centrifugal force, two-body problem, orbit and Kepler's laws of planetary motion. Those problems fall in the general field of analytical dynamics, the determination of orbits from given laws of force.[6] This article is focused more on the kinematical issues surrounding planar motion, that is, determination of the forces necessary to result in a certain trajectory *given* the particle trajectory. General results presented in fictitious forces here are applied to observations of a moving particle as seen from several specific non-inertial frames, for example, a *local* frame (one tied to the moving particle so it appears stationary), and a *co-rotating* frame (one with an arbitrarily located but fixed axis and a rate of rotation that makes the particle appear to have only radial motion and zero azimuthal motion). The Lagrangian approach to fictitious forces is introduced.

Unlike real forces such as electromagnetic forces, fictitious forces do not originate from physical interactions between objects.

21.1 Analysis using fictitious forces

The appearance of fictitious forces normally is associated with use of a non-inertial frame of reference, and their absence with use of an inertial frame of reference. The connection between inertial frames and fictitious forces (also called *inertial forces* or *pseudo-forces*), is expressed, for example, by Arnol'd:[7]

> The equations of motion in an non-inertial system differ from the equations in an inertial system by additional terms called inertial forces. This allows us to detect experimentally the non-inertial nature of a system.
> — V. I. Arnol'd: *Mathematical Methods of Classical Mechanics* Second Edition, p. 129

A slightly different tack on the subject is provided by Iro:[8]

> An additional force due to nonuniform relative motion of two reference frames is called a *pseudo-force*.
> — H Iro in *A Modern Approach to Classical Mechanics* p. 180

Fictitious forces do not appear in the equations of motion in an inertial frame of reference: in an inertial frame, the motion of an object is explained by the real impressed forces. In a non-inertial frame such as a rotating frame, however, Newton's first and second laws still can be used to make accurate physical predictions provided fictitious forces are included along with the real forces. For solving problems of mechanics in non-inertial reference frames, the advice given in textbooks is to treat the fictitious forces like real forces and to pretend you are in an inertial frame.[9] [10]

> Treat the fictitious forces like real forces, and pretend you are in an inertial frame.
> — Louis N. Hand, Janet D. Finch *Analytical Mechanics*, p. 267

It should be mentioned that "treating the fictitious forces like real forces" means, in particular, that fictitious forces as seen in a particular non-inertial frame transform as *vectors* under coordinate transformations made within that frame, that is, like real forces.

21.2   Moving objects and observational frames of reference

Next, it is observed that time varying coordinates are used in both inertial and non-inertial frames of reference, so the use of time varying coordinates should not be confounded with a change of observer, but is only a change of the observer's choice of description. Elaboration of this point and some citations on the subject follow.

21.2.1   Frame of reference and coordinate system

The term frame of reference is used often in a very broad sense, but for the present discussion its meaning is restricted to refer to an observer's *state of motion*, that is, to either an inertial frame of reference or a non-inertial frame of reference.

The term coordinate system is used to differentiate between different possible choices for a set of variables to describe motion, choices available to any observer, regardless of their state of motion. Examples are Cartesian coordinates, polar coordinates and (more generally) curvilinear coordinates.

Here are two quotes relating "state of motion" and "coordinate system":[11][12]

> We first introduce the notion of *reference frame*, itself related to the idea of *observer*: the reference frame is, in some sense, the "Euclidean space carried by the observer". Let us give a more mathematical definition:... the reference frame is... the set of all points in the Euclidean space with the rigid body motion of the observer. The frame, denoted $\mathfrak{R}$, is said to move with the observer.... The spatial positions of particles are labelled relative to a frame $\mathfrak{R}$ by establishing a *coordinate system R* with origin *O*. The corresponding set of axes, sharing the rigid body motion of the frame $\mathfrak{R}$, can be considered to give a physical realization of $\mathfrak{R}$. In a frame $\mathfrak{R}$, coordinates are changed from *R* to *R'* by carrying out, at each instant of time, the same coordinate transformation on the components of *intrinsic* objects (vectors and tensors) introduced to represent physical quantities *in this frame*.
> — Jean Salençon, Stephen Lyle.   (2001). *Handbook of Continuum Mechanics: General Concepts, Thermoelasticity* p. 9

In traditional developments of special and general relativity it has been customary not to distinguish between two quite distinct ideas. The first is the notion of a coordinate system, understood simply as the smooth, invertible assignment of four numbers to events in space-time neighborhoods. The second, the frame of reference, refers to an idealized system used to assign such numbers ... To avoid unnecessary restrictions, we can divorce this arrangement from metrical notions. ... Of special importance for our purposes is that each frame of reference has a definite state of motion at each event of spacetime....Within the context of special relativity and as long as we restrict ourselves to frames of reference in inertial motion, then little of importance depends on the difference between an inertial frame of reference and the inertial coordinate system it induces. This comfortable circumstance ceases immediately once we begin to consider frames of reference in nonuniform motion even within special relativity....the notion of frame of reference has reappeared as a structure distinct from a coordinate system.
— John D. Norton: *General Covariance and the Foundations of General Relativity: eight decades of dispute, Rep. Prog. Phys.*, **56**, pp. 835-7.

21.2.2   Time varying coordinate systems

In a general coordinate system, the basis vectors for the co-ordinates may vary in time at fixed positions, or they may vary with position at fixed times, or both. It may be noted that coordinate systems attached to both inertial frames and non-inertial frames can have basis vectors that vary in time, space or both, for example the description of a trajectory in polar coordinates as seen from an inertial frame.[13] or as seen from a rotating frame.[14] A time-dependent *description* of observations does not change the frame of reference in which the observations are made and recorded.

21.3   Fictitious forces in a local coordinate system

See also: Generalized forces, Curvilinear coordinates, Generalized coordinates, and Frenet-Serret formulas

In discussion of a particle moving in a circular orbit,[15] in an inertial frame of reference one can identify the centripetal and tangential forces. It then seems to be no prob-

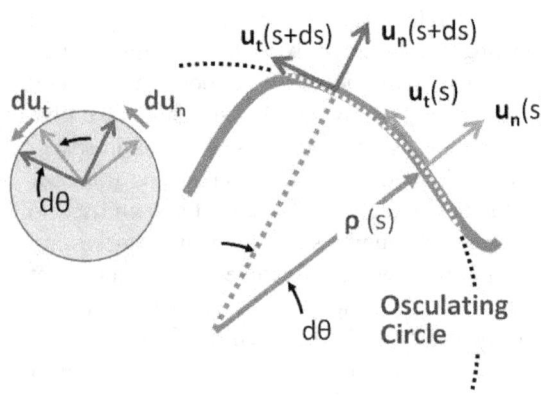

Figure 1: *Local coordinate system for planar motion on a curve. Two different positions are shown for distances s and s + ds along the curve. At each position s, unit vector* **un** *points along the outward normal to the curve and unit vector* **ut** *is tangential to the path. The radius of curvature of the path is ϱ as found from the rate of rotation of the tangent to the curve with respect to arc length, and is the radius of the osculating circle at position s. The unit circle on the left shows the rotation of the unit vectors with s.*

lem to switch hats, change perspective, and talk about the fictitious forces commonly called the centrifugal and Euler force. But what underlies this switch in vocabulary is a change of observational frame of reference from the inertial frame where we started, where centripetal and tangential forces make sense, to a rotating frame of reference where the particle appears motionless and fictitious centrifugal and Euler forces have to be brought into play. That switch is unconscious, but real.

Suppose we sit on a particle in general planar motion (not just a circular orbit). What analysis underlies a switch of hats to introduce fictitious centrifugal and Euler forces?

To explore that question, begin in an inertial frame of reference. By using a coordinate system commonly used in planar motion, the so-called *local* coordinate system,[16] as shown in Figure 1, it becomes easy to identify formulas for the centripetal inward force normal to the trajectory (in direction opposite to **un** in Figure 1), and the tangential force parallel to the trajectory (in direction **ut**), as shown next.

To introduce the unit vectors of the local coordinate system shown in Figure 1, one approach is to begin in Cartesian coordinates in an inertial framework and describe the local coordinates in terms of these Cartesian coordinates. In Figure 1, the arc length *s* is the distance the particle has traveled along its path in time *t*. The path **r** (*t*) with components *x(t)*, *y(t)* in Cartesian coordinates is described using arc length *s(t)* as:[17]

$$\mathbf{r}(s) = [x(s),\ y(s)] \ .$$

One way to look at the use of *s* is to think of the path of the

The arc length s(t) *measures distance along the skywriter's trail. Image from NASA ASRS*

particle as sitting in space, like the trail left by a skywriter, independent of time. Any position on this path is described by stating its distance *s* from some starting point on the path. Then an incremental displacement along the path *ds* is described by:

$$d\mathbf{r}(s) = [dx(s),\ dy(s)] = [x'(s),\ y'(s)]\,ds \ ,$$

where primes are introduced to denote derivatives with respect to *s*. The magnitude of this displacement is *ds*, showing that:[18]

$$[x'(s)^2 + y'(s)^2] = 1 \ . \text{(Eq. 1)}$$

This displacement is necessarily tangent to the curve at *s*, showing that the unit vector tangent to the curve is:

$$\mathbf{u}_t(s) = [x'(s),\ y'(s)] \ ,$$

while the outward unit vector normal to the curve is

$$\mathbf{u}_n(s) = [y'(s),\ -x'(s)] \ ,$$

Orthogonality can be verified by showing the vector dot product is zero. The unit magnitude of these vectors is a consequence of Eq. 1.

As an aside, notice that the use of unit vectors that are not aligned along the Cartesian *xy*-axes does not mean we are no longer in an inertial frame. All it means is that we are using unit vectors that vary with *s* to describe the path, but still observe the motion from the inertial frame.

Using the tangent vector, the angle of the tangent to the curve, say θ, is given by:

$$\sin \theta = \frac{y'(s)}{\sqrt{x'(s)^2+y'(s)^2}} = y'(s) \; ; \text{ and } \cos \theta = \frac{x'(s)}{\sqrt{x'(s)^2+y'(s)^2}} = x'(s) \; .$$

The radius of curvature is introduced completely formally (without need for geometric interpretation) as:

$$\frac{1}{\rho} = \frac{d\theta}{ds} \; .$$

The derivative of θ can be found from that for $\sin \theta$:

$$\frac{d \sin \theta}{ds} = \cos \theta \frac{d\theta}{ds} = \frac{1}{\rho} \cos \theta$$

$$= \frac{1}{\rho} x'(s) \; .$$

Now:

$$\frac{d \sin \theta}{ds} = \frac{d}{ds} \frac{y'(s)}{\sqrt{x'(s)^2+y'(s)^2}} = \frac{y''(s)x'(s)^2 - y'(s)x'(s)x''(s)}{(x'(s)^2+y'(s)^2)^{3/2}} \; ,$$

in which the denominator is unity according to Eq. 1. With this formula for the derivative of the sine, the radius of curvature becomes:

$$\frac{d\theta}{ds} = \frac{1}{\rho} = y''(s)x'(s) - y'(s)x''(s) = \frac{y''(s)}{x'(s)} = -\frac{x''(s)}{y'(s)} \; ,$$

where the equivalence of the forms stems from differentiation of Eq. 1:

$$x'(s)x''(s) + y'(s)y''(s) = 0 \; .$$

Having set up the description of any position on the path in terms of its associated value for s, and having found the properties of the path in terms of this description, motion of the particle is introduced by stating the particle position at any time t as the corresponding value $s\,(t)$.

Using the above results for the path properties in terms of s, the acceleration in the inertial reference frame as described in terms of the components normal and tangential to the path of the particle can be found in terms of the function $s(t)$ and its various time derivatives (as before, *primes* indicate differentiation with respect to s):

$$\mathbf{a}(s) = \frac{d}{dt}\mathbf{v}(s) = \frac{d}{dt} \left[\frac{ds}{dt} \left(x'(s), \; y'(s) \right) \right]$$

$$= \left(\frac{d^2 s}{dt^2} \right) \mathbf{u}_t(s) + \left(\frac{ds}{dt} \right)^2 \left(x''(s), \; y''(s) \right)$$

$$= \left(\frac{d^2 s}{dt^2} \right) \mathbf{u}_t(s) - \left(\frac{ds}{dt} \right)^2 \frac{1}{\rho} \mathbf{u}_n(s) \; ,$$

as can be verified by taking the dot product with the unit vectors $\mathbf{u}_t(s)$ and $\mathbf{u}_n(s)$. This result for acceleration is the same as that for circular motion based on the radius ρ. Using this coordinate system in the inertial frame, it is easy to identify the force normal to the trajectory as the centripetal force and that parallel to the trajectory as the tangential force.

Next, we change observational frames. Sitting on the particle, we adopt a non-inertial frame where the particle is at rest (zero velocity). This frame has a continuously changing origin, which at time t is the center of curvature (the center of the osculating circle in Figure 1) of the path at time t, and whose rate of rotation is the angular rate of motion of the particle about that origin at time t. This non-inertial frame also employs unit vectors normal to the trajectory and parallel to it.

The angular velocity of this frame is the angular velocity of the particle about the center of curvature at time t. The centripetal force of the inertial frame is interpreted in the non-inertial frame where the body is at rest as a force necessary to overcome the centrifugal force. Likewise, the force causing any acceleration of speed along the path seen in the inertial frame becomes the force necessary to overcome the Euler force in the non-inertial frame where the particle is at rest. There is zero Coriolis force in the frame, because the particle has zero velocity in this frame. For a pilot in an airplane, for example, these fictitious forces are a matter of direct experience.[19] However, these fictitious forces cannot be related to a simple observational frame of reference other than the particle itself, unless it is in a particularly simple path, like a circle.

That said, from a qualitative standpoint, the path of an airplane can be approximated by an arc of a circle for a limited time, and for the limited time a particular radius of curvature applies, the centrifugal and Euler forces can be analyzed on the basis of circular motion with that radius. See article discussing turning an airplane.

Next, reference frames rotating about a fixed axis are discussed in more detail.

21.4 Fictitious forces in polar coordinates

Main article: polar coordinates

Description of particle motion often is simpler in non-Cartesian coordinate systems, for example, polar coordinates. When equations of motion are expressed in terms of any curvilinear coordinate system, extra terms appear that represent how the basis vectors change as the coordinates change. These terms arise automatically on transformation to polar (or cylindrical) coordinates and are thus not fictitious *forces*, but rather are simply added *terms* in the acceleration in polar coordinates.[20]

21.4.1 Two terminologies

In a purely mathematical treatment, regardless of the frame that the coordinate system is associated with (inertial or non-inertial), extra terms appear in the acceleration of an observed particle when using curvilinear coordinates. For example, in polar coordinates the acceleration is given by (see below for details):

$$\boldsymbol{a} = \frac{d\boldsymbol{v}}{dt} = \frac{d^2\mathbf{r}}{dt^2} = (\ddot{r} - r\dot{\theta}^2)\hat{\boldsymbol{r}} + (r\ddot{\theta} + 2\dot{r}\dot{\theta})\hat{\boldsymbol{\theta}} \,,$$

which contains not just double time derivatives of the coordinates but added terms. This example employs polar coordinates, but more generally the added terms depend upon which coordinate system is chosen (that is, polar, elliptic, or whatever). Sometimes these coordinate-system dependent *terms* also are referred to as "fictitious forces", introducing a second meaning for "fictitious forces", despite the fact that these terms do not have the vector transformation properties expected of forces. For example, see Shankar[21] and Hildebrand.[22] According to this terminology, fictitious forces are determined in part by the coordinate system itself, regardless of the frame it is attached to, that is, regardless of whether the coordinate system is attached to an inertial or a non-inertial frame of reference. In contrast, the fictitious forces defined in terms of the *state of motion of the observer* vanish in inertial frames of reference. To distinguish these two terminologies, the fictitious forces that vanish in an inertial frame of reference, the inertial forces of Newtonian mechanics, are called in this article the "state-of-motion" fictitious forces and those that originate in the interpretation of time derivatives in particular coordinate systems are called "coordinate" fictitious forces.[23]

Assuming it is clear that "state of motion" and "coordinate system" are *different*, it follows that the dependence of centrifugal force (as in this article) upon "state of motion" and its independence from "coordinate system", which contrasts with the "coordinate" version with exactly the opposite dependencies, indicates that two different ideas are referred to by the terminology "fictitious force". The present article emphasizes one of these two ideas ("state-of-motion"), although the other also is described.

Below, polar coordinates are introduced for use in (first) an inertial frame of reference and then (second) in a rotating frame of reference. The two different uses of the term

"fictitious force" are pointed out. First, however, follows a brief digression to explain further how the "coordinate" terminology for fictitious force has arisen.

Lagrangian approach

See also: Lagrangian mechanics

To motivate the introduction of "coordinate" inertial forces by more than a reference to "mathematical convenience", what follows is a digression to show these forces correspond to what are called by some authors "generalized" fictitious forces or "generalized inertia forces".[24][25][26][27] These forces are introduced via the Lagrangian mechanics approach to mechanics based upon describing a system by *generalized coordinates* usually denoted as $\{qk\}$. The only requirement on these coordinates is that they are necessary and sufficient to uniquely characterize the state of the system: they need not be (although they could be) the coordinates of the particles in the system. Instead, they could be the angles and extensions of links in a robot arm, for instance. If a mechanical system consists of N particles and there are m independent kinematical conditions imposed, it is possible to characterize the system uniquely by $n = 3N - m$ independent generalized coordinates $\{qk\}$.[28]

In classical mechanics, the Lagrangian is defined as the kinetic energy, T, of the system minus its potential energy, U.[29] In symbols,

$$L = T - U.$$

Under conditions that are given in Lagrangian mechanics, if the Lagrangian of a system is known, then the equations of motion of the system may be obtained by a direct substitution of the expression for the Lagrangian into the Euler–Lagrange equation, a particular family of partial differential equations.

Here are some definitions:[30]

> **Definition**:
>
> $$L(\boldsymbol{q}, \, \dot{\boldsymbol{q}}, \, t) = T - U$$
>
> is the *Lagrange function* or *Lagrangian*, qi are the *generalized coordinates*, $\dot{q}_i$ are *generalized velocities*,
>
> $\partial L / \partial \dot{q}_i$ are *generalized momenta*,
>
> $\partial L / \partial q_i$ are *generalized forces*,
>
> $\frac{d}{dt}\frac{\partial L}{\partial \dot{q}_i} - \frac{\partial L}{\partial q_i} = 0$ are *Lagrange's equations*.

It is not the purpose here to outline how Lagrangian mechanics works. The interested reader can look at other articles explaining this approach. For the moment, the goal is simply to show that the Lagrangian approach can lead to "generalized fictitious forces" that *do not vanish in inertial frames*. What is pertinent here is that in the case of a single particle, the Lagrangian approach can be arranged to capture exactly the "coordinate" fictitious forces just introduced.

To proceed, consider a single particle, and introduce the generalized coordinates as $\{qk\} = (r, \theta)$. Then Hildebrand [22] shows in polar coordinates with the $qk = (r, \theta)$ the "generalized momenta" are:

$$p_r = m\dot{r} \,, \ p_\theta = mr^2\dot{\theta} \,,$$

leading, for example, to the generalized force:

$$\frac{d}{dt}p_r = Q_r + mr\dot{\theta}^2 \,,$$

with Q_r the impressed radial force. The connection between "generalized forces" and Newtonian forces varies with the choice of coordinates. This Lagrangian formulation introduces exactly the "coordinate" form of fictitious forces mentioned above that allows "fictitious" (generalized) forces in inertial frames, for example, the term $mr\dot{\theta}^2$. Careful reading of Hildebrand shows he doesn't discuss the role of "inertial frames of reference", and in fact, says "[The] presence or absence [of inertia forces] depends, not upon the particular problem at hand but *upon the coordinate system chosen*." By coordinate system presumably is meant the choice of $\{qk\}$. Later he says "If *accelerations* associated with generalized coordinates are to be of prime interest (as is usually the case), the [nonaccelerational] terms may be conveniently transferred to the right ... and considered as additional (generalized) inertia forces. Such inertia forces are often said to be of the *Coriolis* type."

In short, the emphasis of some authors upon coordinates and their derivatives and their introduction of (generalized) fictitious forces that do not vanish in inertial frames of reference is an outgrowth of the use of generalized coordinates in Lagrangian mechanics. For example, see McQuarrie[31] Hildebrand,[22] and von Schwerin.[32] Below is an example of this usage as employed in the design of robotic manipulators:[33][34][35]

In the above [Lagrange-Euler] equations, there are three types of terms. The first involves the second derivative of the generalized coordinates. The second is quadratic in $\mathbf{q}$ where the coefficients may depend on $\mathbf{q}$. These are further classified into two types. Terms involving a product of the type $\dot{q}_i^2$ are called *centrifugal forces* while those involving a product of the type $\dot{q}_i\dot{q}_j$ for $i \neq j$ are called *Coriolis forces*. The third type is functions of $\mathbf{q}$ only and are called *gravitational forces*.
— Shuzhi S. Ge, Tong Heng Lee & Christopher John Harris: *Adaptive Neural Network Control of Robotic Manipulators*, pp. 47-48

For a robot manipulator, the equations may be written in a form using Christoffel symbols Γijk (discussed further below) as:[36][37]

$$\sum_{j=1}^{n} M_{ij}(\boldsymbol{q})\ddot{q}_j + \sum_{j,k=1}^{n} \Gamma_{ijk}\dot{q}_j\dot{q}_k + \frac{\partial V}{\partial q_i} = \Upsilon_i \,; i = 1, ..., n \,,$$

where M is the "manipulator inertia matrix" and V is the potential energy due to gravity (for example), and Υ_i are the generalized forces on joint i. The terms involving Christoffel symbols therefore determine the "generalized centrifugal" and "generalized Coriolis" terms.

The introduction of *generalized* fictitious forces often is done without notification and without specifying the word "generalized". This sloppy use of terminology leads to endless confusion because these *generalized* fictitious forces, unlike the standard "state-of-motion" fictitious forces, do not vanish in inertial frames of reference.

21.4.2 Polar coordinates in an inertial frame of reference

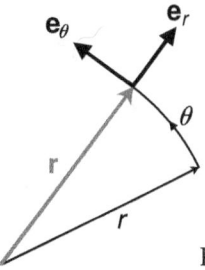

Position vector **r**, always points radially from the origin.

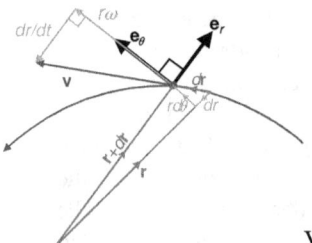

Velocity vector **v**, always tangent to the path of motion.

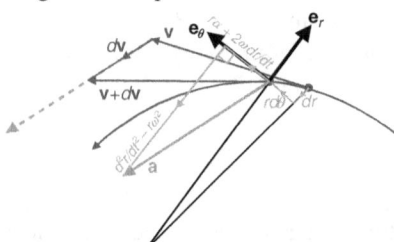

Acceleration vector **a**, not parallel to the radial motion but offset by the angular and Coriolis accelerations, nor tangent to the path but offset by the centripetal and radial accelerations.

Kinematic vectors in plane polar coordinates. Notice the setup is not restricted to 2d space, but a plane in any higher dimension.

Below, the acceleration of a particle is derived as seen in an inertial frame using polar coordinates. There are no "state-of-motion" fictitious forces in an inertial frame, by definition. Following that presentation, the contrasting terminology of "coordinate" fictitious forces is presented and critiqued on the basis of the non-vectorial transformation behavior of these "forces".

In an inertial frame, let **r** be the position vector of a moving particle. Its Cartesian components (x, y) are:

$$\mathbf{r} = (r\cos\theta,\ r\sin\theta)\ ,$$

with polar coordinates r and θ depending on time t.

Unit vectors are defined in the radially outward direction **r** :

$$\hat{r} = \frac{\partial \mathbf{r}}{\partial r} = (\cos\theta,\ \sin\theta)$$

and in the direction at right angles to **r** :

$$\hat{\boldsymbol{\theta}} = \frac{\partial^2 \mathbf{r}}{\partial r\, \partial\theta} = (-\sin\theta,\ \cos\theta)\ .$$

These unit vectors vary in direction with time:

$$\frac{d}{dt}\hat{r} = (-\sin\theta,\ \cos\theta)\frac{d\theta}{dt} = \frac{d\theta}{dt}\hat{\boldsymbol{\theta}},$$

and:

$$\frac{d}{dt}\hat{\boldsymbol{\theta}} = (-\cos\theta,\ -\sin\theta)\frac{d\theta}{dt} = -\frac{d\theta}{dt}\hat{r}.$$

Using these derivatives, the first and second derivatives of position are:

$$\boldsymbol{v} = \frac{d\mathbf{r}}{dt} = \dot{r}\hat{r} + r\dot{\theta}\hat{\boldsymbol{\theta}},$$

$$\boldsymbol{a} = \frac{d\boldsymbol{v}}{dt} = \frac{d^2\mathbf{r}}{dt^2} = (\ddot{r} - r\dot{\theta}^2)\hat{r} + (r\ddot{\theta} + 2\dot{r}\dot{\theta})\hat{\boldsymbol{\theta}}\ ,$$

where dot-overmarkings indicate time differentiation. With this form for the acceleration $\boldsymbol{a}$, in an inertial frame of reference Newton's second law expressed in polar coordinates is:

$$\boldsymbol{F} = m\boldsymbol{a} = m(\ddot{r} - r\dot{\theta}^2)\hat{r} + m(r\ddot{\theta} + 2\dot{r}\dot{\theta})\hat{\boldsymbol{\theta}}\ ,$$

where $\boldsymbol{F}$ is the net real force on the particle. No fictitious forces appear because all fictitious forces are zero by definition in an inertial frame.

From a mathematical standpoint, however, it sometimes is handy to put only the second-order derivatives on the right side of this equation; that is we write the above equation by rearrangement of terms as:

$$\boldsymbol{F} + mr\dot{\theta}^2\hat{r} - m2\dot{r}\dot{\theta}\hat{\boldsymbol{\theta}} = m\tilde{\boldsymbol{a}} = m\ddot{r}\hat{r} + mr\ddot{\theta}\hat{\boldsymbol{\theta}}\ ,$$

where a "coordinate" version of the "acceleration" is introduced:

$$\tilde{\boldsymbol{a}} = \ddot{r}\hat{r} + r\ddot{\theta}\hat{\boldsymbol{\theta}}\ ,$$

consisting of only second-order time derivatives of the coordinates r and θ. The terms moved to the force-side of the equation are now treated as *extra* "fictitious forces" and, confusingly, the resulting forces also are called the "centrifugal" and "Coriolis" force.

These newly defined "forces" are non-zero in an *inertial frame*, and so certainly are not the same as the previously identified fictitious forces that are zero in an inertial frame and non-zero only in a non-inertial frame.[38] In this article, these newly defined forces are called the "coordinate" centrifugal force and the "coordinate" Coriolis force to separate them from the "state-of-motion" forces.

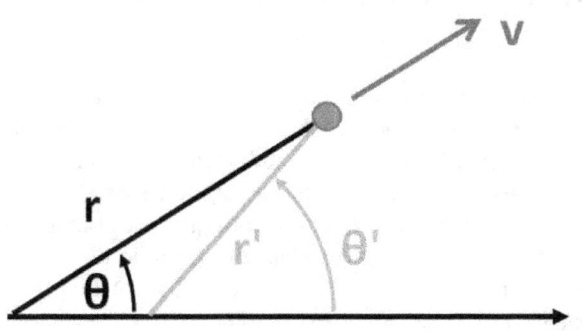

*Figure 2: Two coordinate systems differing by a displacement of origin. Radial motion with constant velocity **v** in one frame is not radial in the other frame. Angular rate $\dot{\theta} = 0$, but $\dot{\theta}' \neq 0$.*

Change of origin

Here is an illustration showing the so-called "centrifugal term" $r\dot{\theta}^2$ does not transform as a true force, putting any reference to this term not just as a "term", but as a centrifugal *force*, in a dubious light. Suppose in frame S a particle moves radially away from the origin at a constant velocity. See Figure 2. The force on the particle is zero by Newton's first law. Now we look at the same thing from frame S', which is the same, but displaced in origin. In S' the particle still is in straight line motion at constant speed, so again the force is zero.

What if we use polar coordinates in the two frames? In frame S the radial motion is constant and there is no angular motion. Hence, the acceleration is:

$$a = \left(\ddot{r} - r\dot{\theta}^2 \right)\hat{r} + \left(r\ddot{\theta} + 2\dot{r}\dot{\theta} \right)\hat{\theta} = 0 ,$$

and each term individually is zero because $\dot{\theta} = 0$, $\ddot{\theta} = 0$ and $\ddot{r} = 0$. There is no force, including no $r\dot{\theta}^2$ "force" in frame S. In frame S', however, we have:

$$a' = \left(\ddot{r}' - r'\dot{\theta}'^2 \right)\hat{r}' + \left(r'\ddot{\theta}' + 2\dot{r}'\dot{\theta}' \right)\hat{\theta}'$$

In this case the azimuthal term is zero, being the rate of change of angular momentum. To obtain zero acceleration in the radial direction, however, we require:

$$\ddot{r}' = r'\dot{\theta}'^2 .$$

The right-hand side is non-zero, inasmuch as neither r' nor $\dot{\theta}'$ is zero. That is, we cannot obtain zero force (zero a') if we retain only $\ddot{r}'$ as the acceleration; we need both terms.

Despite the above facts, suppose we adopt polar coordinates, and wish to say that $r\dot{\theta}^2$ is "centrifugal force", and reinterpret $\ddot{r}$ as "acceleration" (without dwelling upon any possible justification). How does this decision fare when we consider that a proper formulation of physics is geometry and coordinate-independent? See the article on general covariance.[39] To attempt to form a covariant expression, this so-called centrifugal "force" can be put into vector notation as:

$$F_{\dot{\theta}} = -\omega \times (\omega \times r) ,$$

with:

$$\omega = \dot{\theta}\hat{k} ,$$

and $\hat{k}$ a unit vector normal to the plane of motion. Unfortunately, although this expression formally looks like a vector, when an observer changes origin the value of $\dot{\theta}$ changes (see Figure 2), so observers in the same frame of reference standing on different street corners see different "forces" even though the actual events they witness are identical. How can a physical force (be it fictitious or real) be zero in one frame S, but non-zero in another frame S' identical, but a few feet away? Even for exactly the same particle behavior the expression $r\dot{\theta}^2$ is different in every frame of reference, even for very trivial distinctions between frames. In short, if we take $r\dot{\theta}^2$ as "centrifugal force", it does not have a universal significance: it is *unphysical*.

Beyond this problem, the real impressed net force is zero. (There is no real impressed force in straight-line motion at constant speed). If we adopt polar coordinates, and wish to say that $r\dot{\theta}^2$ is "centrifugal force", and reinterpret $\ddot{r}$ as "acceleration", the oddity results in frame S' that straight-line motion at constant speed requires a net force in polar coordinates, but not in Cartesian coordinates. Moreover, this perplexity applies in frame S', but not in frame S.

The absurdity of the behavior of $r\dot{\theta}^2$ indicates that one must say that $r\dot{\theta}^2$ is *not* centrifugal *force*, but simply one of two *terms* in the acceleration. This view, that the acceleration is composed of two terms, is frame-independent: there is zero centrifugal force in any and every inertial frame. It also is coordinate-system independent: we can use Cartesian, polar, or any other curvilinear system: they all produce zero.

Apart from the above physical arguments, of course, the derivation above, based upon application of the mathematical rules of differentiation, shows the radial acceleration does indeed consist of the two terms $\ddot{r} - r\dot{\theta}^2$.

That said, the next subsection shows there is a connection between these centrifugal and Coriolis *terms* and the fictitious *forces* that pertain to a particular *rotating* frame of reference (as distinct from an inertial frame).

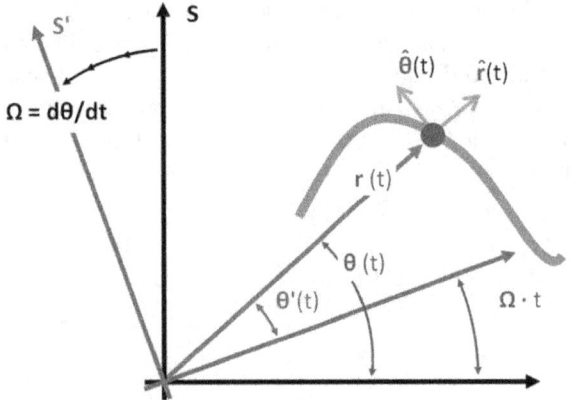

Figure 3: Inertial frame of reference S and instantaneous non-inertial co-rotating frame of reference S'. The co-rotating frame rotates at angular rate Ω equal to the rate of rotation of the particle about the origin of S' at the particular moment t. Particle is located at vector position r(t) and unit vectors are shown in the radial direction to the particle from the origin, and also in the direction of increasing angle θ normal to the radial direction. These unit vectors need not be related to the tangent and normal to the path. Also, the radial distance r need not be related to the radius of curvature of the path.

21.4.3   Co-rotating frame

In the case of planar motion of a particle, the "coordinate" centrifugal and Coriolis acceleration terms found above to be non-zero in an inertial frame can be shown to be the negatives of the "state-of-motion" centrifugal and Coriolis terms that appear in a very particular non-inertial *co-rotating* frame (see next subsection).[40] See Figure 3. To define a co-rotating frame, first an origin is selected from which the distance $r(t)$ to the particle is defined. An axis of rotation is set up that is perpendicular to the plane of motion of the particle, and passing through this origin. Then, at the selected moment t, the rate of rotation of the co-rotating frame Ω is made to match the rate of rotation of the particle about this axis, $d\theta/dt$. The co-rotating frame applies only for a moment, and must be continuously re-selected as the particle moves. For more detail, see Polar coordinates, centrifugal and Coriolis terms.

21.4.4   Polar coordinates in a rotating frame of reference

Next, the same approach is used to find the fictitious forces of a (non-inertial) rotating frame. For example, if a rotating polar coordinate system is adopted for use in a rotating frame of observation, both rotating at the same constant counterclockwise rate Ω, we find the equations of motion in this frame as follows: the radial coordinate in the rotating frame is taken as r, but the angle θ' in the rotating frame changes with time:

$$\theta' = \theta - \Omega t .$$

Consequently,

$$\dot{\theta}' = \dot{\theta} - \Omega .$$

Plugging this result into the acceleration using the unit vectors of the previous section:

$$\frac{d^2\mathbf{r}}{dt^2} = \left[\ddot{r} - r\left(\dot{\theta}' + \Omega\right)^2\right]\hat{\mathbf{r}} + \left[r\ddot{\theta}' + 2\dot{r}\left(\dot{\theta}' + \Omega\right)\right]\hat{\boldsymbol{\theta}}$$

$$= (\ddot{r} - r\dot{\theta}'^2)\hat{\mathbf{r}} + (r\ddot{\theta}' + 2\dot{r}\dot{\theta}')\hat{\boldsymbol{\theta}} - \left(2r\Omega\dot{\theta}' + r\Omega^2\right)\hat{\mathbf{r}} + (2\dot{r}\Omega)\hat{\boldsymbol{\theta}} .$$

The leading two terms are the same form as those in the inertial frame, and they are the only terms if the frame is *not* rotating, that is, if $\Omega=0$. However, in this rotating frame we have the extra terms:[41]

$$-\left(2r\Omega\dot{\theta}' + r\Omega^2\right)\hat{\mathbf{r}} + (2\dot{r}\Omega)\hat{\boldsymbol{\theta}}$$

The radial term $\Omega^2 r$ is the centrifugal force per unit mass due to the system's rotation at rate Ω and the radial term $2r\Omega\dot{\theta}'$ is the radial component of the Coriolis force per unit mass, where $r\dot{\theta}'$ is the tangential component of the particle velocity as seen in the rotating frame. The term $-(2\dot{r}\Omega)\hat{\boldsymbol{\theta}}$ is the so-called *azimuthal* component of the Coriolis force per unit mass. In fact, these extra terms can be used to *measure* Ω and provide a test to see whether or not the frame is rotating, just as explained in the example of rotating identical spheres. If the particle's motion can be described by the observer using Newton's laws of motion *without* these Ω-dependent terms, the observer is in an inertial frame of reference where $\Omega=0$.

These "extra terms" in the acceleration of the particle are the "state of motion" fictitious forces for this rotating frame, the forces introduced by rotation of the frame at angular rate Ω.[42]

In this rotating frame, what are the "coordinate" fictitious forces? As before, suppose we choose to put only the second-order time derivatives on the right side of Newton's law:

$$\boldsymbol{F} + mr\dot{\theta}'^2\hat{\mathbf{r}} - m2\dot{r}\dot{\theta}'\hat{\boldsymbol{\theta}} + m\left(2r\Omega\dot{\theta}' + r\Omega^2\right)\hat{\mathbf{r}} - m\left(2\dot{r}\Omega\right)\hat{\boldsymbol{\theta}} = m\ddot{r}\hat{\mathbf{r}} + mr\ddot{\theta}'\,\hat{\boldsymbol{\theta}} = m\tilde{\boldsymbol{a}}$$

If we choose for convenience to treat $\tilde{a}$ as some so-called "acceleration", then the terms $(mr\dot{\theta}'^2\hat{\mathbf{r}} - m2\dot{r}\dot{\theta}'\hat{\boldsymbol{\theta}})$ are added to the so-called "fictitious force", which are not "state-of-motion" fictitious forces, but are actually components of force that persist even when $\Omega=0$, that is, they persist even in an inertial frame of reference. Because these extra terms are added, the "coordinate" fictitious force is not the same as the "state-of-motion" fictitious force. Because of these extra terms, the "coordinate" fictitious force is not zero even in an inertial frame of reference.

More on the co-rotating frame

Notice however, the case of a rotating frame that happens to have the same angular rate as the particle, so that $\Omega = d\theta/dt$ at some particular moment (that is, the polar coordinates are set up in the instantaneous, non-inertial co-rotating frame of Figure 3). In this case, at this moment, $d\theta'/dt = 0$. In this co-rotating non-inertial frame at this moment the "coordinate" fictitious forces are only those due to the motion of the frame, that is, they are the same as the "state-of-motion" fictitious forces, as discussed in the remarks about the co-rotating frame of Figure 3 in the previous section.

21.5   Fictitious forces in curvilinear coordinates

See also: Curvilinear coordinate system and Covariant derivative

To quote Bullo and Lewis: "Only in exceptional circumstances can the configuration of Lagrangian system be described by a vector in a vector space. In the natural mathematical setting, the system's configuration space is described loosely as a curved space, or more accurately as a differentiable manifold."[43]

Instead of Cartesian coordinates, when equations of motion are expressed in a curvilinear coordinate system, Christoffel symbols appear in the acceleration of a particle expressed in this coordinate system, as described below in more detail.

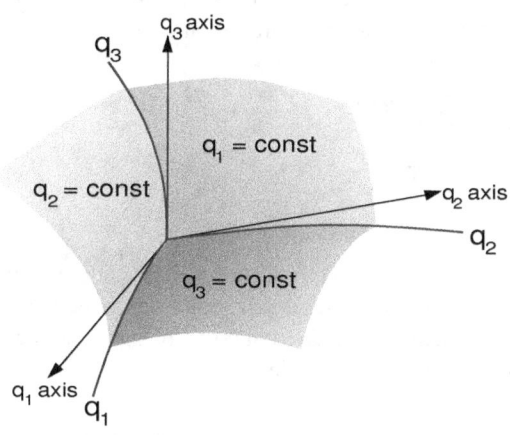

Figure 4: Coordinate surfaces, coordinate lines, and coordinate axes of general curvilinear coordinates.

Consider description of a particle motion from the viewpoint of an *inertial frame of reference* in curvilinear coordinates. Suppose the position of a point P in Cartesian coordinates is (x, y, z) and in curvilinear coordinates is (q_1, q_2, q_3). Then functions exist that relate these descriptions:

$$x = x(q_1, q_2, q_3)\,;\ q_1 = q_1(x, y, z)\,,$$

and so forth. (The number of dimensions may be larger than three.) An important aspect of such coordinate systems is the element of arc length that allows distances to be determined. If the curvilinear coordinates form an orthogonal coordinate system, the element of arc length ds is expressed as:

$$ds^2 = \sum_{k=1}^{d} (h_k)^2 (dq_k)^2\,,$$

where the quantities hk are called *scale factors*.[44] A change dqk in qk causes a displacement $hk\,dqk$ along the coordinate line for qk. At a point P, we place unit vectors $\mathbf{e_k}$ each tangent to a coordinate line of a variable qk. Then any vector can be expressed in terms of these basis vectors, for example, from an inertial frame of reference, the position vector of a moving particle $\mathbf{r}$ located at time t at position P becomes:

$$\boldsymbol{r} = \sum_{k=1}^{d} q_k\,\boldsymbol{e_k}$$

where qk is the vector dot product of $\mathbf{r}$ and $\mathbf{e_k}$. The velocity $\mathbf{v}$ of a particle at P, can be expressed at P as:

$$\boldsymbol{v} = \textstyle\sum_{k=1}^{d} v_k\, \boldsymbol{e_k}$$

$$= \frac{d}{dt}\boldsymbol{r} = \sum_{k=1}^{d} \dot{q}_k\, \boldsymbol{e_k} + \sum_{k=1}^{d} q_k\, \dot{\boldsymbol{e}}_{\boldsymbol{k}}$$

where vk is the vector dot product of $\mathbf{v}$ and $\mathbf{e_k}$, and over dots indicate time differentiation. The time derivatives of the basis vectors can be expressed in terms of the scale factors introduced above. for example:

$$\frac{\partial}{\partial q_2}\boldsymbol{e}_1 = -\boldsymbol{e}_2 \frac{1}{h_2}\frac{\partial h_1}{\partial q_2} - \boldsymbol{e}_3 \frac{1}{h_3}\frac{\partial h_1}{\partial q_3}\ , \text{ or, in general}$$
$$\frac{\partial \boldsymbol{e_j}}{\partial q_k} = \sum_{n=1}^{d} \Gamma^n{}_{kj}\boldsymbol{e_n}\ ,$$

in which the coefficients of the unit vectors are the Christoffel symbols for the coordinate system. The general notation and formulas for the Christoffel symbols are:[45][46]

$$\Gamma^i{}_{ii} = \left\{ \begin{matrix} i \\ i\ i \end{matrix} \right\} = \frac{1}{h_i}\frac{\partial h_i}{\partial q_i}\ ;\ \ \Gamma^i{}_{ij} = \left\{ \begin{matrix} i \\ i\ j \end{matrix} \right\} =$$
$$\frac{1}{h_i}\frac{\partial h_i}{\partial q_j} = \left\{ \begin{matrix} i \\ j\ i \end{matrix} \right\};\ \ \Gamma^j{}_{ii} = \left\{ \begin{matrix} j \\ i\ i \end{matrix} \right\} = -\frac{h_i}{h_j{}^2}\frac{\partial h_i}{\partial q_j}\ ,$$

and the symbol is zero when all the indices are different. Despite appearances to the contrary, the Christoffel symbols *do not form the components of a tensor*. For example, they are zero in Cartesian coordinates, but not in polar coordinates.[47]

Using relations like this one,[48]

$$\dot{\boldsymbol{e}}_{\boldsymbol{j}} = \sum_{k=1}^{d} \frac{\partial}{\partial q_k}\boldsymbol{e}_j \dot{q}_k$$

$$= \sum_{k=1}^{d}\sum_{i=1}^{d} \Gamma^k{}_{ij}\dot{q}_i \boldsymbol{e_k}\ ,$$

which allows all the time derivatives to be evaluated. For example, for the velocity:

$$\boldsymbol{v} = \frac{d}{dt}\boldsymbol{r} = \textstyle\sum_{k=1}^{d} \dot{q}_k\, \boldsymbol{e_k} + \sum_{k=1}^{d} q_k\, \dot{\boldsymbol{e}}_{\boldsymbol{k}}$$

$$= \sum_{k=1}^{d} \dot{q}_k\, \boldsymbol{e_k} + \sum_{j=1}^{d} q_j\, \dot{\boldsymbol{e}}_{\boldsymbol{j}},$$

$$= \sum_{k=1}^{d} \dot{q}_k\, \boldsymbol{e_k} + \sum_{k=1}^{d}\sum_{j=1}^{d}\sum_{i=1}^{d} q_j\, \Gamma^k{}_{ij}\boldsymbol{e_k}\dot{q}_i$$

$$= \sum_{k=1}^{d}\left(\dot{q}_k + \sum_{j=1}^{d}\sum_{i=1}^{d} q_j\, \Gamma^k{}_{ij}\dot{q}_i \right)\boldsymbol{e_k}\ ,$$

with the Γ-notation for the Christoffel symbols replacing the curly bracket notation. Using the same approach, the acceleration is then

$$\boldsymbol{a} = \frac{d}{dt}\boldsymbol{v} = \textstyle\sum_{k=1}^{d} \dot{v}_k\, \boldsymbol{e_k} + \sum_{k=1}^{d} v_k\, \dot{\boldsymbol{e}}_{\boldsymbol{k}}\ .$$

$$= \sum_{k=1}^{d}\left(\dot{v}_k + \sum_{j=1}^{d}\sum_{i=1}^{d} v_j \Gamma^k{}_{ij}\dot{q}_i \right)\boldsymbol{e_k}\ .$$

Looking at the relation for acceleration, the first summation contains the time derivatives of velocity, which would be associated with acceleration if these were Cartesian coordinates, and the second summation (the one with Christoffel symbols) contains terms related to the way the unit vectors change with time.[49]

21.5.1 "State-of-motion" versus "coordinate" fictitious forces

Earlier in this article a distinction was introduced between two terminologies, the fictitious forces that vanish in an inertial frame of reference are called in this article the "state-of-motion" fictitious forces and those that originate from differentiation in a particular coordinate system are called "coordinate" fictitious forces. Using the expression for the acceleration above, Newton's law of motion in the inertial frame of reference becomes:

$$\boldsymbol{F} = m\boldsymbol{a} = m\sum_{k=1}^{d}\left(\dot{v}_k + \sum_{j=1}^{d}\sum_{i=1}^{d} v_j\Gamma^k{}_{ij}\dot{q}_i \right)\boldsymbol{e_k}\ ,$$

where $\boldsymbol{F}$ is the net real force on the particle. No "state-of-motion" fictitious forces are present because the frame is inertial, and "state-of-motion" fictitious forces are zero in an inertial frame, by definition.

The "coordinate" approach to Newton's law above is to retain the second-order time derivatives of the coordinates $\{qk\}$ as the only terms on the right side of this equation, motivated more by mathematical convenience than by physics. To that end, the force law can be rewritten, taking the second summation to the force-side of the equation as:

$$\boldsymbol{F} - m\sum_{j=1}^{d}\sum_{i=1}^{d} v_j\Gamma^k{}_{ij}\dot{q}_i\boldsymbol{e_k} = m\tilde{\boldsymbol{a}}\ ,$$

with the convention that the "acceleration" $\tilde{\boldsymbol{a}}$ is now:

$$\tilde{\boldsymbol{a}} = \sum_{k=1}^{d} \dot{v}_k\boldsymbol{e_k}\ .$$

In the expression above, the summation added to the force-side of the equation now is treated as if added "forces" were present. These summation terms are customarily called fictitious forces within this "coordinate" approach, although in this inertial frame of reference all "state-of-motion" fictitious forces are identically zero. Moreover, these "forces" do not transform under coordinate transformations as *vectors*. Thus, the designation of the terms of the summation as "fictitious forces" uses this terminology for contributions that are completely different from any real force, and from the "state-of-motion" fictitious forces. What adds to this confusion is that these "coordinate" fictitious forces are divided into two groups and given the *same names* as the "state-of-motion" fictitious forces, that is, they are divided into "centrifugal" and "Coriolis" terms, despite their inclusion of terms that are not the "state-of-motion" centrifugal and Coriolis terms. For example, these "coordinate" centrifugal and Coriolis terms can be nonzero *even in an inertial frame of reference* where the "state-of-motion" centrifugal force (the subject of this article) and Coriolis force always are zero.[50]

If the frame is not inertial, for example, in a rotating frame of reference, the "state-of-motion" fictitious forces are included in the above "coordinate" fictitious force expression.[51] Also, if the "acceleration" expressed in terms of first-order time derivatives of the velocity happens to result in terms that are *not* simply second-order derivatives of the coordinates $\{qk\}$ in time, then these terms that are not second-order also are brought to the force-side of the equation and included with the fictitious forces. From the standpoint of a Lagrangian formulation, they can be called *generalized* fictitious forces. See Hildebrand,[22] for example.

Formulation of dynamics in terms of Christoffel symbols and the "coordinate" version of fictitious forces is used often in the design of robots in connection with a Lagrangian formulation of the equations of motion.[35][52]

21.6   Notes and references

[1] See for example, John Joseph Uicker; Gordon R. Pennock; Joseph Edward Shigley (2003). *Theory of Machines and Mechanisms*. Oxford University Press. p. 10. ISBN 0-19-515598-X., Harald Iro (2002). *A Modern Approach to Classical Mechanics*. World Scientific. p. Chapter 3 and Chapter 4. ISBN 981-238-213-5.

[2] *Fictitious forces* (also known as a *pseudo forces*, *inertial forces* or *d'Alembert forces*), exist for observers in a non-inertial reference frames. See, for example, Max Born & Günther Leibfried (1962). *Einstein's Theory of Relativity*. New York: Courier Dover Publications. pp. 76–78. ISBN 0-486-60769-0., NASA: *Accelerated Frames of Reference:*

Inertial Forces, Science Joy Wagon: *Centrifugal force - the false force*

[3] Jerrold E. Marsden; Tudor S. Ratiu (1999). *Introduction to Mechanics and Symmetry: A Basic Exposition of Classical Mechanical Systems*. Springer. p. 251. ISBN 0-387-98643-X.

[4] John Robert Taylor (2004). *Classical Mechanics*. Sausalito CA: University Science Books. p. Chapter 9, pp. 327 ff. ISBN 1-891389-22-X.

[5] Florian Scheck (2005). *Mechanics* (4th ed.). Birkhäuser. p. 13. ISBN 3-540-21925-0.

[6] Edmund Taylor Whittaker (1988). *A Treatise on the Analytical Dynamics of Particles and Rigid Bodies: With an Introduction to the Problem of Three Bodies* (Fourth edition of 1936 with foreword by Sir William McCrea ed.). Cambridge University Press. p. Chapter 1, p. 1. ISBN 0-521-35883-3.

[7] V. I. Arnol'd (1989). *Mathematical Methods of Classical Mechanics*. Springer. p. 129. ISBN 978-0-387-96890-2.

[8] Harald Iroh (2002). *A Modern Approach to Classical Mechanics*. World Scientific. p. 180. ISBN 981-238-213-5.

[9] Louis N. Hand; Janet D. Finch (1998). *Analytical Mechanics*. Cambridge University Press. p. 267. ISBN 0-521-57572-9.

[10] K.S. Rao (2003). *Classical Mechanics*. Orient Longman. p. 162. ISBN 81-7371-436-3.

[11] Jean Salençon; Stephen Lyle (2001). *Handbook of Continuum Mechanics: General Concepts, Thermoelasticity*. Springer. p. 9. ISBN 3-540-41443-6.

[12] John D. Norton (1993). *General covariance and the foundations of general relativity: eight decades of dispute, Rep. Prog. Phys.*, **56**, pp. 835-6.

[13] See Moore and Stommel, Chapter 2, p. 26, which deals with polar coordinates in an inertial frame of reference (what these authors call a "Newtonian frame of reference"), Henry Stommel & Dennis W. Moore (1989). *An Introduction to the Coriolis Force*. Columbia University Press. p. 26. ISBN 0-231-06636-8.

[14] For example, Moore and Stommel point our that in a *rotating* polar coordinate system, the acceleration terms include reference to the rate of rotation of the *rotating frame*. Henry Stommel & Dennis W. Moore. *An Introduction to the Coriolis Force*. p. 55.

[15] The term *particle* is used in mechanics to describe an object without reference to its orientation. The term rigid body is used when orientation is also a factor. Thus, the center of mass of a rigid body is a "particle".

[16] Observational frames of reference and coordinate systems are independent ideas. A frame of reference is a physical notion related to the observer's state of motion. A coordinate system is a mathematical description, which can be chosen to suit the observations. A change to a coordinate system that moves in time affects the description of the particle motion, but does not change the observer's state of motion. For more discussion, see Frame of reference

[17] The article on curvature treats a more general case where the curve is parametrized by an arbitrary variable (denoted t), rather than by the arc length s.

[18] Ahmed A. Shabana; Khaled E. Zaazaa; Hiroyuki Sugiyama (2007). *Railroad Vehicle Dynamics: A Computational Approach*. CRC Press. p. 91. ISBN 1-4200-4581-4.

[19] However, the pilot also will experience Coriolis force, because the pilot is not a *particle*. When the pilot's head moves, for example, the head has a velocity in the non-inertial frame, and becomes subject to Coriolis force. This force causes pilot disorientation in a turn. See Coriolis effect (perception), Arnauld E. Nicogossian (1996). *Space biology and medicine*. Reston, Virginia: American Institute of Aeronautics and Astronautics, Inc. p. 337. ISBN 1-56347-180-9., and Gilles Clément (2003). *Fundamentals of Space Medicine*. Springer. p. 41. ISBN 1-4020-1598-4..

[20] Hugo A Jakobsen (2007). *Chemical Reactor Modeling*. Springer. p. 724. ISBN 3-540-25197-9.

[21] Ramamurti Shankar (1994). *Principles of Quantum Mechanics* (2nd ed.). Springer. p. 81. ISBN 0-306-44790-8.

[22] Francis Begnaud Hildebrand (1992). *Methods of Applied Mathematics* (Reprint of 2nd Edition of 1965 ed.). Courier Dover Publications. p. 156. ISBN 0-486-67002-3.

[23] Although used in this article, these names are not in common use. Alternative names sometimes found are "Newtonian fictitious force" instead of "state-of-motion" fictitious force, and "generalized fictitious force" instead of "coordinate fictitious force". This last term originates in the Lagrangian formulation for mechanics using generalized coordinates. See Francis Begnaud Hildebrand (1992). *Methods of Applied Mathematics* (Reprint of 2nd Edition of 1965 ed.). Courier Dover Publications. p. 156. ISBN 0-486-67002-3.

[24] Donald T. Greenwood (2003). *Advanced Dynamics*. Cambridge University Press. p. 77. ISBN 0-521-82612-8.

[25] Farid M. L. Amirouche (2006). *Fundamentals of Multibody Dynamics: Theory and Applications*. Springer. p. 207. ISBN 0-8176-4236-6.

[26] Harold Josephs; Ronald L. Huston (2002). *Dynamics of Mechanical Systems*. CRC Press. p. 377. ISBN 0-8493-0593-4.

[27] Ahmed A. Shabana (2001). *Computational Dynamics*. Wiley. p. 217. ISBN 0-471-37144-0.

[28] Cornelius Lanczos (1986). *The Variational Principles of Mechanics* (1970 reprint of 4th ed.). Dover Publications. p. 10. ISBN 0-486-65067-7.

[29] Cornelius Lanczos (1986). *The Variational Principles of Mechanics* (Reprint of 1970 4th ed.). Dover Publications. pp. 112–113. ISBN 0-486-65067-7.

[30] Vladimir Igorevich Arnol'd (1989). *Mathematical Methods of Classical Mechanics*. Springer. p. 60. ISBN 0-387-96890-3.

[31] Donald Allan McQuarrie (2000). *Statistical Mechanics*. University Science Books. pp. 5–6. ISBN 1-891389-15-7.

[32] Reinhold von Schwerin (1999). *Multibody system simulation: numerical methods, algorithms, and software*. Springer. p. 24. ISBN 3-540-65662-6.

[33] George F. Corliss, Christele Faure, Andreas Griewank, Laurent Hascoet (editors) (2002). *Automatic Differentiation of Algorithms: From Simulation to Optimization*. Springer. p. 131. ISBN 0-387-95305-1.

[34] Jorge A. C. Ambrósio (editor) (2003). *Advances in Computational Multibody Systems*. Springer. p. 322. ISBN 1-4020-3392-3.

[35] Shuzhi S. Ge; Tong Heng Lee; Christopher John Harris (1998). *Adaptive Neural Network Control of Robotic Manipulators*. World Scientific. pp. 47–48. ISBN 981-02-3452-X.

[36] Richard M. Murray; Zexiang Li; S. Shankar Sastry (1994). *A mathematical introduction to robotic manipulation*. CRC Press. p. 170. ISBN 0-8493-7981-4.

[37] Lorenzo Sciavicco; Bruno Siciliano (2000). *Modelling and control of robot manipulators* (2 ed.). Springer. pp. 142 *ff*. ISBN 1-85233-221-2.

[38] For a treatment using these terms as fictitious forces, see Henry Stommel; Dennis W. Moore. *An Introduction to the Coriolis Force*. p. 36. ISBN 0-231-06636-8.

[39] For a rather abstract but complete discussion, see Harald Atmanspacher & Hans Primas (2008). *Recasting Reality: Wolfgang Pauli's Philosophical Ideas and Contemporary Science*. Springer. p. §2.2, p. 42 *ff*. ISBN 3-540-85197-6.

[40] For the following discussion, see John R Taylor (2005). *Classical Mechanics*. University Science Books. p. §9.10, pp. 358–359. ISBN 1-891389-22-X. At the chosen instant t_0, the frame S' and the particle are rotating at the same rate....In the inertial frame, the forces are simpler (no "fictitious" forces) but the accelerations are more complicated.; in the rotating frame, it is the other way round.

[41] Henry Stommel & Dennis W. Moore (1989). *An Introduction to the Coriolis Force*. p. 55. ISBN 0-231-06636-8.

[42] This derivation can be found in Henry Stommel; Dennis W. Moore. *An Introduction to the Coriolis Force*. p. Chapter III, pp. 54 *ff*.

[43] Francesco Bullo; Andrew D. Lewis (2005). *Geometric Control of Mechanical Systems*. Springer. p. 3. ISBN 0-387-22195-6.

[44] PM Morse & H Feshbach (1953). *Methods of Mathematical Physics* (First ed.). McGraw Hill. p. 25.

[45] PM Morse & H Feshbach (1953). *Methods of Mathematical Physics* (First ed.). McGraw Hill. pp. 47–48.

[46] I-Shih Liu (2002). *Continuum mechanics*. Springer. p. Appendix A2. ISBN 3-540-43019-9.

[47] K. F. Riley; M. P. Hobson; S. J. Bence (2006). *Mathematical Methods for Physics and Engineering*. Cambridge University Press. p. 965. ISBN 0-521-86153-5.

[48] JL Synge & A Schild (1978). *Tensor Calculus* (Reprint of 1969 ed.). Courier Dover Publications. p. 52. ISBN 0-486-63612-7.

[49] For application of the Christoffel symbols formalism to a rotating coordinate system, see Ludwik Silberstein (1922). *The Theory of General Relativity and Gravitation*. D. Van Nostrand. pp. 30–32.

[50] For a more extensive criticism of lumping together the two types of fictitious force, see Ludwik Silberstein (1922). *The Theory of General Relativity and Gravitation*. D. Van Nostrand. p. 29.

[51] See Silberstein.

[52] See R. Kelly; V. Santibáñez; Antonio Loría (2005). *Control of robot manipulators in joint space*. Springer. p. 72. ISBN 1-85233-994-2.

21.7   Further reading

- Newton's description in Principia

- Centrifugal reaction force - Columbia electronic encyclopedia

- M. Alonso and E.J. Finn, *Fundamental university physics*, Addison-Wesley

- Centripetal force vs. Centrifugal force - from an online Regents Exam physics tutorial by the Oswego City School District

- Centrifugal force acts inwards near a black hole

- Centrifugal force at the HyperPhysics concepts site

- A list of interesting links

• Kenneth Franklin Riley; Michael Paul Hobson; Stephen John Bence (2002). "Derivatives of basis vectors and Christoffel symbols". *Mathematical methods for physics and engineering: A comprehensive guide* (2 ed.). Cambridge University Press. pp. 814 *ff*. ISBN 0-521-89067-5.

21.8   External links

- Motion over a flat surface Java physlet by Brian Fiedler (from School of Meteorology at the University of Oklahoma) illustrating fictitious forces. The physlet shows both the perspective as seen from a rotating and from a non-rotating point of view.

- Motion over a parabolic surface Java physlet by Brian Fiedler (from School of Meteorology at the University of Oklahoma) illustrating fictitious forces. The physlet shows both the perspective as seen from a rotating and as seen from a non-rotating point of view.

- Animation clip showing scenes as viewed from both an inertial frame and a rotating frame of reference, visualizing the Coriolis and centrifugal forces.

- Centripetal and Centrifugal Forces at MathPages

- Centrifugal Force at h2g2

- John Baez: *Does centrifugal force hold the Moon up?*

21.9   See also

Chapter 22

Centrifugal force

Not to be confused with Centripetal force.

In Newtonian mechanics, the term **centrifugal force** is used to refer to an inertial force (also called a 'fictitious' force) directed away from the axis of rotation that appears to act on all objects when viewed in a rotating reference frame.

The concept of centrifugal force can be applied in rotating devices such as centrifuges, centrifugal pumps, centrifugal governors, centrifugal clutches, etc., as well as in centrifugal railways, planetary orbits, banked curves, etc. when they are analyzed in a rotating coordinate system.

The name has historically sometimes also been used to refer to the reaction force to the centripetal force.

22.1   Introduction

Centrifugal force is an outward force apparent in a rotating reference frame; it does not exist when measurements are made in an inertial frame of reference.[1]

All measurements of position and velocity must be made relative to some frame of reference. For example, if we are studying the motion of an object in an airliner traveling at great speed, we could calculate the motion of the object with respect to the interior of the airliner, or to the surface of the Earth.[2] An inertial frame of reference is one that is not accelerating (including rotation). The use of an inertial frame of reference, which will be the case for all elementary calculations, is often not explicitly stated but may generally be assumed unless stated otherwise.

In terms of an inertial frame of reference, centrifugal force does not exist. All calculations can be performed using only Newton's laws of motion and the real forces. In its current usage the term 'centrifugal force' has no meaning in an inertial frame.

In an inertial frame, an object that has no forces acting on it travels in a straight line, according to Newton's first law. When measurements are made with respect to a rotating reference frame, however, the same object would have a curved path, because the frame of reference is rotating. If it is desired to apply Newton's laws in the rotating frame, it is necessary to introduce new, fictitious, forces to account for this curved motion.

In the rotating reference frame, all objects, regardless of their state of motion, appear to be under the influence of a radially (from the axis of rotation) outward force that is proportional to their mass, the distance from the axis of rotation of the frame, and to the square of the angular velocity of the frame.[3][4] This is the centrifugal force.

Motion relative to a rotating frame results in another fictitious force, the Coriolis force; and if the rate of rotation of the frame is changing, a third fictitious force, the Euler force is required. Together, these three fictitious forces are necessary for the formulation of correct equations of motion in a rotating reference frame[5][6] and allow Newton's Laws to be used in their normal form in such a frame.[5]

22.2   Examples

22.2.1   A stone on a string

Consider a stone being whirled round on a string. The only real force acting on the stone is the tension in the string. There are no other forces acting on the stone so there is a net force on the stone.

In an inertial frame of reference, were it not for this net force acting on the stone, the stone would travel in a straight line, according to Newton's first law of motion. In order to keep the stone moving in a circular path, this force, known as the centripetal force, must be continuously applied to the stone. As soon as it is removed (for example if the string breaks) the stone moves in a straight line. In this inertial frame, the concept of centrifugal force is not required as all motion can be properly described using only real forces and Newton's laws of motion.

In a frame of reference rotating with the stone around the same axis as the stone, the stone is stationary. However, the tension in the string is still acting on the stone. If Newton's laws were applied in their usual form, the stone would accelerate in the direction of the net applied force; towards the axis of rotation, which it does not do. To use Newton's laws of motion, unchanged, in a rotating frame it is necessary to invent a new force that acts on the stone and is equal and opposite to the tension in the string; this new force acts in the outward direction; it is the centrifugal force. With this new (inertial or fictitious force) the net force on the stone is zero and the stone remains stationary in the rotating frame of reference. With the addition of this extra inertial or fictitious force Newton's laws can be applied in the rotating frame as if it were an inertial (non-rotating) frame.

22.2.2 Weighing an object at the Earth's poles and on the equator

Consider an object that is being weighed with a simple spring balance at one of the Earth's poles. There are only two forces acting on the object, the Earth's gravity, which acts in a downward direction, and the equal and opposite tension in the spring, acting upward. There is no net force acting on the object and the spring balance so the object does not accelerate and remains stationary. The balance shows the value of the force of gravity on the object.

When the same object is weighed on the equator the same two real forces act upon the object. However, the object is moving in a circular path as the Earth rotates. When considered in an inertial frame (that is to say, one that is not rotating with the Earth), some of the force of gravity is expended just to keep the object in its circular path (centripetal force). As such, less tension in the spring is required to counteract the 'remaining' force of gravity. Less tension in the spring would be reflected on a scale as less weight — about 0.3% less at the equator than at the poles.[7] The concept of centrifugal force is not required. However, the Earth is not a perfect sphere, so an object at the poles is slightly closer to the center of the Earth than one at the equator; after accounting for both effects, the actual measured weight of the object is about 0.53% less on the equator.[8]

It is generally more convenient to take measurements in a frame of reference rotating with the Earth. In this reference frame the object is stationary and to account for the loss in measured weight when the object is measured at the equator it is necessary to include the upward acting (inertial or fictitious) centrifugal force. In practice, this is often observed as a reduction in the force of gravity.

22.2.3 An equatorial railway

This thought experiment is more complicated than the previous two examples in that it requires the use of the Coriolis force as well as the centrifugal force.

Imagine a railway line running round the Earth's equator, with a train running at high speed in the opposite direction to the Earth's rotation. The train runs at such a speed that, in an inertial (nonrotating) frame centered on the Earth, it remains stationary as the Earth spins beneath it. In this inertial frame the situation is easy to analyze. The only forces acting on the train are its gravity (downward) and the equal and opposite (upward) reaction force from the track. There is no net force on the train and it therefore remains stationary.

In a frame rotating with the Earth the train is moving in a circular orbit as it travels round the Earth. In this frame, the upward reaction force from the track and the force of gravity on the train remain the same, as they are real forces. However, in the Earth's (rotating) frame, the train is traveling in a circular path and therefore requires a centripetal (downward) force to keep it on this path. Because we are using a rotating frame, we must, as always, apply the (fictitious) centrifugal force to the train. This is equal in value to the required centripetal force but acts in an upward direction—opposite direction to that required. It would therefore seem that there is a net upward force on the train and it should therefore accelerate upward.

In order to explain this paradox we must note that the train is in motion with respect to the rotating frame and we must therefore, in addition to the centrifugal force, add the Coriolis force. In this particular example, this acts in a downward direction and is equal in value to twice the centrifugal force thus canceling out the centrifugal force and supplying the necessary centripetal force to keep the train in its circular path.

22.3 Derivation

Main article: Rotating reference frame
See also: Fictitious force and Mechanics of planar particle motion

For the following formalism, the rotating frame of reference is regarded as a special case of a non-inertial reference frame that is rotating relative to an inertial reference frame denoted the stationary frame.

22.3.1 Velocity

In a rotating frame of reference, the time derivatives of the position vector r, such as velocity and acceleration vectors, of an object will differ from the time derivatives in the stationary frame according to the frame's rotation. The first time derivative $[dr/dt]$ evaluated within a reference frame with a coincident origin at $r = 0$ but rotating with the absolute angular velocity ω is:[9]

$$\frac{dr}{dt} = \left[\frac{dr}{dt}\right] + \omega \times r \,,$$

where $\times$ denotes the vector cross product and square brackets [...] denote evaluation in the rotating frame of reference. In other words, the apparent velocity in the rotating frame is altered by the amount of the apparent rotation $\omega \times r$ at each point, which is perpendicular to both the vector from the origin r and the axis of rotation ω and directly proportional in magnitude to each of them. The vector ω has magnitude ω equal to the rate of rotation and is directed along the axis of rotation according to the right-hand rule.

22.3.2 Acceleration

Newton's law of motion for a particle of mass m written in vector form is:

$$F = ma \,,$$

where F is the vector sum of the physical forces applied to the particle and a is the absolute acceleration (that is, acceleration in an inertial frame) of the particle, given by:

$$a = \frac{d^2 r}{dt^2} \,,$$

where r is the position vector of the particle.

By twice applying the transformation above from the stationary to the rotating frame, the absolute acceleration of the particle can be written as:

$$a = \frac{d^2 r}{dt^2} = \frac{d}{dt}\frac{dr}{dt} = \frac{d}{dt}\left(\left[\frac{dr}{dt}\right] + \omega \times r\right)$$
$$= \left[\frac{d^2 r}{dt^2}\right] + \frac{d\omega}{dt} \times r + 2\omega \times \left[\frac{dr}{dt}\right] + \omega \times (\omega \times r) \,.$$

22.3.3 Force

The apparent acceleration in the rotating frame is $[d^2r/dt^2]$. An observer unaware of the rotation would expect this to be zero in the absence of outside forces. However Newton's laws of motion apply only in the inertial frame and describe dynamics in terms of the absolute acceleration d^2r/dt^2. Therefore, the observer perceives the extra terms as contributions due to fictitious forces. These terms in the apparent acceleration are independent of mass; so it appears that each of these fictitious forces, like gravity, pulls on an object in proportion to its mass. When these forces are added, the equation of motion has the form:[10][11][12]

$$F - m\frac{d\omega}{dt} \times r - 2m\omega \times \left[\frac{dr}{dt}\right] - m\omega \times (\omega \times r)$$
$$= m\left[\frac{d^2 r}{dt^2}\right] \,.$$

From the perspective of the rotating frame, the additional force terms are experienced just like the real external forces and contribute to the apparent acceleration.[13][14] The additional terms on the force side of the equation can be recognized as, reading from left to right, the Euler force $m\,d\omega/dt \times r$, the Coriolis force $2m\omega \times [dr/dt]$, and the centrifugal force $m\omega \times (\omega \times r)$, respectively.[15] Unlike the other two fictitious forces, the centrifugal force always points radially outward from the axis of rotation of the rotating frame, with magnitude $m\omega^2 r$, and unlike the Coriolis force in particular, it is independent of the motion of the particle in the rotating frame. As expected, for a non-rotating inertial frame of reference ($\omega = 0$) the centrifugal force and all other fictitious forces disappear.[16]

22.4 Absolute rotation

Main article: Absolute rotation

Three scenarios were suggested by Newton to answer

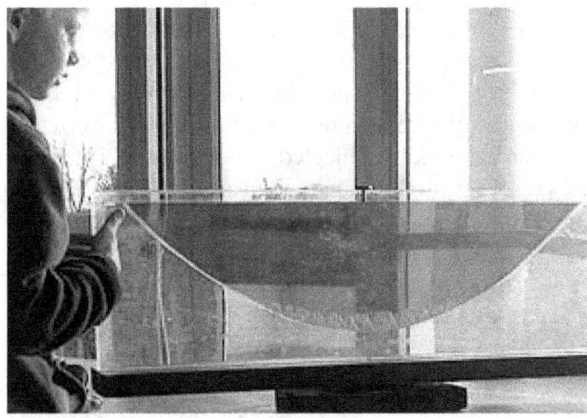

The interface of two immiscible liquids rotating around a vertical axis is an upward-opening circular paraboloid.

the question of whether the absolute rotation of a local frame can be detected; that is, if an observer can decide

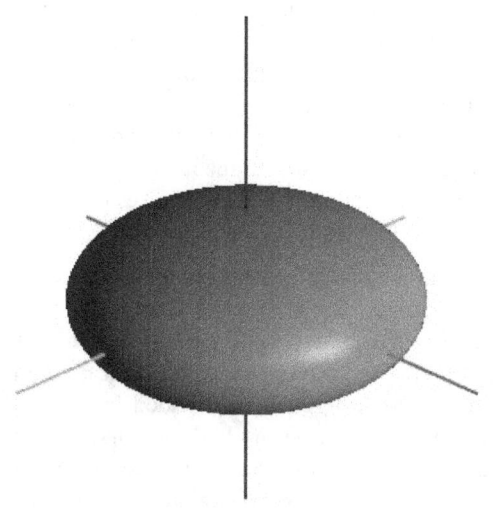

When analysed in a rotating reference frame of the planet, centrifugal force causes rotating planets to assume the shape of an oblate spheroid

whether an observed object is rotating or if the observer is rotating.[17][18]

- The shape of the surface of water rotating in a bucket. The shape of the surface becomes concave to balance the centrifugal force against the other forces upon the liquid.

- The tension in a string joining two spheres rotating about their center of mass. The tension in the string will be proportional to the centrifugal force on each sphere as it rotates around the common center of mass.

In these scenarios, the effects attributed to centrifugal force are only observed in the local frame (the frame in which the object is stationary) if the object is undergoing absolute rotation relative to an inertial frame. By contrast, in an inertial frame, the observed effects arise as a consequence of the inertia and the known forces without the need to introduce a centrifugal force. Based on this argument, the privileged frame, wherein the laws of physics take on the simplest form, is a stationary frame in which no fictitious forces need to be invoked.

Within this view of physics, any other phenomenon that is usually attributed to centrifugal force can be used to identify absolute rotation. For example, the oblateness of a sphere of freely flowing material is often explained in terms of centrifugal force. The oblate spheroid shape reflects, following Clairaut's theorem, the balance between containment by gravitational attraction and dispersal by centrifugal force. That the Earth is itself an oblate spheroid, bulging at the

equator where the radial distance and hence the centrifugal force is larger, is taken as one of the evidences for its absolute rotation.[19]

22.5 Applications

The operations of numerous common rotating mechanical systems are most easily conceptualized in terms of centrifugal force. For example:

- A centrifugal governor regulates the speed of an engine by using spinning masses that move radially, adjusting the throttle, as the engine changes speed. In the reference frame of the spinning masses, centrifugal force causes the radial movement.

- A centrifugal clutch is used in small engine-powered devices such as chain saws, go-karts and model helicopters. It allows the engine to start and idle without driving the device but automatically and smoothly engages the drive as the engine speed rises. Inertial drum brake ascenders used in rock climbing and the inertia reels used in many automobile seat belts operate on the same principle.

- Centrifugal forces can be used to generate artificial gravity, as in proposed designs for rotating space stations. The Mars Gravity Biosatellite would have studied the effects of Mars-level gravity on mice with gravity simulated in this way.

- Spin casting and centrifugal casting are production methods that uses centrifugal force to disperse liquid metal or plastic throughout the negative space of a mold.

- Centrifuges are used in science and industry to separate substances. In the reference frame spinning with the centrifuge, the centrifugal force induces a hydrostatic pressure gradient in fluid-filled tubes oriented perpendicular to the axis of rotation, giving rise to large buoyant forces which push low-density particles inward. Elements or particles denser than the fluid move outward under the influence of the centrifugal force. This is effectively Archimedes' principle as generated by centrifugal force as opposed to being generated by gravity.

- Some amusement rides make use of centrifugal forces. For instance, a Gravitron's spin forces riders against a wall and allows riders to be elevated above the machine's floor in defiance of Earth's gravity.[20]

Nevertheless, all of these systems can also be described without requiring the concept of centrifugal force, in terms of motions and forces in a stationary frame, at the cost of taking somewhat more care in the consideration of forces and motions within the system.

22.6 History of conceptions of centrifugal and centripetal forces

Main article: History of centrifugal and centripetal forces

The conception of centrifugal force has evolved since the time of Huygens, Newton, Leibniz, and Hooke who expressed early conceptions of it. Its modern conception as a fictitious force arising in a rotating reference frame evolved in the eighteenth and nineteenth centuries.

Centrifugal force has also played a role in debates in classical mechanics about detection of absolute motion. Newton suggested two arguments to answer the question of whether absolute rotation can be detected: the rotating bucket argument, and the rotating spheres argument.[21] According to Newton, in each scenario the centrifugal force would be observed in the object's local frame (the frame where the object is stationary) only if the frame were rotating with respect to absolute space. Nearly two centuries later, Mach's principle was proposed where, instead of absolute rotation, the motion of the distant stars relative to the local inertial frame gives rise through some (hypothetical) physical law to the centrifugal force and other inertia effects. Today's view is based upon the idea of an inertial frame of reference, which privileges observers for which the laws of physics take on their simplest form, and in particular, frames that do not use centrifugal forces in their equations of motion in order to describe motions correctly.

The analogy between centrifugal force (sometimes used to create artificial gravity) and gravitational forces led to the equivalence principle of general relativity.[22][23]

22.7 Other uses of the term

While the majority of the scientific literature uses the term *centrifugal force* to refer to the particular fictitious force that arises in rotating frames, there are a few limited instances in the literature of the term applied to other distinct physical concepts. One of these instances occurs in Lagrangian mechanics. Lagrangian mechanics formulates mechanics in terms of generalized coordinates $\{q_k\}$, which can be as simple as the usual polar coordinates (r, θ) or a much more extensive list of variables.[24][25] Within this formulation the motion is described in terms of *generalized forces*, using in place of Newton's laws the Euler–Lagrange equations.

Among the generalized forces, those involving the square of the time derivatives $\{(dq_k/dt)^2\}$ are sometimes called **centrifugal forces**.[26][27][28][29] In the case of motion in a central potential the Lagrangian centrifugal force has the same form as the fictitious centrifugal force derived in a co-rotating frame.[30] However, the Lagrangian use of "centrifugal force" in other, more general cases has only a limited connection to the Newtonian definition.

In another instance the term refers to the reaction force to a centripetal force, or reactive centrifugal force. A body undergoing curved motion, such as circular motion, is accelerating toward a center at any particular point in time. This centripetal acceleration is provided by a centripetal force, which is exerted on the body in curved motion by some other body. In accordance with Newton's third law of motion, the body in curved motion exerts an equal and opposite force on the other body. This reactive force is exerted *by* the body in curved motion *on* the other body that provides the centripetal force and its direction is from that other body toward the body in curved motion.[31][32] [33][34]

This reaction force is sometimes described as a *centrifugal inertial reaction*,[35][36] that is, a force that is centrifugally directed, which is a reactive force equal and opposite to the centripetal force that is curving the path of the mass.

The concept of the reactive centrifugal force is sometimes used in mechanics and engineering. It is sometimes referred to as just *centrifugal force* rather than as *reactive* centrifugal force[37][38] although this usage is deprecated in elementary mechanics.[39]

22.8 See also

- Centrifugal mechanism of acceleration
- Equivalence principle
- Folk physics
- Lamm equation
- Lagrangian point
- Balancing of rotating masses

22.9 References

[1]

[2] http://www-spof.gsfc.nasa.gov/stargaze/Sframes1.htm

[3] Encyclopaedia Britannica, article on Centrifuge

[4] Feynman lectures on physics, Book 1 12-11

[5] Alexander L. Fetter; John Dirk Walecka (2003). *Theoretical Mechanics of Particles and Continua*. Courier Dover Publications. pp. 38–39. ISBN 0-486-43261-0.

[6] Jerrold E. Marsden; Tudor S. Ratiu (1999). *Introduction to Mechanics and Symmetry: A Basic Exposition of Classical Mechanical Systems*. Springer. p. 251. ISBN 0-387-98643-X.

[7] "Curious About Astronomy?" Archived January 17, 2015, at the Wayback Machine., Cornell University, retrieved June 2007

[8] Boynton, Richard (2001). *"Precise Measurement of Mass"* (PDF). *Sawe Paper No. 3147*. Arlington, Texas: S.A.W.E., Inc. Retrieved 2007-01-21.

[9] John L. Synge (2007). *Principles of Mechanics* (Reprint of Second Edition of 1942 ed.). Read Books. p. 347. ISBN 1-4067-4670-3.

[10] Taylor (2005). p. 342.

[11] LD Landau; LM Lifshitz (1976). *Mechanics* (Third ed.). Oxford: Butterworth-Heinemann. p. 128. ISBN 978-0-7506-2896-9.

[12] Louis N. Hand; Janet D. Finch (1998). *Analytical Mechanics*. Cambridge University Press. p. 267. ISBN 0-521-57572-9.

[13] Mark P Silverman (2002). *A universe of atoms, an atom in the universe* (2 ed.). Springer. p. 249. ISBN 0-387-95437-6.

[14] Taylor (2005). p. 329.

[15] Cornelius Lanczos (1986). *The Variational Principles of Mechanics* (Reprint of Fourth Edition of 1970 ed.). Dover Publications. Chapter 4, §5. ISBN 0-486-65067-7.

[16] Morton Tavel (2002). *Contemporary Physics and the Limits of Knowledge*. Rutgers University Press. p. 93. ISBN 0-8135-3077-6. Noninertial forces, like centrifugal and Coriolis forces, can be eliminated by jumping into a reference frame that moves with constant velocity, the frame that Newton called inertial.

[17] Louis N. Hand; Janet D. Finch (1998). *Analytical Mechanics*. Cambridge University Press. p. 324. ISBN 0-521-57572-9.

[18] I. Bernard Cohen; George Edwin Smith (2002). *The Cambridge companion to Newton*. Cambridge University Press. p. 43. ISBN 0-521-65696-6.

[19] Simon Newcomb (1878). *Popular astronomy*. Harper & Brothers. pp. 86–88.

[20] Myers, Rusty L. (2006). *The basics of physics*. Greenwood Publishing Group. p. 57. ISBN 0-313-32857-9.

[21] An English translation is found at Isaac Newton (1934). *Philosophiae naturalis principia mathematica* (Andrew Motte translation of 1729, revised by Florian Cajori ed.). University of California Press. pp. 10–12.

[22] Barbour, Julian B. and Herbert Pfister (1995). *Mach's principle: from Newton's bucket to quantum gravity*. Birkhäuser. ISBN 0-8176-3823-7, p. 69.

[23] Eriksson, Ingrid V. (2008). *Science education in the 21st century*. Nova Books. ISBN 1-60021-951-9, p. 194.

[24] For an introduction, see for example Cornelius Lanczos (1986). *The variational principles of mechanics* (Reprint of 1970 University of Toronto ed.). Dover. p. 1. ISBN 0-486-65067-7.

[25] For a description of generalized coordinates, see Ahmed A. Shabana (2003). "Generalized coordinates and kinematic constraints". *Dynamics of Multibody Systems* (2 ed.). Cambridge University Press. p. 90 *ff*. ISBN 0-521-54411-4.

[26] Christian Ott (2008). *Cartesian Impedance Control of Redundant and Flexible-Joint Robots*. Springer. p. 23. ISBN 3-540-69253-3.

[27] Shuzhi S. Ge; Tong Heng Lee; Christopher John Harris (1998). *Adaptive Neural Network Control of Robotic Manipulators*. World Scientific. pp. 47–48. ISBN 981-02-3452-X. In the above Euler–Lagrange equations, there are three types of terms. The first involves the second derivative of the generalized co-ordinates. The second is quadratic in $\dot{q}$ where the coefficients may depend on q . These are further classified into two types. Terms involving a product of the type $\dot{q}_i^2$ are called *centrifugal forces* while those involving a product of the type $\dot{q}_i\dot{q}_j$ for $i \neq j$ are called *Coriolis forces*. The third type is functions of q only and are called *gravitational forces*.

[28] R. K. Mittal; I. J. Nagrath (2003). *Robotics and Control*. Tata McGraw-Hill. p. 202. ISBN 0-07-048293-4.

[29] T Yanao; K Takatsuka (2005). "Effects of an intrinsic metric of molecular internal space". In Mikito Toda; Tamiki Komatsuzaki; Stuart A. Rice; Tetsuro Konishi; R. Stephen Berry. *Geometrical Structures Of Phase Space In Multidimensional Chaos: Applications to chemical reaction dynamics in complex systems*. Wiley. p. 98. ISBN 0-471-71157-8. As is evident from the first terms ..., which are proportional to the square of $\dot{\phi}$, a kind of "centrifugal force" arises ... We call this force "democratic centrifugal force". Of course, DCF is different from the ordinary centrifugal force, and it arises even in a system of zero angular momentum.

[30] See p. 5 in Donato Bini; Paolo Carini; Robert T Jantzen (1997). "The intrinsic derivative and centrifugal forces in general relativity: I. Theoretical foundations". *International Journal of Modern Physics D*. **6** (1).. The companion paper is Donato Bini; Paolo Carini; Robert T Jantzen (1997). "The intrinsic derivative and centrifugal forces in general relativity: II. Applications to circular orbits in some stationary

axisymmetric spacetimes". *International Journal of Modern Physics D.* **6** (1).

[31] Mook, Delo E. & Thomas Vargish (1987). *Inside relativity.* Princeton NJ: Princeton University Press. ISBN 0-691-02520-7, p. 47.

[32] G. David Scott (1957). "Centrifugal Forces and Newton's Laws of Motion". **25**. American Journal of Physics. p. 325.

[33] Signell, Peter (2002). "Acceleration and force in circular motion" *Physnet.* Michigan State University, "Acceleration and force in circular motion", §5b, p. 7.

[34] Mohanty, A. K. (2004). *Fluid Mechanics.* PHI Learning Pvt. Ltd. ISBN 81-203-0894-8, p. 121.

[35] Roche, John (September 2001). "Introducing motion in a circle". *Physics Education* **43** (5), pp. 399-405, "Introducing motion in a circle". Retrieved 2009-05-07.

[36] Lloyd William Taylor (1959). *Physics, the pioneer science.* **1**. Dover Publications. p. 173.

[37] Edward Albert Bowser (1920). *An elementary treatise on analytic mechanics: with numerous examples* (25th ed.). D. Van Nostrand Company. p. 357.

[38] Joseph A. Angelo (2007). *Robotics: a reference guide to the new technology.* Greenwood Press. p. 267. ISBN 1-57356-337-4.

[39] Eric M Rogers (1960). *Physics for the Inquiring Mind.* Princeton University Press. p. 302.

Chapter 23

Centripetal force

Not to be confused with centrifugal force. For other meanings of "centripetal", see Centripetal (disambiguation).

A **centripetal force** (from Latin *centrum*, "center" and *petere*, "to seek"[1]) is a force that makes a body follow a curved path. Its direction is always orthogonal to the motion of the body and towards the fixed point of the instantaneous center of curvature of the path. Isaac Newton described it as "a force by which bodies are drawn or impelled, or in any way tend, towards a point as to a centre".[2] In Newtonian mechanics, gravity provides the centripetal force responsible for astronomical orbits.

One common example involving centripetal force is the case in which a body moves with uniform speed along a circular path. The centripetal force is directed at right angles to the motion and also along the radius towards the centre of the circular path.[3][4] The mathematical description was derived in 1659 by the Dutch physicist Christiaan Huygens.[5]

23.1 Formula

The magnitude of the centripetal force on an object of mass m moving at tangential speed v along a path with radius of curvature r is:[6]

$$F = ma_c = \frac{mv^2}{r}$$

where a_c is the centripetal acceleration. The direction of the force is toward the center of the circle in which the object is moving, or the osculating circle (the circle that best fits the local path of the object, if the path is not circular).[7] The speed in the formula is squared, so twice the speed needs four times the force. The inverse relationship with the radius of curvature shows that half the radial distance requires twice the force. This force is also sometimes written in terms of the angular velocity ω of the object about the center of the circle, related to the tangential velocity by the formula

$$v = \omega r$$

so that

$$F = mr\omega^2 .$$

Expressed using the orbital period T for one revolution of the circle,

$$\omega = \frac{2\pi}{T}$$

the equation becomes

$$F = mr \left(\tfrac{2\pi}{T} \right)^2 . \text{[8]}$$

In particle accelerators, velocity can be very high (close to the speed of light in vacuum) so the same rest mass now exerts greater inertia (relativistic mass) thereby requiring greater force for the same centripetal acceleration, so the equation becomes:

$$F = \frac{\gamma mv^2}{r}$$

where

$$\gamma = \frac{1}{\sqrt{1 - v^2/c^2}}$$

is called the Lorentz factor.

More intuitively:

$$F = \gamma mv\omega$$

which is the rate of change of relativistic momentum (γmv)

23.2 Sources of centripetal force

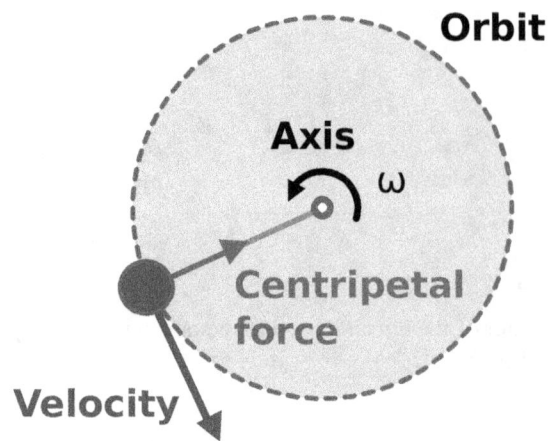

A body experiencing uniform circular motion requires a centripetal force, towards the axis as shown, to maintain its circular path.

In the case of an object that is swinging around on the end of a rope in a horizontal plane, the centripetal force on the object is supplied by the tension of the rope. The rope example is an example involving a 'pull' force. The centripetal force can also be supplied as a 'push' force, such as in the case where the normal reaction of a wall supplies the centripetal force for a wall of death rider.

Newton's idea of a centripetal force corresponds to what is nowadays referred to as a central force. When a satellite is in orbit around a planet, gravity is considered to be a centripetal force even though in the case of eccentric orbits, the gravitational force is directed towards the focus, and not towards the instantaneous center of curvature.[9]

Another example of centripetal force arises in the helix that is traced out when a charged particle moves in a uniform magnetic field in the absence of other external forces. In this case, the magnetic force is the centripetal force that acts towards the helix axis.

23.3 Analysis of several cases

Below are three examples of increasing complexity, with derivations of the formulas governing velocity and acceleration.

23.3.1 Uniform circular motion

See also: Uniform circular motion

Uniform circular motion refers to the case of constant rate of rotation. Here are two approaches to describing this case.

23.3.2 Calculus derivation

In two dimensions, the position vector $\mathbf{r}$, which has magnitude (length) r and directed at an angle θ above the x-axis, can be expressed in Cartesian coordinates using the unit vectors $\hat{x}$ and $\hat{y}$:[10]

$$\mathbf{r} = r\cos(\theta)\hat{x} + r\sin(\theta)\hat{y}.$$

Assume uniform circular motion, which requires three things.

1. The object moves only on a circle.

2. The radius of the circle r does not change in time.

3. The object moves with constant angular velocity ω around the circle. Therefore, $\theta = \omega t$ where t is time.

Now find the velocity $\mathbf{v}$ and acceleration $\mathbf{a}$ of the motion by taking derivatives of position with respect to time.

$$\mathbf{r} = r\cos(\omega t)\hat{x} + r\sin(\omega t)\hat{y}$$

$$\dot{\mathbf{r}} = \mathbf{v} = -r\omega\sin(\omega t)\hat{x} + r\omega\cos(\omega t)\hat{y}$$

$$\ddot{\mathbf{r}} = \mathbf{a} = -r\omega^2\cos(\omega t)\hat{x} - r\omega^2\sin(\omega t)\hat{y}$$

$$\mathbf{a} = -\omega^2(r\cos(\omega t)\hat{x} + r\sin(\omega t)\hat{y})$$

Notice that the term in parenthesis is the original expression of $\mathbf{r}$ in Cartesian coordinates. Consequently,

$$\mathbf{a} = -\omega^2\mathbf{r}.$$

negative shows that the acceleration is pointed towards the center of the circle (opposite the radius), hence it is called "centripetal" (i.e. "center-seeking"). While objects naturally follow a straight path (due to inertia), this centripetal acceleration describes the circular motion path caused by a centripetal force.

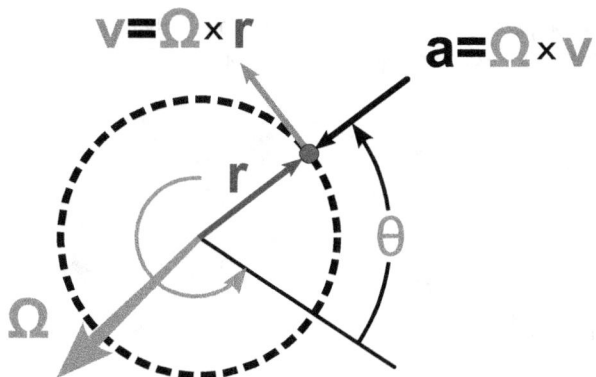

*Vector relationships for uniform circular motion; vector **Ω** representing the rotation is normal to the plane of the orbit with polarity determined by the right-hand rule and magnitude dθ /dt.*

Derivation using vectors

The image at right shows the vector relationships for uniform circular motion. The rotation itself is represented by the angular velocity vector **Ω**, which is normal to the plane of the orbit (using the right-hand rule) and has magnitude given by:

$$|\mathbf{\Omega}| = \frac{d\theta}{dt} = \omega \,,$$

with θ the angular position at time t. In this subsection, $d\theta/dt$ is assumed constant, independent of time. The distance traveled $\mathbf{d\ell}$ of the particle in time dt along the circular path is

$$\mathbf{d\ell} = \mathbf{\Omega} \times \mathbf{r}(t)dt \,,$$

which, by properties of the vector cross product, has magnitude $rd\theta$ and is in the direction tangent to the circular path.

Consequently,

$$\frac{d\mathbf{r}}{dt} = \lim_{\Delta t \to 0} \frac{\mathbf{r}(t + \Delta t) - \mathbf{r}(t)}{\Delta t} = \frac{d\mathbf{\ell}}{dt} \,.$$

In other words,

$$\mathbf{v} \overset{\text{def}}{=} \frac{d\mathbf{r}}{dt} = \frac{d\mathbf{\ell}}{dt} = \mathbf{\Omega} \times \mathbf{r}(t) \,.$$

Differentiating with respect to time,

$$\mathbf{a} \overset{\text{def}}{=} \frac{d\mathbf{v}}{dt} = \mathbf{\Omega} \times \frac{d\mathbf{r}(t)}{dt} = \mathbf{\Omega} \times [\mathbf{\Omega} \times \mathbf{r}(t)] \,.$$

Lagrange's formula states:

$$\mathbf{a} \times (\mathbf{b} \times \mathbf{c}) = \mathbf{b}\,(\mathbf{a} \cdot \mathbf{c}) - \mathbf{c}\,(\mathbf{a} \cdot \mathbf{b}) \,.$$

Applying Lagrange's formula with the observation that $\mathbf{\Omega} \bullet \mathbf{r}(t) = 0$ at all times,

$$\mathbf{a} = -|\mathbf{\Omega}|^2 \mathbf{r}(t) \,.$$

In words, the acceleration is pointing directly opposite to the radial displacement **r** at all times, and has a magnitude:

$$|\mathbf{a}| = |\mathbf{r}(t)| \left(\frac{d\theta}{dt}\right)^2 = r\omega^2$$

where vertical bars |...| denote the vector magnitude, which in the case of $\mathbf{r}(t)$ is simply the radius r of the path. This result agrees with the previous section, though the notation is slightly different.

When the rate of rotation is made constant in the analysis of nonuniform circular motion, that analysis agrees with this one.

A merit of the vector approach is that it is manifestly independent of any coordinate system.

Example: The banked turn

Main article: Banked turn
See also: Reactive centrifugal force
The upper panel in the image at right shows a ball in circular motion on a banked curve. The curve is banked at an angle θ from the horizontal, and the surface of the road is considered to be slippery. The objective is to find what angle the bank must have so the ball does not slide off the road.[11] Intuition tells us that, on a flat curve with no banking at all, the ball will simply slide off the road; while with a very steep banking, the ball will slide to the center unless it travels the curve rapidly.

Apart from any acceleration that might occur in the direction of the path, the lower panel of the image above indicates the forces on the ball. There are *two* forces; one is the force of gravity vertically downward through the center of mass of the ball $m\mathbf{g}$, where m is the mass of the ball and **g** is the gravitational acceleration; the second is the upward normal force exerted by the road perpendicular to the road surface $m\mathbf{a_n}$. The centripetal force demanded by the curved motion is also shown above. This centripetal force

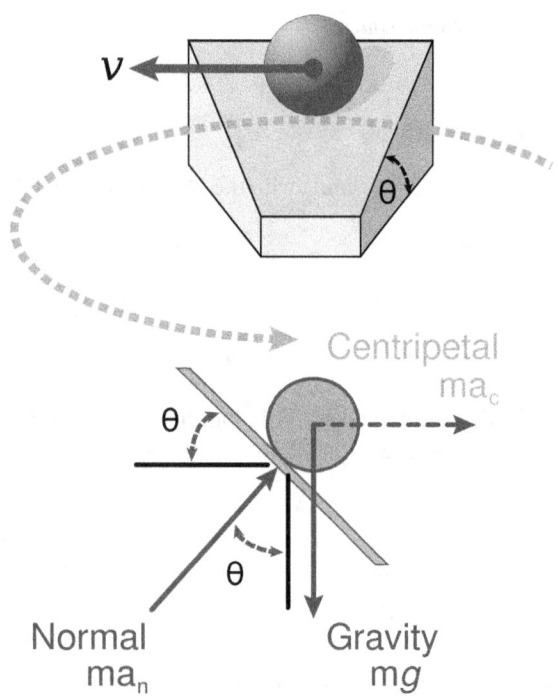

Upper panel: Ball on a banked circular track moving with constant speed v; Lower panel: Forces on the ball

is not a third force applied to the ball, but rather must be provided by the net force on the ball resulting from vector addition of the normal force and the force of gravity. The resultant or net force on the ball found by vector addition of the normal force exerted by the road and vertical force due to gravity must equal the centripetal force dictated by the need to travel a circular path. The curved motion is maintained so long as this net force provides the centripetal force requisite to the motion.

The horizontal net force on the ball is the horizontal component of the force from the road, which has magnitude $|\mathbf{F}_h| = m|\mathbf{a}_n|\sin\theta$. The vertical component of the force from the road must counteract the gravitational force: $|\mathbf{F}_v| = m|\mathbf{a}_n|\cos\theta = m|\mathbf{g}|$, which implies $|\mathbf{a}_n| = |\mathbf{g}| / \cos\theta$. Substituting into the above formula for $|\mathbf{F}_h|$ yields a horizontal force to be:

$$|\mathbf{F}_h| = m|\mathbf{g}|\frac{\sin\theta}{\cos\theta} = m|\mathbf{g}|\tan\theta .$$

On the other hand, at velocity $|\mathbf{v}|$ on a circular path of radius r, kinematics says that the force needed to turn the ball continuously into the turn is the radially inward centripetal force $\mathbf{F}_c$ of magnitude:

$$|\mathbf{F}_c| = m|\mathbf{a}_c| = \frac{m|\mathbf{v}|^2}{r} .$$

Consequently, the ball is in a stable path when the angle of the road is set to satisfy the condition:

$$m|\mathbf{g}|\tan\theta = \frac{m|\mathbf{v}|^2}{r} ,$$

or,

$$\tan\theta = \frac{|\mathbf{v}|^2}{|\mathbf{g}|r} .$$

As the angle of bank θ approaches 90°, the tangent function approaches infinity, allowing larger values for $|\mathbf{v}|^2/r$. In words, this equation states that for faster speeds (bigger $|\mathbf{v}|$) the road must be banked more steeply (a larger value for θ), and for sharper turns (smaller r) the road also must be banked more steeply, which accords with intuition. When the angle θ does not satisfy the above condition, the horizontal component of force exerted by the road does not provide the correct centripetal force, and an additional frictional force tangential to the road surface is called upon to provide the difference. If friction cannot do this (that is, the coefficient of friction is exceeded), the ball slides to a different radius where the balance can be realized.[12][13]

These ideas apply to air flight as well. See the FAA pilot's manual.[14]

23.3.3 Nonuniform circular motion

See also: Circular motion and Non-uniform circular motion
As a generalization of the uniform circular motion case,

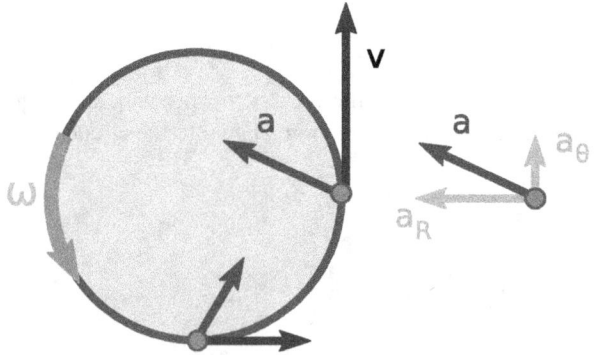

Velocity and acceleration for nonuniform circular motion: the velocity vector is tangential to the orbit, but the acceleration vector is not radially inward because of its tangential component $a\theta$ that increases the rate of rotation: $d\omega / dt = | a\theta | / R$.

suppose the angular rate of rotation is not constant. The acceleration now has a tangential component, as shown the image at right. This case is used to demonstrate a derivation strategy based on a polar coordinate system.

Let $\mathbf{r}(t)$ be a vector that describes the position of a point mass as a function of time. Since we are assuming circular motion, let $\mathbf{r}(t) = R \cdot \mathbf{u}_r$, where R is a constant (the radius of the circle) and $\mathbf{u}_r$ is the unit vector pointing from the origin to the point mass. The direction of $\mathbf{u}_r$ is described by θ, the angle between the x-axis and the unit vector, measured counterclockwise from the x-axis. The other unit vector for polar coordinates, $\mathbf{u}\theta$ is perpendicular to $\mathbf{u}_r$ and points in the direction of increasing θ. These polar unit vectors can be expressed in terms of Cartesian unit vectors in the x and y directions, denoted $\mathbf{i}$ and $\mathbf{j}$ respectively:[15]

$$\mathbf{u}_r = \cos\theta\, \mathbf{i} + \sin\theta\, \mathbf{j}$$

and

$$\mathbf{u}\theta = -\sin\theta\, \mathbf{i} + \cos\theta\, \mathbf{j}.$$

One can differentiate to find velocity:

$$\mathbf{v} = r\frac{d\mathbf{u}_r}{dt} = r\frac{d}{dt}\left(\cos\theta\, \mathbf{i} + \sin\theta\, \mathbf{j}\right)$$

$$= r\frac{d\theta}{dt}\left(-\sin\theta\, \mathbf{i} + \cos\theta\, \mathbf{j}\right)$$

$$= r\frac{d\theta}{dt}\mathbf{u}_\theta$$

$$= \omega r\mathbf{u}_\theta$$

where ω is the angular velocity $d\theta/dt$.

This result for the velocity matches expectations that the velocity should be directed tangentially to the circle, and that the magnitude of the velocity should be $r\omega$. Differentiating again, and noting that

$$\frac{d\mathbf{u}_\theta}{dt} = -\frac{d\theta}{dt}\mathbf{u}_r = -\omega\mathbf{u}_r\,,$$

we find that the acceleration, $\mathbf{a}$ is:

$$\mathbf{a} = r\left(\frac{d\omega}{dt}\mathbf{u}_\theta - \omega^2\mathbf{u}_r\right)\,.$$

Thus, the radial and tangential components of the acceleration are:

$$\mathbf{a}_r = -\omega^2 r\, \mathbf{u}_r = -\frac{|\mathbf{v}|^2}{r}\, \mathbf{u}_r \text{ and } \mathbf{a}_\theta = r\frac{d\omega}{dt}\, \mathbf{u}_\theta = \frac{d|\mathbf{v}|}{dt}\, \mathbf{u}_\theta\,,$$

where $|\mathbf{v}| = r\,\omega$ is the magnitude of the velocity (the speed).

These equations express mathematically that, in the case of an object that moves along a circular path with a changing speed, the acceleration of the body may be decomposed into a perpendicular component that changes the direction of motion (the centripetal acceleration), and a parallel, or tangential component, that changes the speed.

23.3.4 General planar motion

See also: Generalized forces, Generalized force, Curvilinear coordinates, Generalized coordinates, and Orthogonal coordinates

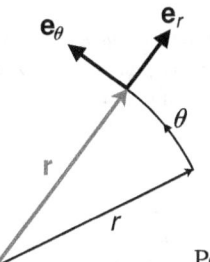

Position vector $\mathbf{r}$, always points radially from the origin.

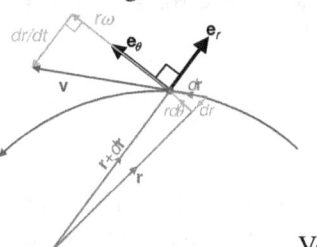

Velocity vector $\mathbf{v}$, always tangent to the path of motion.

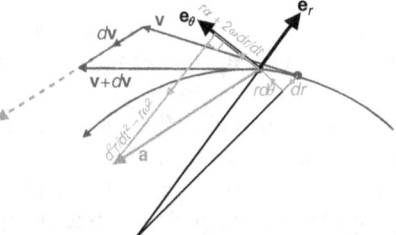

Acceleration vector $\mathbf{a}$, not parallel to the radial motion but offset by the angular and Coriolis accelerations, nor tangent to the path but offset by the centripetal and radial accelerations.

Kinematic vectors in plane polar coordinates. Notice the setup is not restricted to 2d space, but a plane in any higher dimension.

Polar coordinates

The above results can be derived perhaps more simply in polar coordinates, and at the same time extended to general

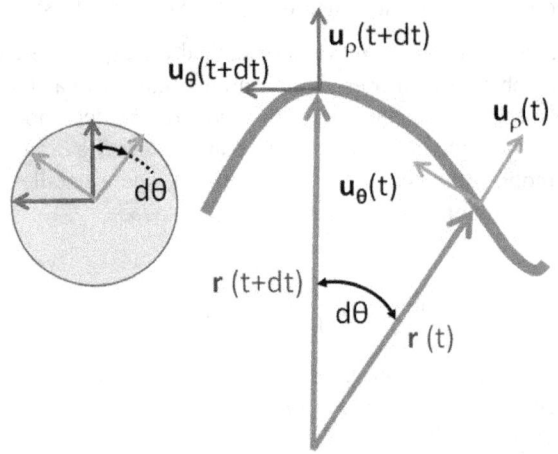

Polar unit vectors at two times t *and* t + dt *for a particle with trajectory* r *(* t *); on the left the unit vectors* uₚ *and* uθ *at the two times are moved so their tails all meet, and are shown to trace an arc of a unit radius circle. Their rotation in time* dt *is* dθ*, just the same angle as the rotation of the trajectory* r *(* t *).*

motion within a plane, as shown next. Polar coordinates in the plane employ a radial unit vector $\mathbf{u}_\rho$ and an angular unit vector $\mathbf{u}\theta$, as shown above.[16] A particle at position $\mathbf{r}$ is described by:

$$\mathbf{r} = \rho\mathbf{u}_\rho ,$$

where the notation ϱ is used to describe the distance of the path from the origin instead of R to emphasize that this distance is not fixed, but varies with time. The unit vector $\mathbf{u}_\rho$ travels with the particle and always points in the same direction as $\mathbf{r}(t)$. Unit vector $\mathbf{u}\theta$ also travels with the particle and stays orthogonal to $\mathbf{u}_\rho$. Thus, $\mathbf{u}_\rho$ and $\mathbf{u}\theta$ form a local Cartesian coordinate system attached to the particle, and tied to the path traveled by the particle.[17] By moving the unit vectors so their tails coincide, as seen in the circle at the left of the image above, it is seen that $\mathbf{u}_\rho$ and $\mathbf{u}\theta$ form a right-angled pair with tips on the unit circle that trace back and forth on the perimeter of this circle with the same angle $\theta(t)$ as $\mathbf{r}(t)$.

When the particle moves, its velocity is

$$\mathbf{v} = \frac{d\rho}{dt}\mathbf{u}_\rho + \rho\frac{d\mathbf{u}_\rho}{dt} .$$

To evaluate the velocity, the derivative of the unit vector $\mathbf{u}_\rho$ is needed. Because $\mathbf{u}_\rho$ is a unit vector, its magnitude is fixed, and it can change only in direction, that is, its change $d\mathbf{u}_\rho$ has a component only perpendicular to $\mathbf{u}_\rho$. When the trajectory $\mathbf{r}(t)$ rotates an amount $d\theta$, $\mathbf{u}_\rho$, which points in the

same direction as $\mathbf{r}(t)$, also rotates by $d\theta$. See image above. Therefore, the change in $\mathbf{u}_\rho$ is

$$d\mathbf{u}_\rho = \mathbf{u}_\theta d\theta ,$$

or

$$\frac{d\mathbf{u}_\rho}{dt} = \mathbf{u}_\theta\frac{d\theta}{dt} .$$

In a similar fashion, the rate of change of $\mathbf{u}\theta$ is found. As with $\mathbf{u}_\rho$, $\mathbf{u}\theta$ is a unit vector and can only rotate without changing size. To remain orthogonal to $\mathbf{u}_\rho$ while the trajectory $\mathbf{r}(t)$ rotates an amount $d\theta$, $\mathbf{u}\theta$, which is orthogonal to $\mathbf{r}(t)$, also rotates by $d\theta$. See image above. Therefore, the change $d\mathbf{u}\theta$ is orthogonal to $\mathbf{u}\theta$ and proportional to $d\theta$ (see image above):

$$\frac{d\mathbf{u}_\theta}{dt} = -\frac{d\theta}{dt}\mathbf{u}_\rho .$$

The image above shows the sign to be negative: to maintain orthogonality, if $d\mathbf{u}_\rho$ is positive with $d\theta$, then $d\mathbf{u}\theta$ must decrease.

Substituting the derivative of $\mathbf{u}_\rho$ into the expression for velocity:

$$\mathbf{v} = \frac{d\rho}{dt}\mathbf{u}_\rho + \rho\mathbf{u}_\theta\frac{d\theta}{dt} = v_\rho\mathbf{u}_\rho + v_\theta\mathbf{u}_\theta = \mathbf{v}_\rho + \mathbf{v}_\theta .$$

To obtain the acceleration, another time differentiation is done:

$$\mathbf{a} = \frac{d^2\rho}{dt^2}\mathbf{u}_\rho + \frac{d\rho}{dt}\frac{d\mathbf{u}_\rho}{dt} + \frac{d\rho}{dt}\mathbf{u}_\theta\frac{d\theta}{dt} + \rho\frac{d\mathbf{u}_\theta}{dt}\frac{d\theta}{dt} + \rho\mathbf{u}_\theta\frac{d^2\theta}{dt^2} .$$

Substituting the derivatives of $\mathbf{u}_\rho$ and $\mathbf{u}\theta$, the acceleration of the particle is:[18]

$$\mathbf{a} = \frac{d^2\rho}{dt^2}\mathbf{u}_\rho + 2\frac{d\rho}{dt}\mathbf{u}_\theta\frac{d\theta}{dt} - \rho\mathbf{u}_\rho\left(\frac{d\theta}{dt}\right)^2 + \rho\mathbf{u}_\theta\frac{d^2\theta}{dt^2} ,$$

$$= \mathbf{u}_\rho\left[\frac{d^2\rho}{dt^2} - \rho\left(\frac{d\theta}{dt}\right)^2\right] + \mathbf{u}_\theta\left[2\frac{d\rho}{dt}\frac{d\theta}{dt} + \rho\frac{d^2\theta}{dt^2}\right]$$

$$= \mathbf{u}_\rho\left[\frac{dv_\rho}{dt} - \frac{v_\theta^2}{\rho}\right] + \mathbf{u}_\theta\left[\frac{2}{\rho}v_\rho v_\theta + \rho\frac{d}{dt}\frac{v_\theta}{\rho}\right] .$$

As a particular example, if the particle moves in a circle of constant radius R, then $d\varrho/dt = 0$, $\mathbf{v} = \mathbf{v}\theta$, and:

$$\mathbf{a} = \mathbf{u}_\rho \left[-\rho \left(\frac{d\theta}{dt} \right)^2 \right] + \mathbf{u}_\theta \left[\rho \frac{d^2\theta}{dt^2} \right]$$

$$= \mathbf{u}_\rho \left[-\frac{v^2}{r} \right] + \mathbf{u}_\theta \left[\frac{dv}{dt} \right]$$

where $v = v_\theta$.

These results agree with those above for nonuniform circular motion. See also the article on non-uniform circular motion. If this acceleration is multiplied by the particle mass, the leading term is the centripetal force and the negative of the second term related to angular acceleration is sometimes called the Euler force.[19]

For trajectories other than circular motion, for example, the more general trajectory envisioned in the image above, the instantaneous center of rotation and radius of curvature of the trajectory are related only indirectly to the coordinate system defined by $\mathbf{u}_\rho$ and $\mathbf{u}\theta$ and to the length $|\mathbf{r}(t)|$ = ϱ. Consequently, in the general case, it is not straightforward to disentangle the centripetal and Euler terms from the above general acceleration equation.[20] [21] To deal directly with this issue, local coordinates are preferable, as discussed next.

Local coordinates

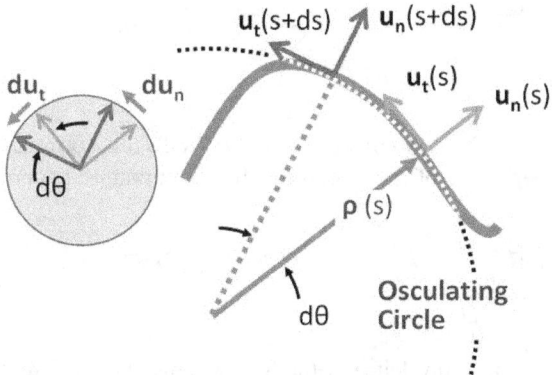

Local coordinate system for planar motion on a curve. Two different positions are shown for distances s *and* s + ds *along the curve. At each position* s, *unit vector* **u**n *points along the outward normal to the curve and unit vector* **u**t *is tangential to the path. The radius of curvature of the path is* ρ *as found from the rate of rotation of the tangent to the curve with respect to arc length, and is the radius of the osculating circle at position* s. *The unit circle on the left shows the rotation of the unit vectors with* s.

Local coordinates mean a set of coordinates that travel with the particle,[22] and have orientation determined by the path of the particle.[23] Unit vectors are formed as shown in the image at right, both tangential and normal to the path. This coordinate system sometimes is referred to as intrin-

sic or *path coordinates*[24][25] or *nt-coordinates*, for *normal-tangential*, referring to these unit vectors. These coordinates are a very special example of a more general concept of local coordinates from the theory of differential forms.[26]

Distance along the path of the particle is the arc length s, considered to be a known function of time.

$$s = s(t) .$$

A center of curvature is defined at each position s located a distance ϱ (the radius of curvature) from the curve on a line along the normal $\mathbf{u}_n$ (s). The required distance $\varrho(s)$ at arc length s is defined in terms of the rate of rotation of the tangent to the curve, which in turn is determined by the path itself. If the orientation of the tangent relative to some starting position is $\theta(s)$, then $\varrho(s)$ is defined by the derivative $d\theta/ds$:

$$\frac{1}{\rho(s)} = \kappa(s) = \frac{d\theta}{ds} .$$

The radius of curvature usually is taken as positive (that is, as an absolute value), while the *curvature* κ is a signed quantity.

A geometric approach to finding the center of curvature and the radius of curvature uses a limiting process leading to the osculating circle.[27][28] See image above.

Using these coordinates, the motion along the path is viewed as a succession of circular paths of ever-changing center, and at each position s constitutes non-uniform circular motion at that position with radius ϱ. The local value of the angular rate of rotation then is given by:

$$\omega(s) = \frac{d\theta}{dt} = \frac{d\theta}{ds}\frac{ds}{dt} = \frac{1}{\rho(s)}\frac{ds}{dt} = \frac{v(s)}{\rho(s)} ,$$

with the local speed v given by:

$$v(s) = \frac{ds}{dt} .$$

As for the other examples above, because unit vectors cannot change magnitude, their rate of change is always perpendicular to their direction (see the left-hand insert in the image above):[29]

$$\frac{d\mathbf{u}_n(s)}{ds} = \mathbf{u}_t(s)\frac{d\theta}{ds} = \mathbf{u}_t(s)\frac{1}{\rho} ; \frac{d\mathbf{u}_t(s)}{ds} = -\mathbf{u}_n(s)\frac{d\theta}{ds} = -\mathbf{u}_n(s)\frac{1}{\rho} .$$

Consequently, the velocity and acceleration are:[28][30][31]

$$\mathbf{v}(t) = v\mathbf{u}_t(s) \; ;$$

and using the chain-rule of differentiation:

$$\mathbf{a}(t) = \frac{dv}{dt}\mathbf{u}_t(s) - \frac{v^2}{\rho}\mathbf{u}_n(s) \; ; \text{ with the tangential}$$

acceleration $\frac{dv}{dt} = \frac{dv}{ds}\frac{ds}{dt} = \frac{dv}{ds}\,v$.

In this local coordinate system, the acceleration resembles the expression for nonuniform circular motion with the local radius $\varrho(s)$, and the centripetal acceleration is identified as the second term.[32]

Extending this approach to three dimensional space curves leads to the Frenet–Serret formulas.[33][34]

Alternative approach   Looking at the image above, one might wonder whether adequate account has been taken of the difference in curvature between $\varrho(s)$ and $\varrho(s + ds)$ in computing the arc length as $ds = \varrho(s)d\theta$. Reassurance on this point can be found using a more formal approach outlined below. This approach also makes connection with the article on curvature.

To introduce the unit vectors of the local coordinate system, one approach is to begin in Cartesian coordinates and describe the local coordinates in terms of these Cartesian coordinates. In terms of arc length s, let the path be described as:[35]

$$\mathbf{r}(s) = [x(s), y(s)] \; .$$

Then an incremental displacement along the path ds is described by:

$$d\mathbf{r}(s) = [dx(s), dy(s)] = [x'(s), y'(s)]\,ds \; ,$$

where primes are introduced to denote derivatives with respect to s. The magnitude of this displacement is ds, showing that:[36]

$$\left[x'(s)^2 + y'(s)^2\right] = 1 \; . \text{ (Eq. 1)}$$

This displacement is necessarily a tangent to the curve at s, showing that the unit vector tangent to the curve is:

$$\mathbf{u}_t(s) = [x'(s), y'(s)] \; ,$$

while the outward unit vector normal to the curve is

$$\mathbf{u}_n(s) = [y'(s), -x'(s)] \; ,$$

Orthogonality can be verified by showing that the vector dot product is zero. The unit magnitude of these vectors is a consequence of Eq. 1. Using the tangent vector, the angle θ of the tangent to the curve is given by:

$$\sin\theta = \frac{y'(s)}{\sqrt{x'(s)^2 + y'(s)^2}} = y'(s) \; ; \text{ and } \cos\theta = \frac{x'(s)}{\sqrt{x'(s)^2 + y'(s)^2}} = x'(s) \; .$$

The radius of curvature is introduced completely formally (without need for geometric interpretation) as:

$$\frac{1}{\rho} = \frac{d\theta}{ds} \; .$$

The derivative of θ can be found from that for $\sin\theta$:

$$\frac{d\sin\theta}{ds} = \cos\theta\frac{d\theta}{ds} = \frac{1}{\rho}\cos\theta = \frac{1}{\rho}x'(s) \; .$$

Now:

$$\frac{d\sin\theta}{ds} = \frac{d}{ds}\frac{y'(s)}{\sqrt{x'(s)^2 + y'(s)^2}} = \frac{y''(s)x'(s)^2 - y'(s)x'(s)x''(s)}{(x'(s)^2 + y'(s)^2)^{3/2}} \; ,$$

in which the denominator is unity. With this formula for the derivative of the sine, the radius of curvature becomes:

$$\frac{d\theta}{ds} = \frac{1}{\rho} = y''(s)x'(s) - y'(s)x''(s) = \frac{y''(s)}{x'(s)} = -\frac{x''(s)}{y'(s)} \; ,$$

where the equivalence of the forms stems from differentiation of Eq. 1:

$$x'(s)x''(s) + y'(s)y''(s) = 0 \; .$$

With these results, the acceleration can be found:

$$\mathbf{a}(s) = \frac{d}{dt}\mathbf{v}(s) = \frac{d}{dt}\left[\frac{ds}{dt}(x'(s), y'(s))\right]$$

$$= \left(\frac{d^2 s}{dt^2}\right)\mathbf{u}_t(s) + \left(\frac{ds}{dt}\right)^2 (x''(s), y''(s))$$

$$= \left(\frac{d^2 s}{dt^2}\right)\mathbf{u}_t(s) - \left(\frac{ds}{dt}\right)^2\frac{1}{\rho}\mathbf{u}_n(s) \; ,$$

as can be verified by taking the dot product with the unit vectors $\mathbf{u}_t(s)$ and $\mathbf{u}_n(s)$. This result for acceleration is the same as that for circular motion based on the radius ϱ. Using this coordinate system in the inertial frame, it is easy to identify the force normal to the trajectory as the centripetal force and that parallel to the trajectory as the tangential force. From a qualitative standpoint, the path can be approximated by an arc of a circle for a limited time, and for the limited time a particular radius of curvature applies, the centrifugal and Euler forces can be analyzed on the basis of circular motion with that radius.

This result for acceleration agrees with that found earlier. However, in this approach, the question of the change in radius of curvature with s is handled completely formally, consistent with a geometric interpretation, but not relying upon it, thereby avoiding any questions the image above might suggest about neglecting the variation in ϱ.

Example: circular motion   To illustrate the above formulas, let x, y be given as:

$$x = \alpha \cos \frac{s}{\alpha} \; ; \; y = \alpha \sin \frac{s}{\alpha} \; .$$

Then:

$$x^2 + y^2 = \alpha^2 \; ,$$

which can be recognized as a circular path around the origin with radius α. The position $s = 0$ corresponds to $[\alpha, 0]$, or 3 o'clock. To use the above formalism, the derivatives are needed:

$$y'(s) = \cos \frac{s}{\alpha} \; ; \; x'(s) = -\sin \frac{s}{\alpha} \; ,$$

$$y''(s) = -\frac{1}{\alpha} \sin \frac{s}{\alpha} \; ; \; x''(s) = -\frac{1}{\alpha} \cos \frac{s}{\alpha} \; .$$

With these results, one can verify that:

$$x'(s)^2 + y'(s)^2 = 1 \; ; \; \frac{1}{\rho} = y''(s)x'(s) - y'(s)x''(s) = \frac{1}{\alpha} \; .$$

The unit vectors can also be found:

$$\mathbf{u}_t(s) = \left[-\sin \frac{s}{\alpha} \; , \; \cos \frac{s}{\alpha}\right] \; ; \; \mathbf{u}_n(s) = \left[\cos \frac{s}{\alpha} \; , \; \sin \frac{s}{\alpha}\right] \; ,$$

which serve to show that $s = 0$ is located at position $[\varrho, 0]$ and $s = \varrho\pi/2$ at $[0, \varrho]$, which agrees with the original expressions for x and y. In other words, s is measured counterclockwise around the circle from 3 o'clock. Also, the derivatives of these vectors can be found:

$$\frac{d}{ds}\mathbf{u}_t(s) = -\frac{1}{\alpha}\left[\cos \frac{s}{\alpha} \; , \; \sin \frac{s}{\alpha}\right] = -\frac{1}{\alpha}\mathbf{u}_n(s) \; ;$$

$$\frac{d}{ds}\mathbf{u}_n(s) = \frac{1}{\alpha}\left[-\sin \frac{s}{\alpha} \; , \; \cos \frac{s}{\alpha}\right] = \frac{1}{\alpha}\mathbf{u}_t(s) \; .$$

To obtain velocity and acceleration, a time-dependence for s is necessary. For counterclockwise motion at variable speed $v(t)$:

$$s(t) = \int_0^t dt' \, v(t') \; ,$$

where $v(t)$ is the speed and t is time, and $s(t = 0) = 0$. Then:

$$\mathbf{v} = v(t)\mathbf{u}_t(s) \; ,$$

$$\mathbf{a} = \frac{dv}{dt}\mathbf{u}_t(s) + v\frac{d}{dt}\mathbf{u}_t(s) = \frac{dv}{dt}\mathbf{u}_t(s) - v\frac{1}{\alpha}\mathbf{u}_n(s)\frac{ds}{dt}$$

$$\mathbf{a} = \frac{dv}{dt}\mathbf{u}_t(s) - \frac{v^2}{\alpha}\mathbf{u}_n(s) \; ,$$

where it already is established that $\alpha = \rho$. This acceleration is the standard result for non-uniform circular motion.

23.4   See also

23.5   Notes and references

[1] Craig, John (1849). *A new universal etymological, technological and pronouncing dictionary of the English language: embracing all terms used in art, science, and literature, Volume 1*. Harvard University. p. 291. Extract of page 291

[2] Newton, Isaac (2010). *The principia : mathematical principles of natural philosophy*. [S.l.]: Snowball Pub. p. 10. ISBN 978-1-60796-240-3.

[3] Russelkl C Hibbeler (2009). "Equations of Motion: Normal and tangential coordinates". *Engineering Mechanics: Dynamics* (12 ed.). Prentice Hall. p. 131. ISBN 0-13-607791-9.

[4] Paul Allen Tipler; Gene Mosca (2003). *Physics for scientists and engineers* (5th ed.). Macmillan. p. 129. ISBN 0-7167-8339-8.

[5] P. Germain; M. Piau; D. Caillerie, eds. (2012). *Theoretical and Applied Mechanics*. Elsevier. ISBN 9780444600202.

[6] Chris Carter (2001). *Facts and Practice for A-Level: Physics*. S.l.: Oxford University Press. p. 30. ISBN 978-0-19-914768-7.

[7] Eugene Lommel; George William Myers (1900). *Experimental physics*. K. Paul, Trench, Trübner & Co. p. 63.

[8] Colwell, Catharine H. "A Derivation of the Formulas for Centripetal Acceleration". *PhysicsLAB*. Retrieved 31 July 2011.

[9] Theo Koupelis (2010). *In Quest of the Universe* (6th ed.). Jones & Bartlett Learning. p. 83. ISBN 978-0-7637-6858-4.

[10] A. V. Durrant (1996). *Vectors in physics and engineering*. CRC Press. p. 103. ISBN 978-0-412-62710-1.

[11] Lawrence S. Lerner (1997). *Physics for Scientists and Engineers*. Boston: Jones & Bartlett Publishers. p. 128. ISBN 0-86720-479-6.

[12] Arthur Beiser (2004). *Schaum's Outline of Applied Physics*. New York: McGraw-Hill Professional. p. 103. ISBN 0-07-142611-6.

[13] Alan Darbyshire (2003). *Mechanical Engineering: BTEC National Option Units*. Oxford: Newnes. p. 56. ISBN 0-7506-5761-8.

[14] Federal Aviation Administration (2007). *Pilot's Encyclopedia of Aeronautical Knowledge*. Oklahoma City OK: Skyhorse Publishing Inc. Figure 3–21. ISBN 1-60239-034-7.

[15] Note: unlike the Cartesian unit vectors **i** and **j**, which are constant, in polar coordinates the direction of the unit vectors $\mathbf{u_r}$ and $\mathbf{u\theta}$ depend on θ, and so in general have non-zero time derivatives.

[16] Although the polar coordinate system moves with the particle, the observer does not. The description of the particle motion remains a description from the stationary observer's point of view.

[17] Notice that this local coordinate system is not autonomous; for example, its rotation in time is dictated by the trajectory traced by the particle. Note also that the radial vector $\mathbf{r}(t)$ does not represent the radius of curvature of the path.

[18] John Robert Taylor (2005). *Classical Mechanics*. Sausalito CA: University Science Books. pp. 28–29. ISBN 1-891389-22-X.

[19] Cornelius Lanczos (1986). *The Variational Principles of Mechanics*. New York: Courier Dover Publications. p. 103. ISBN 0-486-65067-7.

[20] See, for example, Howard D. Curtis (2005). *Orbital Mechanics for Engineering Students*. Butterworth-Heinemann. p. 5. ISBN 0-7506-6169-0.

[21] S. Y. Lee (2004). *Accelerator physics* (2nd ed.). Hackensack NJ: World Scientific. p. 37. ISBN 981-256-182-X.

[22] The *observer* of the motion along the curve is using these local coordinates to describe the motion from the observer's *frame of reference*, that is, from a stationary point of view. In other words, although the local coordinate system moves with the particle, the observer does not. A change in coordinate system used by the observer is only a change in their *description* of observations, and does not mean that the observer has changed their state of motion, and *vice versa*.

[23] Zhilin Li; Kazufumi Ito (2006). *The immersed interface method: numerical solutions of PDEs involving interfaces and irregular domains*. Philadelphia: Society for Industrial and Applied Mathematics. p. 16. ISBN 0-89871-609-8.

[24] K L Kumar (2003). *Engineering Mechanics*. New Delhi: Tata McGraw-Hill. p. 339. ISBN 0-07-049473-8.

[25] Lakshmana C. Rao; J. Lakshminarasimhan; Raju Sethuraman; SM Sivakuma (2004). *Engineering Dynamics: Statics and Dynamics*. Prentice Hall of India. p. 133. ISBN 81-203-2189-8.

[26] Shigeyuki Morita (2001). *Geometry of Differential Forms*. American Mathematical Society. p. 1. ISBN 0-8218-1045-6.

[27] The osculating circle at a given point P on a curve is the limiting circle of a sequence of circles that pass through P and two other points on the curve, Q and R, on either side of P, as Q and R approach P. See the online text by Lamb: Horace Lamb (1897). *An Elementary Course of Infinitesimal Calculus*. University Press. p. 406. ISBN 1-108-00534-9.

[28] Guang Chen; Fook Fah Yap (2003). *An Introduction to Planar Dynamics* (3rd ed.). Central Learning Asia/Thomson Learning Asia. p. 34. ISBN 981-243-568-9.

[29] R. Douglas Gregory (2006). *Classical Mechanics: An Undergraduate Text*. Cambridge University Press. p. 20. ISBN 0-521-82678-0.

[30] Edmund Taylor Whittaker; William McCrea (1988). *A Treatise on the Analytical Dynamics of Particles and Rigid Bodies: with an introduction to the problem of three bodies* (4th ed.). Cambridge University Press. p. 20. ISBN 0-521-35883-3.

[31] Jerry H. Ginsberg (2007). *Engineering Dynamics*. Cambridge University Press. p. 33. ISBN 0-521-88303-2.

[32] Joseph F. Shelley (1990). *800 solved problems in vector mechanics for engineers: Dynamics*. McGraw-Hill Professional. p. 47. ISBN 0-07-056687-9.

[33] Larry C. Andrews; Ronald L. Phillips (2003). *Mathematical Techniques for Engineers and Scientists*. SPIE Press. p. 164. ISBN 0-8194-4506-1.

[34] Ch V Ramana Murthy; NC Srinivas (2001). *Applied Mathematics*. New Delhi: S. Chand & Co. p. 337. ISBN 81-219-2082-5.

[35] The article on curvature treats a more general case where the curve is parametrized by an arbitrary variable (denoted t), rather than by the arc length s.

[36] Ahmed A. Shabana; Khaled E. Zaazaa; Hiroyuki Sugiyama (2007). *Railroad Vehicle Dynamics: A Computational Approach*. CRC Press. p. 91. ISBN 1-4200-4581-4.

23.6   Further reading

- Serway, Raymond A.; Jewett, John W. (2004). *Physics for Scientists and Engineers* (6th ed.). Brooks/Cole. ISBN 0-534-40842-7.

- Tipler, Paul (2004). *Physics for Scientists and Engineers: Mechanics, Oscillations and Waves, Thermodynamics* (5th ed.). W. H. Freeman. ISBN 0-7167-0809-4.

- Centripetal force vs. Centrifugal force, from an online Regents Exam physics tutorial by the Oswego City School District

23.7   External links

- Notes from University of Winnipeg

- Notes from Physics and Astronomy HyperPhysics at Georgia State University; see also home page

- Notes from Britannica

- Notes from PhysicsNet

- NASA notes by David P. Stern

- Notes from U Texas.

- Analysis of smart yo-yo

- The Inuit yo-yo

- Kinematic Models for Design Digital Library (KMODDL)
  Movies and photos of hundreds of working mechanical-systems models at Cornell University. Also includes an e-book library of classic texts on mechanical design and engineering.

Chapter 24

Coriolis force

For the psychophysical perception effect, see Coriolis effect (perception).

In physics, the **Coriolis force** is an inertial force (also

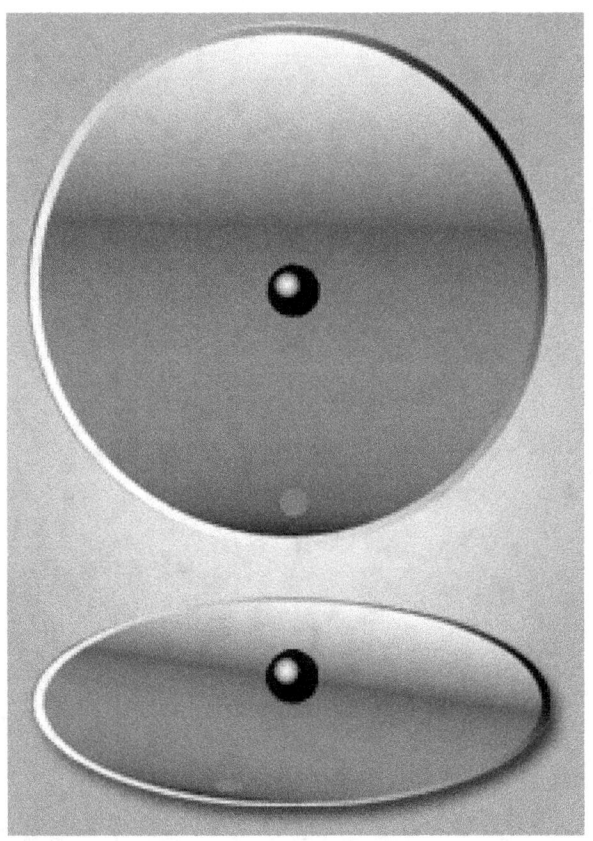

In the inertial frame of reference (upper part of the picture), the black ball moves in a straight line. However, the observer (red dot) who is standing in the rotating/non-inertial frame of reference (lower part of the picture) sees the object as following a curved path due to the Coriolis and centrifugal forces present in this frame.

called a *fictitious force*)[1] that acts on objects that are in motion relative to a rotating reference frame. In a reference frame with clockwise rotation, the force acts to the left of the motion of the object. In one with anticlockwise rotation, the force acts to the right. Though recognized pre-

viously by others, the mathematical expression for the Coriolis force appeared in an 1835 paper by French scientist Gaspard-Gustave de Coriolis, in connection with the theory of water wheels. Early in the 20th century, the term *Coriolis force* began to be used in connection with meteorology. Deflection of an object due to the Coriolis force is called the 'Coriolis effect'.

Newton's laws of motion describe the motion of an object in an inertial (non-accelerating) frame of reference. When Newton's laws are transformed to a rotating frame of reference, the Coriolis force and centrifugal force appear. Both forces are proportional to the mass of the object. The Coriolis force is proportional to the rotation rate and the centrifugal force is proportional to its square. The Coriolis force acts in a direction perpendicular to the rotation axis and to the velocity of the body in the rotating frame and is proportional to the object's speed in the rotating frame. The centrifugal force acts outwards in the radial direction and is proportional to the distance of the body from the axis of the rotating frame. These additional forces are termed inertial forces, fictitious forces or *pseudo forces*.[2] They allow the application of Newton's laws to a rotating system. They are correction factors that do not exist in a non-accelerating or inertial reference frame.

A commonly encountered rotating reference frame is the Earth. The Coriolis effect is caused by the rotation of the Earth and the inertia of the mass experiencing the effect. Because the Earth completes only one rotation per day, the Coriolis force is quite small, and its effects generally become noticeable only for motions occurring over large distances and long periods of time, such as large-scale movement of air in the atmosphere or water in the ocean. Such motions are constrained by the surface of the Earth, so only the horizontal component of the Coriolis force is generally important. This force causes moving objects on the surface of the Earth to be deflected to the right (with respect to the direction of travel) in the Northern Hemisphere and to the left in the Southern Hemisphere. The horizontal deflection effect is greater near the poles and smallest at the equator, since the rate of change in the diameter of the circles of latitude when travelling north or south, increases the closer the object is to the poles.[3] Rather than flowing directly from areas of high pressure to low pressure, as they would in a non-rotating system, winds and currents tend to flow to the right of this direction north of the equator and to the left of this direction south of it. This effect is responsible for the rotation of large cyclones (see Coriolis effects in meteorology). To explain this intuitively, consider how an object that moves northwards from the equator has a tendency to maintain its greater speed at the equator (rotating around towards the right as you look at the sphere of the Earth), where the "horizontal diameter" is larger, and therefore tends to move towards the right as it passed northwards where the "horizontal diameter" of the Earth (the rings of latitude) is smaller, and the linear speed of local objects on the Earth's surface at that latitude is slower.

24.1   History

Italian scientist Giovanni Battista Riccioli and his assistant Francesco Maria Grimaldi described the effect in connection with artillery in the 1651 *Almagestum Novum*, writing that rotation of the Earth should cause a cannonball fired to the north to deflect to the east.[4] The effect was described in the tidal equations of Pierre-Simon Laplace in 1778.

Gaspard-Gustave Coriolis published a paper in 1835 on the energy yield of machines with rotating parts, such as waterwheels.[5] That paper considered the supplementary forces that are detected in a rotating frame of reference. Coriolis divided these supplementary forces into two categories. The second category contained a force that arises from the cross product of the angular velocity of a coordinate system and the projection of a particle's velocity into a plane perpendicular to the system's axis of rotation. Coriolis referred to this force as the "compound centrifugal force" due to its analogies with the centrifugal force already considered in category one.[6][7] The effect was known in the early 20th century as the "acceleration of Coriolis",[8] and by 1920 as "Coriolis force".[9]

In 1856, William Ferrel proposed the existence of a circulation cell in the mid-latitudes with air being deflected by the Coriolis force to create the prevailing westerly winds.[10]

The understanding of the kinematics of how exactly the rotation of the Earth affects airflow was partial at first.[11] Late in the 19th century, the full extent of the large scale interaction of pressure gradient force and deflecting force that in the end causes air masses to move 'along' isobars was understood.

24.2   Formula

See also: Fictitious force

In non-vector terms: at a given rate of rotation of the observer, the magnitude of the Coriolis acceleration of the object is proportional to the velocity of the object and also to the sine of the angle between the direction of movement of the object and the axis of rotation.

The vector formula for the magnitude and direction of the Coriolis acceleration [12] is

$$a_C = -2\,\Omega \times v$$

where (here and below) a_C is the acceleration of the particle in the rotating system, v is the velocity of the particle with respect to the rotating system, and Ω is the angular velocity vector having magnitude equal to the rotation rate ω, with direction along the axis of rotation of the rotating reference frame, and the × symbol represents the cross product operator.

The equation may be multiplied by the mass of the relevant object to produce the Coriolis force:

$$F_C = -2\,m\,\Omega \times v$$

See *fictitious force* for a derivation.

The *Coriolis effect* is the behavior added by the *Coriolis acceleration*. The formula implies that the Coriolis acceleration is perpendicular both to the direction of the velocity of the moving mass and to the frame's rotation axis. So in particular:

- if the velocity is parallel to the rotation axis, the Coriolis acceleration is zero.

- if the velocity is straight inward to the axis, the acceleration is in the direction of local rotation.

- if the velocity is straight outward from the axis, the acceleration is against the direction of local rotation.

- if the velocity is in the direction of local rotation, the acceleration is outward from the axis.

- if the velocity is against the direction of local rotation, the acceleration is inward to the axis.

The vector cross product can be evaluated as the determinant of a matrix:

$$\mathbf{\Omega} \times \mathbf{v} = \begin{vmatrix} \mathbf{i} & \mathbf{j} & \mathbf{k} \\ \Omega_x & \Omega_y & \Omega_z \\ v_x & v_y & v_z \end{vmatrix} = \begin{pmatrix} \Omega_y v_z - \Omega_z v_y \\ \Omega_z v_x - \Omega_x v_z \\ \Omega_x v_y - \Omega_y v_x \end{pmatrix},$$

where the vectors $\mathbf{i}$, $\mathbf{j}$, $\mathbf{k}$ are unit vectors in the x, y and z directions.

24.3 Causes

The Coriolis force exists only when one uses a rotating reference frame. In the rotating frame it behaves exactly like a real force (that is to say, it causes acceleration and has real effects). However, the Coriolis force is a consequence of inertia,[13] and is not attributable to an identifiable originating body, as is the case for electromagnetic or nuclear forces, for example. From an analytical viewpoint, to use Newton's second law in a rotating system, the Coriolis force is mathematically necessary, but it disappears in a non-accelerating, inertial frame of reference. For example, consider two children on opposite sides of a spinning roundabout (Merry-go-round), who are throwing a ball to each other. From the children's point of view, this ball's path is curved sideways by the Coriolis force. Suppose the roundabout spins anticlockwise when viewed from above. From the thrower's perspective, the deflection is to the right.[14] From the non-thrower's perspective, deflection is to the left. *For a mathematical formulation see Mathematical derivation of fictitious forces.*

An observer in a rotating frame, such as an astronaut in a rotating space station, very probably will find the interpretation of everyday life in terms of the Coriolis force accords more simply with intuition and experience than a cerebral reinterpretation of events from an inertial standpoint. For example, nausea due to an experienced push may be more instinctively explained by the Coriolis force than by the law of inertia.[15][16] See also Coriolis effect (perception). In meteorology, a rotating frame (the Earth) with its Coriolis force provides a more natural framework for explanation of air movements than a non-rotating, inertial frame without Coriolis forces.[17] In long-range gunnery, sight corrections for the Earth's rotation are based upon the Coriolis force.[18] These examples are described in more detail below.

The acceleration entering the Coriolis force arises from two sources of change in velocity that result from rotation: the first is the change of the velocity of an object in time. The same velocity (in an inertial frame of reference where the normal laws of physics apply) is seen as different velocities at different times in a rotating frame of reference. The apparent acceleration is proportional to the angular velocity of the reference frame (the rate at which the coordinate axes change direction), and to the component of velocity of

the object in a plane perpendicular to the axis of rotation. This gives a term $-\mathbf{\Omega} \times \mathbf{v}$. The minus sign arises from the traditional definition of the cross product (right hand rule), and from the sign convention for angular velocity vectors.

The second is the change of velocity in space. Different positions in a rotating frame of reference have different velocities (as seen from an inertial frame of reference). For an object to move in a straight line, it must accelerate so that its velocity changes from point to point by the same amount as the velocities of the frame of reference. The force is proportional to the angular velocity (which determines the relative speed of two different points in the rotating frame of reference), and to the component of the velocity of the object in a plane perpendicular to the axis of rotation (which determines how quickly it moves between those points). This also gives a term $-\mathbf{\Omega} \times \mathbf{v}$.

24.4 Length scales and the Rossby number

Further information: Rossby number

The time, space and velocity scales are important in determining the importance of the Coriolis force. Whether rotation is important in a system can be determined by its Rossby number, which is the ratio of the velocity, U, of a system to the product of the Coriolis parameter, $f = 2\omega \sin \varphi$, and the length scale, L, of the motion:

$$Ro = \frac{U}{fL}.$$

The Rossby number is the ratio of inertial to Coriolis forces. A small Rossby number indicates a system is strongly affected by Coriolis forces, and a large Rossby number idicates a system in which inertial forces dominate. For example, in tornadoes, the Rossby number is large, in low-pressure systems it is low, and in oceanic systems it is around 1. As a result, in tornadoes the Coriolis force is negligible, and balance is between pressure and centrifugal forces. In low-pressure systems, centrifugal force is negligible and balance is between Coriolis and pressure forces. In the oceans all three forces are comparable.[19]

An atmospheric system moving at U = 10 m/s (22 mph) occupying a spatial distance of L = 1,000 km (621 mi), has a Rossby number of approximately 0.1.

A baseball pitcher may throw the ball at U = 45 m/s (100 mph) for a distance of L = 18.3 m (60 ft). The Rossby number in this case would be 32,000.

Baseball players don't care about which hemisphere they're

playing in. However, an unguided missile obeys exactly the same physics as a baseball, but can travel far enough and be in the air long enough to experience the effect of Coriolis force. Long-range shells in the Northern Hemisphere landed close to, but to the right of, where they were aimed until this was noted. (Those fired in the Southern Hemisphere landed to the left.) In fact, it was this effect that first got the attention of Coriolis himself.[20][21][22]

24.5   Simple cases

24.5.1   Cannon on turntable

See also: Fictitious force § Crossing a carousel

The animation at the top of this article is a classic illustra-

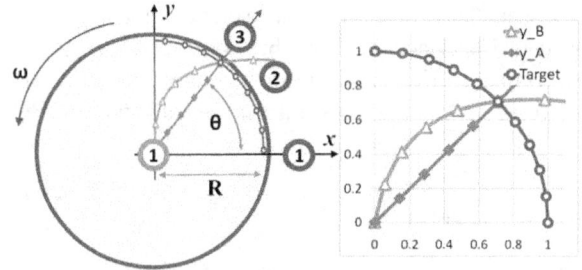

Cannon at the center of a rotating turntable. To hit the target located at position 1 on the perimeter at time t = 0 s, the cannon must be aimed ahead of the target at angle θ. That way, by the time the cannonball reaches position 3 on the periphery, the target is also at that position. In an inertial frame of reference, the cannonball travels a straight radial path to the target (curve yA). However, in the frame of the turntable, the path is arched (curve yB), as also shown in the figure.

tion of Coriolis force. Another visualization of the Coriolis and centrifugal forces is this animation clip.

Given the radius R of the turntable in that animation, the rate of angular rotation ω, and the speed of the cannonball (assumed constant) v, the correct angle θ to aim so as to hit the target at the edge of the turntable can be calculated.

The inertial frame of reference provides one way to handle the question: calculate the time to interception, which is $tf = R / v$. Then, the turntable revolves an angle $\omega\, tf$ in this time. If the cannon is pointed an angle $\theta = \omega\, tf = \omega R / v$, then the cannonball arrives at the periphery at position number 3 at the same time as the target.

No discussion of Coriolis force can arrive at this solution as simply, so the reason to treat this problem is to demonstrate Coriolis formalism in an easily visualized situation.

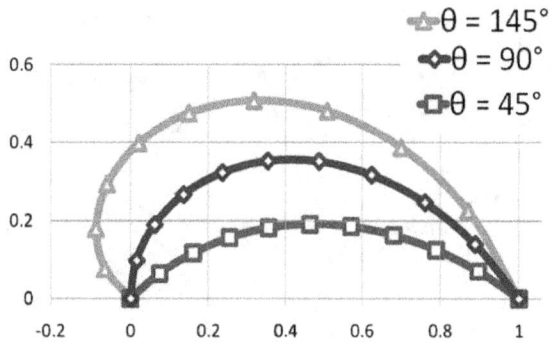

Successful trajectory of cannonball as seen from the turntable for three angles of launch θ. Plotted points are for the same equally spaced times steps on each curve. Cannonball speed v is held constant and angular rate of rotation ω is varied to achieve a successful "hit" for selected θ. For example, for a radius of 1 m and a cannonball speed of 1 m/s, the time of flight $t_f = 1$ s, and $\omega t_f = \theta \rightarrow \omega$ and θ have the same numerical value if θ is expressed in radians. The wider spacing of the plotted points as the target is approached show the speed of the cannonball is accelerating as seen on the turntable, due to fictitious Coriolis and centrifugal forces.

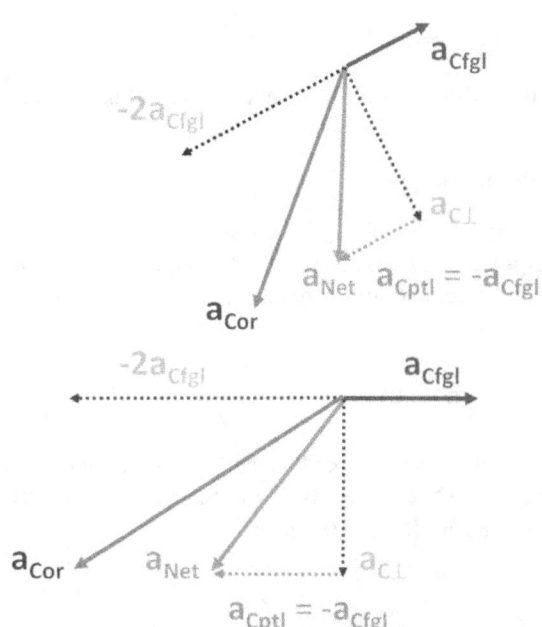

Acceleration components at an earlier time (top) and at arrival time at the target (bottom)

Trajectory in the inertial frame

The trajectory in the inertial frame (denoted A) is a straight line radial path at angle θ. The position of the cannonball in (x, y) coordinates at time t is:

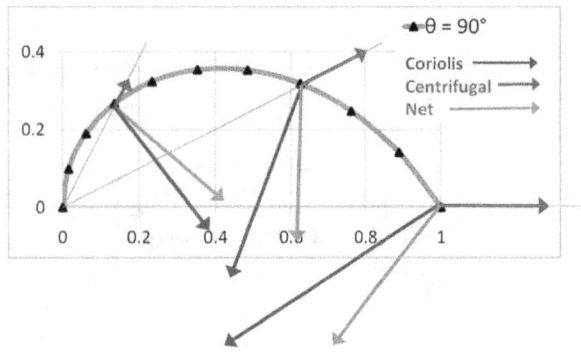

Coriolis acceleration, centrifugal acceleration and net acceleration vectors at three selected points on the trajectory as seen on the turntable.

$$\mathbf{r}_A(t) = vt \ (\cos(\theta), \ \sin(\theta)) \ .$$

In the turntable frame (denoted *B*), the *x*- *y* axes rotate at angular rate ω, so the trajectory becomes:

$$\mathbf{r}_B(t) = vt \ (\cos(\theta - \omega t), \ \sin(\theta - \omega t)) \ ,$$

and three examples of this result are plotted in the figure.

Accelerations

Components of acceleration   To determine the components of acceleration, a general expression is used from the article fictitious force:

$$\mathbf{a}_B = \mathbf{a}_A - 2\mathbf{\Omega} \times \mathbf{v}_B - \mathbf{\Omega} \times (\mathbf{\Omega} \times \mathbf{r}_B) - \frac{d\mathbf{\Omega}}{dt} \times \mathbf{r}_B \ ,$$

in which the term in **Ω × vB** is the Coriolis acceleration and the term in **Ω × (Ω × rB)** is the centrifugal acceleration. The results are (let α = θ − ωt):

$$\mathbf{\Omega} \times \mathbf{r}_B = \begin{vmatrix} \mathbf{i} & \mathbf{j} & \mathbf{k} \\ 0 & 0 & \omega \\ vt\cos\alpha & vt\sin\alpha & 0 \end{vmatrix}$$
$$= \omega tv \left(-\sin\alpha, \cos\alpha \right) \ ,$$

$$\mathbf{\Omega} \times (\mathbf{\Omega} \times \mathbf{r}_B) = \begin{vmatrix} \mathbf{i} & \mathbf{j} & \mathbf{k} \\ 0 & 0 & \omega \\ -\omega tv \sin\alpha & \omega tv \cos\alpha & 0 \end{vmatrix} \ ,$$

Producing accelerations   Producing a centrifugal acceleration:

$$\mathbf{a}_{Cfgl} = \omega^2 vt \left(\cos\alpha, \sin\alpha \right) = \omega^2 \mathbf{r}_B(t) \ .$$

Also:

$$\mathbf{v}_B = \frac{d\mathbf{r}_B(t)}{dt} = (v\cos\alpha + \omega t \ v\sin\alpha, \ v\sin\alpha - \omega t \ v\cos\alpha, \ 0) \ ,$$

$$\mathbf{\Omega} \times \mathbf{v}_B = \begin{vmatrix} \mathbf{i} & \mathbf{j} & \mathbf{k} \\ 0 & 0 & \omega \\ v\cos\alpha & v\sin\alpha & 0 \\ +\omega t \ v\sin\alpha & -\omega t \ v\cos\alpha & 0 \end{vmatrix} \ ,$$

producing a Coriolis acceleration:

$$\mathbf{a}_{Cor} = -2 \left[-\omega v \left(\sin\alpha - \omega t \cos\alpha \right), \ \omega v \left(\cos\alpha + \omega t \sin\alpha \right) \right]$$
$$= 2\omega v \left(\sin\alpha, \ -\cos\alpha \right) - 2\omega^2 \mathbf{r}_B(t) \ .$$

These accelerations are shown in the diagrams for a particular example.

It is seen that the Coriolis acceleration not only cancels the centrifugal acceleration, but together they provide a net "centripetal", radially inward component of acceleration (that is, directed toward the center of rotation):[23]

$$\mathbf{a}_{Cptl} = -\omega^2 \mathbf{r}_B(t) \ ,$$

and an additional component of acceleration perpendicular to **rB** *(t)*:

$$\mathbf{a}_{C\perp} = 2\omega v \left(\sin\alpha, \ -\cos\alpha \right) \ .$$

The "centripetal" component of acceleration resembles that for circular motion at radius *r*B, while the perpendicular component is velocity dependent, increasing with the radial velocity *v* and directed to the right of the velocity. The situation could be described as a circular motion combined with an "apparent Coriolis acceleration" of 2ω*v*. However, this is a rough labelling: a careful designation of the true centripetal force refers to a local reference frame that employs the directions normal and tangential to the path, not coordinates referred to the axis of rotation.

These results also can be obtained directly by two time differentiations of **rB** *(t)*. Agreement of the two approaches demonstrates that one could start from the general expression for fictitious acceleration above and derive the trajectories shown here. However, working from the acceleration to the trajectory is more complicated than the reverse procedure used here, which, of course, is made possible in this example by knowing the answer in advance.

As a result of this analysis an important point appears: *all* the fictitious accelerations must be included to obtain the correct trajectory. In particular, besides the Coriolis acceleration, the centrifugal force plays an essential role. It is easy to get the impression from verbal discussions of the cannonball problem, which focus on displaying the Coriolis effect particularly, that the Coriolis force is the only factor that must be considered,[24] but that is not so.[25] A turntable for which the Coriolis force *is* the only factor is the parabolic turntable. A somewhat more complex situation is the idealized example of flight routes over long distances, where the centrifugal force of the path and aeronautical lift are countered by gravitational attraction.[26][27]

24.5.2   Tossed ball on a rotating carousel

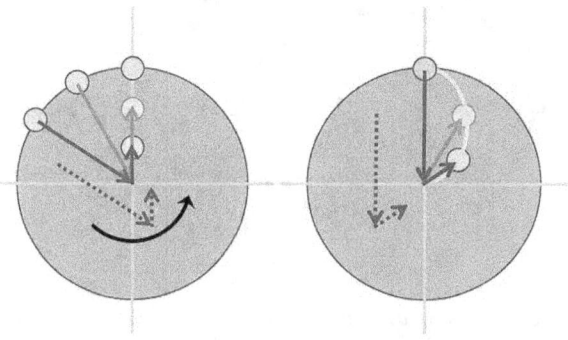

A carousel is rotating counter-clockwise. Left panel: a ball is tossed by a thrower at 12:00 o'clock and travels in a straight line to the center of the carousel. While it travels, the thrower circles in a counter-clockwise direction. Right panel: The ball's motion as seen by the thrower, who now remains at 12:00 o'clock, because there is no rotation from their viewpoint.

The figure illustrates a ball tossed from 12:00 o'clock toward the center of a counter-clockwise rotating carousel. On the left, the ball is seen by a stationary observer above the carousel, and the ball travels in a straight line to the center, while the ball-thrower rotates counter-clockwise with the carousel. On the right the ball is seen by an observer rotating with the carousel, so the ball-thrower appears to stay at 12:00 o'clock. The figure shows how the trajectory of the ball as seen by the rotating observer can be constructed.

On the left, two arrows locate the ball relative to the ball-thrower. One of these arrows is from the thrower to the center of the carousel (providing the ball-thrower's line of sight), and the other points from the center of the carousel to the ball.(This arrow gets shorter as the ball approaches the center.) A shifted version of the two arrows is shown dotted.

On the right is shown this same dotted pair of arrows, but now the pair are rigidly rotated so the arrow corresponding to the line of sight of the ball-thrower toward the center of the carousel is aligned with 12:00 o'clock. The other

arrow of the pair locates the ball relative to the center of the carousel, providing the position of the ball as seen by the rotating observer. By following this procedure for several positions, the trajectory in the rotating frame of reference is established as shown by the curved path in the right-hand panel.

The ball travels in the air, and there is no net force upon it. To the stationary observer the ball follows a straight-line path, so there is no problem squaring this trajectory with zero net force. However, the rotating observer sees a *curved* path. Kinematics insists that a force (pushing to the *right* of the instantaneous direction of travel for a *counter-clockwise* rotation) must be present to cause this curvature, so the rotating observer is forced to invoke a combination of centrifugal and Coriolis forces to provide the net force required to cause the curved trajectory.

24.5.3   Bounced ball

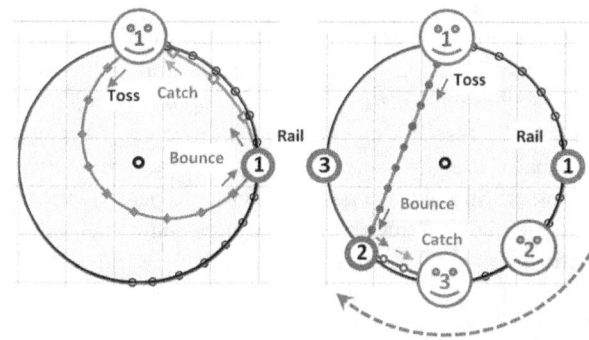

Bird's-eye view of carousel. The carousel rotates clockwise. Two viewpoints are illustrated: that of the camera at the center of rotation rotating with the carousel (left panel) and that of the inertial (stationary) observer (right panel). Both observers agree at any given time just how far the ball is from the center of the carousel, but not on its orientation. Time intervals are 1/10 of time from launch to bounce.

The figure describes a more complex situation where the tossed ball on a turntable bounces off the edge of the carousel and then returns to the tosser, who catches the ball. The effect of Coriolis force on its trajectory is shown again as seen by two observers: an observer (referred to as the "camera") that rotates with the carousel, and an inertial observer. The figure shows a bird's-eye view based upon the same ball speed on forward and return paths. Within each circle, plotted dots show the same time points. In the left panel, from the camera's viewpoint at the center of rotation, the tosser (smiley face) and the rail both are at fixed locations, and the ball makes a very considerable arc on its travel toward the rail, and takes a more direct route on the way back. From the ball tosser's viewpoint, the ball seems to return more quickly than it went (because the tosser is rotating toward the ball on the return flight).

On the carousel, instead of tossing the ball straight at a rail to bounce back, the tosser must throw the ball toward the right of the target and the ball then seems to the camera to bear continuously to the left of its direction of travel to hit the rail (*left* because the carousel is turning *clockwise*). The ball appears to bear to the left from direction of travel on both inward and return trajectories. The curved path demands this observer to recognize a leftward net force on the ball. (This force is "fictitious" because it disappears for a stationary observer, as is discussed shortly.) For some angles of launch, a path has portions where the trajectory is approximately radial, and Coriolis force is primarily responsible for the apparent deflection of the ball (centrifugal force is radial from the center of rotation, and causes little deflection on these segments). When a path curves away from radial, however, centrifugal force contributes significantly to deflection.

The ball's path through the air is straight when viewed by observers standing on the ground (right panel). In the right panel (stationary observer), the ball tosser (smiley face) is at 12 o'clock and the rail the ball bounces from is at position one (1). From the inertial viewer's standpoint, positions one (1), two (2), three (3) are occupied in sequence. At position 2 the ball strikes the rail, and at position 3 the ball returns to the tosser. Straight-line paths are followed because the ball is in free flight, so this observer requires that no net force is applied.

24.6   Applied to the Earth

An important case where the Coriolis force is observed is the rotating Earth. Unless otherwise stated, directions of forces and motion apply to the Northern Hemisphere.

24.6.1   Intuitive explanation

As the Earth turns around its axis, everything attached to it turns with it (imperceptibly to our senses). An object that is moving without being dragged along with this rotation travels in a straight motion over the turning Earth. From our rotating perspective on the planet, its direction of motion changes as it moves, bending in the opposite direction to our actual motion. When viewed from a stationary point in space above, any land feature in the Northern Hemisphere turns anticlockwise—and, fixing our gaze on that location, any other location in that hemisphere rotates around it the same way. The traced ground path of a freely moving body travelling from one point to another therefore bends the opposite way, clockwise, which is conventionally labeled as

"right," where it will be if the direction of motion is considered "ahead," and "down" is defined naturally.

24.6.2   Rotating sphere

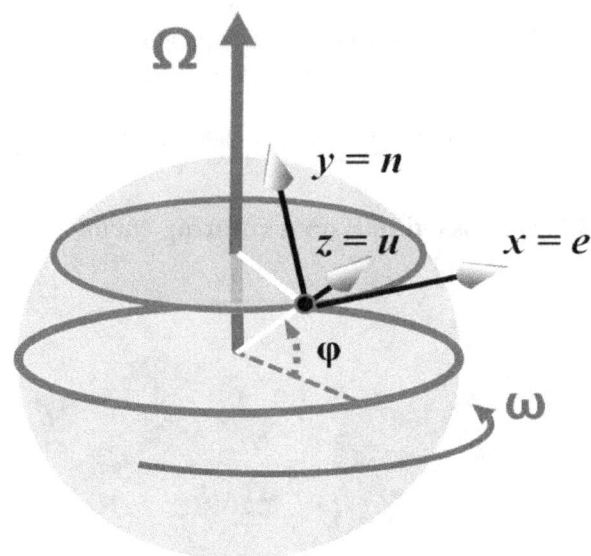

Coordinate system at latitude φ with x-axis east, y-axis north and z-axis upward (that is, radially outward from center of sphere).

Consider a location with latitude φ on a sphere that is rotating around the north-south axis.[28] A local coordinate system is set up with the x axis horizontally due east, the y axis horizontally due north and the z axis vertically upwards. The rotation vector, velocity of movement and Coriolis acceleration expressed in this local coordinate system (listing components in the order east (e), north (n) and upward (u)) are:

$$\boldsymbol{\Omega} = \omega \begin{pmatrix} 0 \\ \cos\varphi \\ \sin\varphi \end{pmatrix}, \quad \boldsymbol{v} = \begin{pmatrix} v_e \\ v_n \\ v_u \end{pmatrix},$$

$$\boldsymbol{a}_C = -2\boldsymbol{\Omega} \times \boldsymbol{v} = 2\omega \begin{pmatrix} v_n \sin\varphi - v_u \cos\varphi \\ -v_e \sin\varphi \\ v_e \cos\varphi \end{pmatrix}.$$

When considering atmospheric or oceanic dynamics, the vertical velocity is small, and the vertical component of the Coriolis acceleration is small compared to gravity. For such cases, only the horizontal (east and north) components matter. The restriction of the above to the horizontal plane is (setting $v_u = 0$):

$$\boldsymbol{v} = \begin{pmatrix} v_e \\ v_n \end{pmatrix}, \quad \boldsymbol{a}_c = \begin{pmatrix} v_n \\ -v_e \end{pmatrix} f,$$

where $f = 2\omega \sin \varphi$ is called the Coriolis parameter.

By setting $vn = 0$, it can be seen immediately that (for positive φ and ω) a movement due east results in an acceleration due south. Similarly, setting $ve = 0$, it is seen that a movement due north results in an acceleration due east. In general, observed horizontally, looking along the direction of the movement causing the acceleration, the acceleration always is turned 90° to the right and of the same size regardless of the horizontal orientation.

As a different case, consider equatorial motion setting $\varphi = 0°$. In this case, $\boldsymbol{\Omega}$ is parallel to the north or n-axis, and:

$$\boldsymbol{\Omega} = \omega \begin{pmatrix} 0 \\ 1 \\ 0 \end{pmatrix}, \boldsymbol{v} = \begin{pmatrix} v_e \\ v_n \\ v_u \end{pmatrix}, \boldsymbol{a}_C = -2\boldsymbol{\Omega} \times \boldsymbol{v} = 2\omega \begin{pmatrix} -v_u \\ 0 \\ v_e \end{pmatrix}.$$

Accordingly, an eastward motion (that is, in the same direction as the rotation of the sphere) provides an upward acceleration known as the Eötvös effect, and an upward motion produces an acceleration due west.

24.6.3   Meteorology

This low-pressure system over Iceland spins counterclockwise due to balance between the Coriolis force and the pressure gradient force.

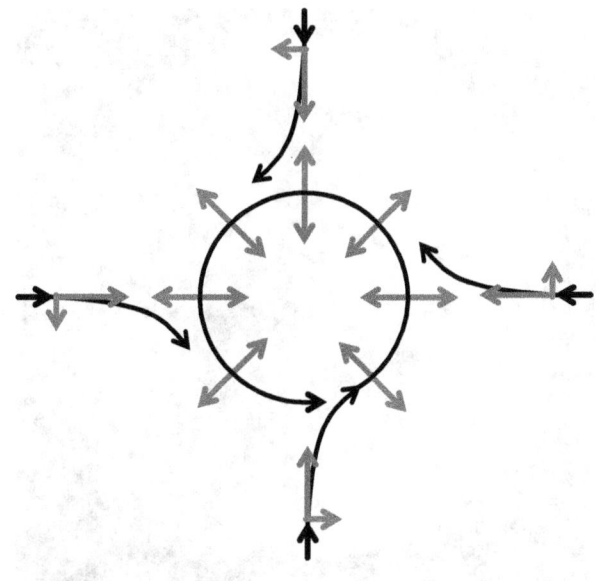

*Schematic representation of flow around a **low**-pressure area in the Northern Hemisphere. The Rossby number is low, so the centrifugal force is virtually negligible. The pressure-gradient force is represented by blue arrows, the Coriolis acceleration (always perpendicular to the velocity) by red arrows*

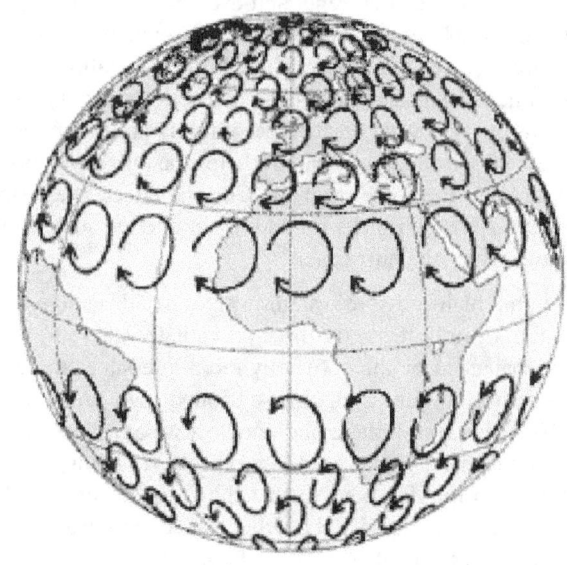

Schematic representation of inertial circles of air masses in the absence of other forces, calculated for a wind speed of approximately 50 to 70 m/s (110 to 160 mph).

Perhaps the most important impact of the Coriolis effect is in the large-scale dynamics of the oceans and the atmosphere. In meteorology and oceanography, it is convenient to postulate a rotating frame of reference wherein the Earth is stationary. In accommodation of that provisional postulation, the centrifugal and Coriolis forces are introduced.

Their relative importance is determined by the applicable Rossby numbers. Tornadoes have high Rossby numbers, so, while tornado-associated centrifugal forces are quite substantial, Coriolis forces associated with tornadoes are for practical purposes negligible.[29]

Because ocean currents are driven by the movement of wind

Cloud formations in a famous image of Earth from Apollo 17, makes similar circulation directly visible

over the water's surface, the Coriolis force also affects the movement of ocean currents and cyclones as well. Many of the ocean's largest currents circulate around warm, high-pressure areas called gyres. Though the circulation is not as significant as that in the air, the deflection caused by the Coriolis effect is what creates the spiraling pattern in these gyres. The spiraling wind pattern helps the hurricane form. The stronger the force from the Coriolis effect, the faster the wind spins and picks up additional energy, increasing the strength of the hurricane.[30]

Air within high-pressure systems rotates in a direction such that the Coriolis force is directed radially inwards, and nearly balanced by the outwardly radial pressure gradient. As a result, air travels clockwise around high pressure in the Northern Hemisphere and anticlockwise in the Southern Hemisphere. Air within low-pressure systems rotates in the opposite direction, so that the Coriolis force is directed radially outward and nearly balances an inwardly radial pressure gradient.

Flow around a low-pressure area

Main article: Low-pressure area

If a low-pressure area forms in the atmosphere, air tends to flow in towards it, but is deflected perpendicular to its velocity by the Coriolis force. A system of equilibrium can then establish itself creating circular movement, or a cyclonic flow. Because the Rossby number is low, the force balance is largely between the pressure gradient force acting

towards the low-pressure area and the Coriolis force acting away from the center of the low pressure.

Instead of flowing down the gradient, large scale motions in the atmosphere and ocean tend to occur perpendicular to the pressure gradient. This is known as geostrophic flow.[31] On a non-rotating planet, fluid would flow along the straightest possible line, quickly eliminating pressure gradients. Note that the geostrophic balance is thus very different from the case of "inertial motions" (see below), which explains why mid-latitude cyclones are larger by an order of magnitude than inertial circle flow would be.

This pattern of deflection, and the direction of movement, is called Buys-Ballot's law. In the atmosphere, the pattern of flow is called a cyclone. In the Northern Hemisphere the direction of movement around a low-pressure area is anticlockwise. In the Southern Hemisphere, the direction of movement is clockwise because the rotational dynamics is a mirror image there. At high altitudes, outward-spreading air rotates in the opposite direction.[32] Cyclones rarely form along the equator due to the weak Coriolis effect present in this region.

Inertial circles

An air or water mass moving with speed v subject only to the Coriolis force travels in a circular trajectory called an 'inertial circle'. Since the force is directed at right angles to the motion of the particle, it moves with a constant speed around a circle whose radius R is given by:

$$R = \frac{v}{f}$$

where f is the Coriolis parameter $2\Omega \sin \varphi$, introduced above (where φ is the latitude). The time taken for the mass to complete a full circle is therefore $2\pi/f$. The Coriolis parameter typically has a mid-latitude value of about 10^{-4} s^{-1}; hence for a typical atmospheric speed of 10 m/s (22 mph) the radius is 100 km (62 mi), with a period of about 17 hours. For an ocean current with a typical speed of 10 cm/s (0.22 mph), the radius of an inertial circle is 1 km (0.6 mi). These inertial circles are clockwise in the Northern Hemisphere (where trajectories are bent to the right) and anticlockwise in the Southern Hemisphere.

If the rotating system is a parabolic turntable, then f is constant and the trajectories are exact circles. On a rotating planet, f varies with latitude and the paths of particles do not form exact circles. Since the parameter f varies as the sine of the latitude, the radius of the oscillations associated with a given speed are smallest at the poles (latitude = ±90°), and increase toward the equator.[33]

Other terrestrial effects

The Coriolis effect strongly affects the large-scale oceanic and atmospheric circulation, leading to the formation of robust features like jet streams and western boundary currents. Such features are in geostrophic balance, meaning that the Coriolis and *pressure gradient* forces balance each other. Coriolis acceleration is also responsible for the propagation of many types of waves in the ocean and atmosphere, including Rossby waves and Kelvin waves. It is also instrumental in the so-called Ekman dynamics in the ocean, and in the establishment of the large-scale ocean flow pattern called the Sverdrup balance.

24.6.4   Eötvös effect

Main article: Eötvös effect

The practical impact of the "Coriolis effect" is mostly caused by the horizontal acceleration component produced by horizontal motion.

There are other components of the Coriolis effect. Westward-travelling objects are deflected downwards (feel heavier), while Eastward-travelling objects are deflected upwards (feel lighter).[34] This is known as the Eötvös effect. This aspect of the Coriolis effect is greatest near the equator. The force produced by this effect is similar to the horizontal component, but the much larger vertical forces due to gravity and pressure mean that it is generally unimportant dynamically.

In addition, objects travelling upwards (*i.e.*, out) or downwards (*i.e.*, in) are deflected to the west or east respectively. This effect is also the greatest near the equator. Since vertical movement is usually of limited extent and duration, the size of the effect is smaller and requires precise instruments to detect. However, in the case of large changes of momentum, such as a spacecraft being launched into orbit, the effect becomes significant. The fastest and most fuel-efficient path to orbit is a launch from the equator that curves to a directly eastward heading.

Intuitive example

Imagine a train that travels through a frictionless railway line along the equator. Assume that, when in motion, it moves at the necessary speed to complete a trip around the world in one day (465 m/s).[35] The Coriolis effect can be considered in three cases: when the train travels west, when it is at rest, and when it travels east. In each case, the Coriolis effect can be calculated from the rotating frame of reference on Earth first, and then checked against a fixed inertial frame. The

image below illustrates the three cases in an inertial frame as observed from a fixed point above Earth along its axis of rotation:

Earth and train

1. The train travels toward the west: In that case, it moves against the direction of rotation. Therefore, on the Earth's rotating frame the Coriolis term is pointed inwards towards the axis of rotation (down). This additional force downwards should cause the train to be heavier while moving in that direction.

   - If one looks at this train from the fixed non-rotating frame on top of the center of the Earth, at that speed it remains stationary as the Earth spins beneath it. Hence, the only force acting on it is gravity and the reaction from the track. This force is greater (by 0.34%)[35] than the force that the passengers and the train experience when at rest (rotating along with Earth). This difference is what the Coriolis effect accounts for in the rotating frame of reference.

2. The train comes to a stop: From the point of view on the Earth's rotating frame, the velocity of the train is zero, thus the Coriolis force is also zero and the train and its passengers recuperate their usual weight.

   - From the fixed inertial frame of reference above Earth, the train now rotates along with the rest of the Earth. 0.34% of the force of gravity provides the centripetal force needed to achieve the circular motion on that frame of reference. The remaining force, as measured by a scale, makes the train and passengers "lighter" than in the previous case.

3. The train travels east. In this case, because it moves in the direction of Earth's rotating frame, the Coriolis term is directed outward from the axis of rotation (up). This upward force makes the train seem lighter still than when at rest.

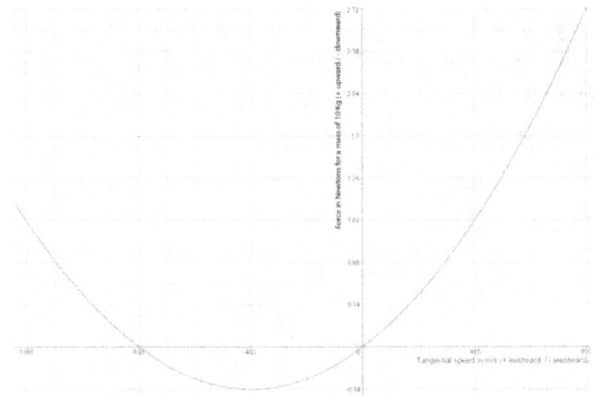

Graph of the force experienced by a 10 kilograms object as a function of its speed moving along Earth's equator (as measured within the rotating frame). (Positive force in the graph is directed upward. Positive speed is directed eastward and negative speed is directed westward).

- From the fixed frame of reference on space, the train travelling east now rotates at twice the rate as when it was at rest—so the amount of centripetal force needed to cause that circular path increases leaving less force from gravity to act on the track. This is what the Coriolis term accounts for on the previous paragraph.

- As a final check one can imagine a frame of reference rotating along with the train. Such frame would be rotating at twice the angular velocity as Earth's rotating frame. The resulting centrifugal force component for that imaginary frame would be greater. Since the train and its passengers are at rest, that would be the only component in that frame explaining again why the train and the passengers are lighter than in the previous two cases.

This also explains why high speed projectiles that travel west are deflected down, and those that travel east are deflected up. This vertical component of the Coriolis effect is called the Eötvös effect.[36]

The above example can be used to explain why the Eötvös effect starts diminishing on an object travelling westward as its tangential speed increases above Earth's rotation (465 m/s). If the westward train in the above example increases speed, part of the force of gravity that pushes against the track accounts for the centripetal force needed to keep it in circular motion on the inertial frame. Once the train doubles its westward speed at 930 m/s that centripetal force becomes equal to the force the train experiences when it stops. From the inertial frame, in both cases it rotates at the same

speed but in the opposite directions. Thus, the force is the same cancelling completely the Eötvös effect. Any object that moves westward at a speed above 930 m/s experiences an upward force instead. In the figure, the Eötvös effect is illustrated for a 10 kilogram object on the train at different speeds. The parabolic shape is because the centripetal force is proportional to the square of the tangential speed. On the inertial frame, the bottom of the parabola is centered at the origin. The offset is because this argument uses the Earth's rotating frame of reference. The graph shows that the Eötvös effect is not symmetrical, and that the resulting downward force experienced by an object that travels west at high velocity is less than the resulting upward force when it travels east at the same speed.

24.6.5 Draining in bathtubs and toilets

Contrary to popular misconception, water rotation in home bathrooms under *normal* circumstances is not related to the Coriolis effect or to the rotation of the Earth, and no consistent difference in rotation direction between toilet drainage in the Northern and Southern Hemispheres can be observed. The formation of a vortex over the plug hole may be explained by the conservation of angular momentum: The radius of rotation decreases as water approaches the plug hole, so the rate of rotation increases, for the same reason that an ice skater's rate of spin increases as they pull their arms in. Any rotation around the plug hole that is initially present accelerates as water moves inward.

Of course, the Coriolis force does still impact the direction of the flow of water, but only minutely. Only if the water is so still that the effective rotation rate of the Earth is faster than that of the water relative to its container, and if externally applied torques (such as might be caused by flow over an uneven bottom surface) are small enough, the Coriolis effect may indeed determine the direction of the vortex. Without such careful preparation, the Coriolis effect is likely to be much smaller than various other influences on drain direction[37] such as any residual rotation of the water[38] and the geometry of the container.[39] Despite this, the idea that toilets and bathtubs drain differently in the Northern and Southern Hemispheres has been popularized by several television programs and films, including *Escape Plan*, *Wedding Crashers*, *The Simpsons* episode "Bart vs. Australia", and *The X-Files* episode "Die Hand Die Verletzt".[40] Several science broadcasts and publications, including at least one college-level physics textbook, have also stated this.[41][42]

In 1908, the Austrian physicist Ottokar Tumlirz described careful and effective experiments that demonstrated the effect of the rotation of the Earth on the outflow of water through a central aperture.[43] The subject was later popularized in a famous 1962 article in the journal *Nature*, which described an experiment in which all other forces to the system were removed by filling a 6 ft (1.8 m) tank with 300 U.S. gal (1,100 L) of water and allowing it to settle for 24 hours (to allow any movement due to filling the tank to die away), in a room where the temperature had stabilized. The drain plug was then very slowly removed, and tiny pieces of floating wood were used to observe rotation. During the first 12 to 15 minutes, no rotation was observed. Then, a vortex appeared and consistently began to rotate in an anticlockwise direction (the experiment was performed in Boston, Massachusetts, in the Northern Hemisphere). This was repeated and the results averaged to make sure the effect was real. The report noted that the vortex rotated, "about 30,000 times faster than the effective rotation of the Earth in 42° North (the experiment's location)". This shows that the small initial rotation due to the Earth is amplified by gravitational draining and conservation of angular momentum to become a rapid vortex and may be observed under carefully controlled laboratory conditions.[44][45]

24.6.6 Ballistic trajectories

The Coriolis force became important in external ballistics for calculating the trajectories of very long-range artillery shells. The most famous historical example was the Paris gun, used by the Germans during World War I to bombard Paris from a range of about 120 km (75 mi). Similarly, a bullet does not fly straight from the barrel to a target. The Coriolis force minutely changes the trajectory of a bullet, curving the path of the projectile into a more arched 'semi-circle' shape. This effect only affects accuracy at extremely long distances and is therefore adjusted for by accurate long-distance shooters, such as snipers and other trained professionals.[18]

The effects of the Coriolis force on ballistic trajectories should not be confused with the curvature of the paths of missiles, satellites, and similar objects when the paths are plotted on two-dimensional (flat) maps, such as the Mercator projection. The projections of the three-dimensional curved surface of the Earth to a two-dimensional surface (the map) necessarily results in distorted features. The apparent curvature of the path is a consequence of the sphericity of the Earth and would occur even in a non-rotating frame.

24.7 Visualization of the Coriolis effect

To demonstrate the Coriolis effect, a parabolic turntable can be used. On a flat turntable, the inertia of a co-rotating

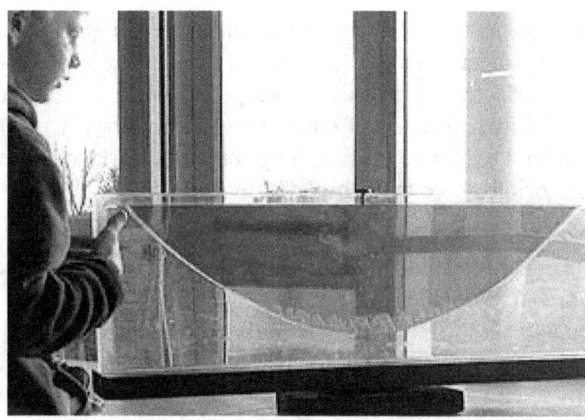

Fluid assuming a parabolic shape as it is rotating

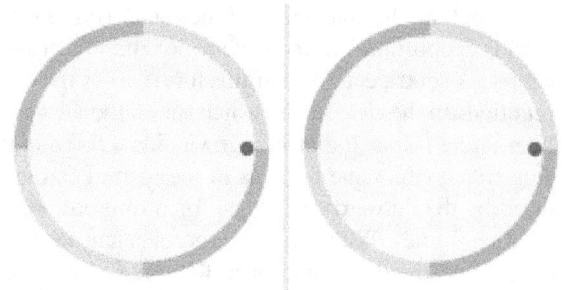

Object moving frictionlessly over the surface of a very shallow parabolic dish. The object has been released in such a way that it follows an elliptical trajectory.
Left: *The inertial point of view.*
Right: *The co-rotating point of view.*

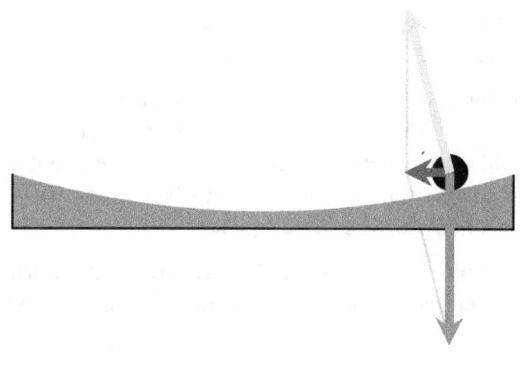

The forces at play in the case of a curved surface.
Red: *gravity*
Green: *the normal force*
Blue: *the net resultant centripetal force.*

object forces it off the edge. However, if the turntable surface has the correct paraboloid (parabolic bowl) shape (see the figure) and rotates at the corresponding rate, the force components shown in the figure make the component of gravity tangential to the bowl surface exactly equal to the centripetal force necessary to keep the object rotating at its velocity and radius of curvature (assuming no friction). (See banked turn.) This carefully contoured surface allows the Coriolis force to be displayed in isolation.[46][47]

Discs cut from cylinders of dry ice can be used as pucks, moving around almost frictionlessly over the surface of the parabolic turntable, allowing effects of Coriolis on dynamic phenomena to show themselves. To get a view of the motions as seen from the reference frame rotating with the turntable, a video camera is attached to the turntable so as to co-rotate with the turntable, with results as shown in the figure. In the left panel of the figure, which is the viewpoint of a stationary observer, the gravitational force in the inertial frame pulling the object toward the center (bottom) of the dish is proportional to the distance of the object from the center. A centripetal force of this form causes the elliptical motion. In the right panel, which shows the viewpoint of the rotating frame, the inward gravitational force in the rotating frame (the same force as in the inertial frame) is balanced by the outward centrifugal force (present only in the rotating frame). With these two forces balanced, in the rotating frame the only unbalanced force is Coriolis (also present only in the rotating frame), and the motion is an *inertial circle*. Analysis and observation of circular motion in the rotating frame is a simplification compared to analysis or observation of elliptical motion in the inertial frame.

Because this reference frame rotates several times a minute rather than only once a day like the Earth, the Coriolis acceleration produced is many times larger and so easier to observe on small time and spatial scales than is the Coriolis acceleration caused by the rotation of the Earth.

In a manner of speaking, the Earth is analogous to such a turntable.[48] The rotation has caused the planet to settle on a spheroid shape, such that the normal force, the gravitational force and the centrifugal force exactly balance each other on a "horizontal" surface. (See equatorial bulge.)

The Coriolis effect caused by the rotation of the Earth can be seen indirectly through the motion of a Foucault pendulum.

24.8 Coriolis effects in other areas

24.8.1 Coriolis flow meter

A practical application of the Coriolis effect is the mass flow meter, an instrument that measures the mass flow rate and density of a fluid flowing through a tube. The operating principle involves inducing a vibration of the tube through which the fluid passes. The vibration, though not completely circular, provides the rotating reference frame that gives rise to the Coriolis effect. While specific methods vary according to the design of the flow meter, sensors monitor and analyze changes in frequency, phase shift, and amplitude of the vibrating flow tubes. The changes observed represent the mass flow rate and density of the fluid.[49]

24.8.2 Molecular physics

In polyatomic molecules, the molecule motion can be described by a rigid body rotation and internal vibration of atoms about their equilibrium position. As a result of the vibrations of the atoms, the atoms are in motion relative to the rotating coordinate system of the molecule. Coriolis effects are therefore present, and make the atoms move in a direction perpendicular to the original oscillations. This leads to a mixing in molecular spectra between the rotational and vibrational levels, from which Coriolis coupling constants can be determined.[50]

24.8.3 Gyroscopic precession

When an external torque is applied to a spinning gyroscope along an axis that is at right angles to the spin axis, the rim velocity that is associated with the spin becomes radially directed in relation to the external torque axis. This causes a Coriolis force to act on the rim in such a way as to tilt the gyroscope at right angles to the direction that the external torque would have tilted it. This tendency has the effect of keeping spinning bodies stably aligned in space.

24.8.4 Insect flight

Flies (Diptera) and some moths (Lepidoptera) exploit the Coriolis effect in flight with specialized appendages and organs that relay information about the angular velocity of their bodies. Coriolis forces resulting from linear motion of these appendages are detected within the rotating frame of reference of the insects' bodies. In the case of flies, their specialized appendages are dumbbell shaped organs located just behind their wings called halteres.[51] The halteres oscillate in a plane at the same beat frequency as the main wings so that any body rotation results in lateral deviation of the halteres from their plane of motion.[52] In moths, their antennae are responsible for the sensing of Coriolis forces in the similar manner as with the halteres in flies.[53] In both flies and moths, a collection of mechanosensors at the base of the appendage are sensitive to deviations at the beat frequency, correlating to rotation in the pitch and roll planes, and at twice the beat frequency, correlating to rotation in the yaw plane.[54][53]

24.9   See also

- Analytical mechanics

- Applied mechanics

- Classical mechanics

- Dynamics (physics)

- Earth's rotation

- Equatorial Rossby wave

- Frenet–Serret formulas

- Gyroscope

- Kinetics (physics)

- Mechanics of planar particle motion

- Reactive centrifugal force

- Secondary flow

- Statics

- Uniform circular motion

24.10   References

[1] Frautschi, Steven C.; Olenick, Richard P.; Apostol, Tom M.; Goodstein, David L. (2007). *The Mechanical Universe: Mechanics and Heat, Advanced Edition* (illustrated ed.). Cambridge University Press. p. 208. ISBN 978-0-521-71590-4. Extract of page 208

[2] Bhatia, V.B. (1997). *Classical Mechanics: With introduction to Nonlinear Oscillations and Chaos*. Narosa Publishing House. p. 201. ISBN 81-7319-105-0.

[3] "Coriolis Effect: Because the Earth turns – Teacher's guide" (PDF). *Project ATMOSPHERE*. American Meteorological Society. Retrieved 10 April 2015.

[4] Graney, Christopher M. (2011). "Coriolis effect, two centuries before Coriolis". *Physics Today*. **64**: 8. Bibcode:2011PhT....64h...8G. doi:10.1063/PT.3.1195.

[5] G-G Coriolis (1835). "Sur les équations du mouvement relatif des systèmes de corps". *J. de l'Ecole royale polytechnique*. **15**: 144–154.

[6] Dugas, René and J. R. Maddox (1988). *A History of Mechanics*. Courier Dover Publications: p. 374. ISBN 0-486-65632-2

[7] Bartholomew Price (1862). *A Treatise on Infinitesimal Calculus : Vol. IV. The dynamics of material systems*. Oxford : University Press. pp. 418–420.

[8] Arthur Gordon Webster (1912). *The Dynamics of Particles and of Rigid, Elastic, and Fluid Bodies*. B. G. Teubner. p. 320. ISBN 1-113-14861-6.

[9] Edwin b. Wilson (1920). James McKeen Cattell, ed. "Space, *Time*, and Gravitation". *The Scientific Monthly*. American Association for the Advancement of Science. **10**: 226.

[10] William Ferrel (November 1856). "An Essay on the Winds and the Currents of the Ocean" (PDF). *Nashville Journal of Medicine and Surgery*. **xi** (4): 7–19. Retrieved on 1 January 2009.

[11] Anders O. Persson. "The Coriolis Effect:Four centuries of conflict between common sense and mathematics, Part I: A history to 1885" (PDF). Swedish Meteorological and Hydrological Institute.

[12] Hestenes, David (1990). *New Foundations for Classical Mechanics*. The Netherlands: Kluwer Academic Publishers. p. 312. ISBN 90-277-2526-8.

[13] Schneider, Stephen H.; Root, Terry L.; Mastrandrea, Michael, eds. (2011). *Encyclopedia of Climate and Weather*. **3**. Oxford University Press. p. 310.

[14] John M. Wallace; Peter V. Hobbs (1977). *Atmospheric Science: An Introductory Survey*. Academic Press, Inc. pp. 368–371. ISBN 0-12-732950-1.

[15] Sheldon M. Ebenholtz (2001). *Oculomotor Systems and Perception*. Cambridge University Press. ISBN 0-521-80459-0.

[16] George Mather (2006). *Foundations of perception*. Taylor & Francis. ISBN 0-86377-835-6.

[17] Roger Graham Barry; Richard J. Chorley (2003). *Atmosphere, Weather and Climate*. Routledge. p. 113. ISBN 0-415-27171-1.

[18] The claim is made that in the Falklands in WW I, the British failed to correct their sights for the southern hemisphere, and so missed their targets. John Edensor Littlewood (1953). *A Mathematician's Miscellany*. Methuen And Company Limited. p. 51. John Robert Taylor (2005). *Classical Mechanics*. University Science Books. p. 364; Problem 9.28. ISBN 1-891389-22-X. For set up of the calculations, see Carlucci & Jacobson (2007), p. 225

[19] Lakshmi H. Kantha; Carol Anne Clayson (2000). *Numerical Models of Oceans and Oceanic Processes*. Academic Press. p. 103. ISBN 0-12-434068-7.

[20] Stephen D. Butz (2002). *Science of Earth Systems*. Thomson Delmar Learning. p. 305. ISBN 0-7668-3391-7.

[21] James R. Holton (2004). *An Introduction to Dynamic Meteorology*. Academic Press. p. 18. ISBN 0-12-354015-1.

[22] Carlucci, Donald E.; Jacobson, Sidney S. (2007). *Ballistics: Theory and Design of Guns and Ammunition*. CRC Press. pp. 224–226. ISBN 1-4200-6618-8.

[23] Here the description "radially inward" means "toward the axis of rotation". That direction is *not* toward the center of curvature of the path, however, which is the direction of the true centripetal force. Hence, the quotation marks on "centripetal".

[24] George E. Owen (2003). *Fundamentals of Scientific Mathematics* (original edition published by Harper & Row, New York, 1964 ed.). Courier Dover Publications. p. 23. ISBN 0-486-42808-7.

[25] Morton Tavel (2002). *Contemporary Physics and the Limits of Knowledge*. Rutgers University Press. p. 88. ISBN 0-8135-3077-6.

[26] James R Ogden; M Fogiel (1995). *High School Earth Science Tutor*. Research & Education Assoc. p. 167. ISBN 0-87891-975-9.

[27] James Greig McCully (2006). *Beyond the moon: A Conversational, Common Sense Guide to Understanding the Tides*. World Scientific. pp. 74–76. ISBN 981-256-643-0.

[28] William Menke; Dallas Abbott (1990). *Geophysical Theory*. Columbia University Press. pp. 124–126. ISBN 0-231-06792-5.

[29] James R. Holton (2004). *An Introduction to Dynamic Meteorology*. Burlington, MA: Elsevier Academic Press. p. 64. ISBN 0-12-354015-1.

[30] Brinney, Amanda. "Coriolis Effect – An Overview of the Coriolis Effect". About.com.

[31] Roger Graham Barry; Richard J. Chorley (2003). *Atmosphere, Weather and Climate*. Routledge. p. 115. ISBN 0-415-27171-1.

[32] Cloud Spirals and Outflow in Tropical Storm Katrina from Earth Observatory (NASA)

[33] John Marshall; R. Alan Plumb (2007). *p. 98*. Amsterdam: Elsevier Academic Press. ISBN 0-12-558691-4.

[34] Lowrie, William (1997). *Fundamentals of Geophysics* (illustrated ed.). Cambridge University Press. p. 45. ISBN 0-521-46728-4. Extract of page 45

[35] Persson, Anders. "The Coriolis Effect – a conflict between common sense and mathematics" (PDF). Norrköping, Sweden: The Swedish Meteorological and Hydrological Institute: 8. Archived from the original (PDF) on 6 September 2005. Retrieved 6 September 2015.

[36] Rugai, Nick (1 December 2012). *Computational Epistemology: From Reality To Wisdom*. Lulu.com. p. 304. ISBN 1300477237. Retrieved 6 September 2015.

[37] Larry D. Kirkpatrick; Gregory E. Francis (2006). *Physics: A World View*. Cengage Learning. pp. 168–9. ISBN 978-0-495-01088-3. Retrieved 1 April 2011.

[38] Y. A. Stepanyants; G. H. Yeoh (2008). "Stationary bathtub vortices and a critical regime of liquid discharge". *Journal of Fluid Mechanics*. **604** (1): 77–98. Bibcode:2008JFM...604...77S. doi:10.1017/S0022112008001080.

[39] Creative Media Applications (2004). *A Student's Guide to Earth Science: Words and terms*. Greenwood Publishing Group. p. 22. ISBN 978-0-313-32902-9.

[40] Emery, C. Eugene, Jr. (May 1, 1995). "*X-Files* coriolis error leaves viewers wondering". *Skeptical Inquirer*

[41] Fraser, Alistair. "Bad Coriolis". *Bad Meteorology*. Pennsylvania State College of Earth and Mineral Science. Retrieved 17 January 2011.

[42] Tipler, Paul (1998). *Physics for Engineers and Scientists* (4th ed.). W.H.Freeman, Worth Publishers. p. 128. ISBN 978-1-57259-616-0. ...on a smaller scale, the coriolis effect causes water draining out a bathtub to rotate anticlockwise in the northern hemisphere...

[43] Tumlirz, Ottokar (1908). "Ein neuer physikalischer Beweis für die Achsendrehung der Erde". *Sitzungsberichte der math.-nat. Klasse der kaiserlichen Akademie der Wissenschaften IIa*. **117**: 819–841.

[44] Shapiro, Ascher H. (1962). "Bath-Tub Vortex". *Nature*. **196** (4859): 1080–1081. Bibcode:1962Natur.196.1080S. doi:10.1038/1961080b0.

[45] (Vorticity, Part 1). Web.mit.edu. Retrieved 8 November 2011.

[46] When a container of fluid is rotating on a turntable, the surface of the fluid naturally assumes the correct parabolic shape. This fact may be exploited to make a parabolic turntable by using a fluid that sets after several hours, such as a synthetic resin. For a video of the Coriolis effect on such a parabolic surface, see Geophysical fluid dynamics lab demonstration John Marshall, Massachusetts Institute of Technology.

[47] For a java applet of the Coriolis effect on such a parabolic surface, see Brian Fiedler School of Meteorology at the University of Oklahoma.

[48] John Marshall; R. Alan Plumb (2007). *Atmosphere, Ocean, and Climate Dynamics: An Introductory Text*. Academic Press. p. 101. ISBN 0-12-558691-4.

[49] Omega Engineering. "Mass Flowmeters".

[50] califano, S (1976). *Vibrational states*. Wiley. pp. 226–227. ISBN 0471129968.

[51] Fraenkel, G.; Pringle, W.S. (21 May 1938). "Halteres of Flies as Gyroscopic Organs of Equilibrium". *Nature*. **141** (141): 919–920. Bibcode:1938Natur.141..919F. doi:10.1038/141919a0.

[52] Dickinson, M. (1999). "Haltere-mediated equilibrium reflexes of the fruit fly, Drosophila melanogaster" (PDF). *Phil. Trans. R. Soc. Lond.* **354** (354): 903–916. doi:10.1098/rstb.1999.0442. PMC 1692594⊚. PMID 10382224.

[53] Sane S., Dieudonné, A., Willis, M., Daniel, T. (February 2007). "Antennal mechanosensors mediate flight control in moths". *Science.* **315**: 863–866. Bibcode:2007Sci...315..863S. doi:10.1126/science.1133598.

[54] Fox, J; Daniel, T (2008). "A neural basis for gyroscopic force measurement in the halteres of Holorusia". *Journal of Comparative Physiology.* **194** (10): 887–897. doi:10.1007/s00359-008-0361-z. PMID 18751714.

24.10.1 Further reading

Physics and meteorology

- Riccioli, G. B., 1651: *Almagestum Novum*, Bologna, pp. 425–427
  (Original book [in Latin], scanned images of complete pages.)

- Coriolis, G. G., 1832: "Mémoire sur le principe des forces vives dans les mouvements relatifs des machines." *Journal de l'école Polytechnique*, Vol 13, pp. 268–302.
  (Original article [in French], PDF file, 1.6 MB, scanned images of complete pages.)

- Coriolis, G. G., 1835: "Mémoire sur les équations du mouvement relatif des systèmes de corps." *Journal de l'école Polytechnique*, Vol 15, pp. 142–154
  (Original article [in French] PDF file, 400 KB, scanned images of complete pages.)

- Gill, A. E. *Atmosphere-Ocean dynamics*, Academic Press, 1982.

- Robert Ehrlich (1990). *Turning the World Inside Out and 174 Other Simple Physics Demonstrations.* Princeton University Press. p. *Rolling a ball on a rotating turntable*; p. 80 ff. ISBN 0-691-02395-6.

- Durran, D. R., 1993: *Is the Coriolis force really responsible for the inertial oscillation?*, Bull. Amer. Meteor. Soc., 74, pp. 2179–2184; Corrigenda. Bulletin of the American Meteorological Society, 75, p. 261

- Durran, D. R., and S. K. Domonkos, 1996: *An apparatus for demonstrating the inertial oscillation*, Bulletin of the American Meteorological Society, 77, pp. 557–559.

- Marion, Jerry B. 1970, *Classical Dynamics of Particles and Systems*, Academic Press.

- Persson, A., 1998 *How do we Understand the Coriolis Force?* Bulletin of the American Meteorological Society 79, pp. 1373–1385.

- Symon, Keith. 1971, *Mechanics*, Addison–Wesley

- Akira Kageyama & Mamoru Hyodo: *Eulerian derivation of the Coriolis force*

- James F. Price: *A Coriolis tutorial* Woods Hole Oceanographic Institute (2003)

- McDonald, James E. (May 1952). "The Coriolis Effect" (PDF). *Scientific American*: 72–78. Retrieved 2016-01-04. Everything that moves over the surface of the Earth – water, air, animals, machines and projectiles – sidles to the right in the Northern Hemisphere and to the left in the Southern. Elementary, non-mathematical; but well written.

Historical

- Grattan-Guinness, I., Ed., 1994: *Companion Encyclopedia of the History and Philosophy of the Mathematical Sciences.* Vols. I and II. Routledge, 1840 pp. 1997: *The Fontana History of the Mathematical Sciences.* Fontana, 817 pp. 710 pp.

- Khrgian, A., 1970: *Meteorology: A Historical Survey.* Vol. 1. Keter Press, 387 pp.

- Kuhn, T. S., 1977: Energy conservation as an example of simultaneous discovery. *The Essential Tension, Selected Studies in Scientific Tradition and Change*, University of Chicago Press, 66–104.

- Kutzbach, G., 1979: *The Thermal Theory of Cyclones. A History of Meteorological Thought in the Nineteenth Century.* Amer. Meteor. Soc., 254 pp.

24.11 External links

- The definition of the Coriolis effect from the Glossary of Meteorology

- The Coriolis Effect — a conflict between common sense and mathematics PDF-file. 20 pages. A general discussion by Anders Persson of various aspects of the coriolis effect, including Foucault's Pendulum and Taylor columns.

- 10 Coriolis Effect Videos and Games- from the About.com Weather Page

- Coriolis Force – from ScienceWorld

- *Coriolis Effect and Drains* An article from the NEW-TON web site hosted by the Argonne National Laboratory.

- Catalog of Coriolis videos

- *Coriolis Effect: A graphical animation*, a visual Earth animation with precise explanation

- *An introduction to fluid dynamics* SPINLab Educational Film explains the Coriolis effect with the aid of lab experiments

- *Do bathtubs drain counterclockwise in the Northern Hemisphere?* by Cecil Adams.

- *Bad Coriolis.* An article uncovering misinformation about the Coriolis effect. By Alistair B. Fraser, Emeritus Professor of Meteorology at Pennsylvania State University

- *The Coriolis Effect: A (Fairly) Simple Explanation*, an explanation for the layperson

- Observe an animation of the Coriolis effect over Earth's surface

- Animation clip showing scenes as viewed from both an inertial frame and a rotating frame of reference, visualizing the Coriolis and centrifugal forces.

- Vincent Mallette *The Coriolis Force* @ INWIT

- NASA notes

- Interactive Coriolis Fountain lets you control rotation speed, droplet speed and frame of reference to explore the Coriolis effect.

Chapter 25

Friction

For other uses, see Friction (disambiguation).

Friction is the force resisting the relative motion of solid surfaces, fluid layers, and material elements sliding against each other.[1] There are several types of friction:

- **Dry friction** resists relative lateral motion of two solid surfaces in contact. Dry friction is subdivided into *static friction* ("stiction") between non-moving surfaces, and *kinetic friction* between moving surfaces.

- **Fluid friction** describes the friction between layers of a viscous fluid that are moving relative to each other.[2][3]

- **Lubricated friction** is a case of fluid friction where a lubricant fluid separates two solid surfaces.[4][5][6]

- **Skin friction** is a component of drag, the force resisting the motion of a fluid across the surface of a body.

- **Internal friction** is the force resisting motion between the elements making up a solid material while it undergoes deformation.[3]

When surfaces in contact move relative to each other, the friction between the two surfaces converts kinetic energy into thermal energy (that is, it converts work to heat). This property can have dramatic consequences, as illustrated by the use of friction created by rubbing pieces of wood together to start a fire. Kinetic energy is converted to thermal energy whenever motion with friction occurs, for example when a viscous fluid is stirred. Another important consequence of many types of friction can be wear, which may lead to performance degradation and/or damage to components. Friction is a component of the science of tribology.

Friction is not itself a fundamental force. Dry friction arises from a combination of inter-surface adhesion, surface roughness, surface deformation, and surface contamination. The complexity of these interactions makes the calculation of friction from first principles impractical and necessitates the use of empirical methods for analysis and the development of theory.

25.1 History

The classic rules of sliding friction were discovered by Leonardo da Vinci in 1493 but remained in his notebooks, unpublished.[7][8][9][10][11][12] These rules were rediscovered by Guillaume Amontons in 1699. Amontons presented the nature of friction in terms of surface irregularities and the force required to raise the weight pressing the surfaces together. This view was further elaborated by Bernard Forest de Bélidor [13] and Leonhard Euler (1750), who derived the angle of repose of a weight on an inclined plane and first distinguished between static and kinetic friction.[14] A different explanation was provided by John Theophilus Desaguliers (1725), who demonstrated the strong cohesion forces between lead spheres of which a small cap is cut off and which were then brought into contact with each other.

The understanding of friction was further developed by Charles-Augustin de Coulomb (1785). Coulomb investigated the influence of four main factors on friction: the nature of the materials in contact and their surface coatings; the extent of the surface area; the normal pressure (or load); and the length of time that the surfaces remained in contact (time of repose).[7] Coulomb further considered the influence of sliding velocity, temperature and humidity, in order to decide between the different explanations on the nature of friction that had been proposed. The distinction between static and dynamic friction is made in Coulomb's friction law (see below), although this distinction was already drawn by Johann Andreas von Segner in 1758.[7] The effect of the time of repose was explained by Pieter van Musschenbroek (1762) by considering the surfaces of fibrous materials, with fibers meshing together, which takes a finite time in which the friction increases.

John Leslie (1766–1832) noted a weakness in the views of Amontons and Coulomb: If friction arises from a weight being drawn up the inclined plane of successive asperities, why then isn't it balanced through descending the opposite slope? Leslie was equally skeptical about the role of adhesion proposed by Desaguliers, which should on the whole have the same tendency to accelerate as to retard the motion.[7] In Leslie's view, friction should be seen as a time-dependent process of flattening, pressing down asperities, which creates new obstacles in what were cavities before.

Arthur Jules Morin (1833) developed the concept of sliding versus rolling friction. Osborne Reynolds (1866) derived the equation of viscous flow. This completed the classic empirical model of friction (static, kinetic, and fluid) commonly used today in engineering.[8] In 1877, Fleeming Jenkin and J. A. Ewing investigated the continuity between static and kinetic friction.[15]

The focus of research during the 20th century has been to understand the physical mechanisms behind friction. Frank Philip Bowden and David Tabor (1950) showed that, at a microscopic level, the actual area of contact between surfaces is a very small fraction of the apparent area.[9] This actual area of contact, caused by "asperities" (roughness) increases with pressure. The development of the atomic force microscope (ca. 1986) enabled scientists to study friction at the atomic scale,[8] showing that, on that scale, dry friction is the product of the inter-surface shear stress and the contact area. These two discoveries explain the macroscopic proportionality between normal force and static frictional force between dry surfaces.

25.2   Laws of dry friction

The elementary property of sliding (kinetic) friction were discovered by experiment in the 15th to 18th centuries and were expressed as three empirical laws:

- **Amontons' First Law**: The force of friction is directly proportional to the applied load.

- **Amontons' Second Law**: The force of friction is independent of the apparent area of contact.

- **Coulomb's Law of Friction**: Kinetic friction is independent of the sliding velocity.

25.3   Dry friction

Dry friction resists relative lateral motion of two solid surfaces in contact. The two regimes of dry friction are 'static friction' ("stiction") between non-moving surfaces, and *kinetic friction* (sometimes called sliding friction or dynamic friction) between moving surfaces.

Coulomb friction, named after Charles-Augustin de Coulomb, is an approximate model used to calculate the force of dry friction. It is governed by the model:

$$F_f \leq \mu F_n,$$

where

- F_f is the force of friction exerted by each surface on the other. It is parallel to the surface, in a direction opposite to the net applied force.

- μ is the coefficient of friction, which is an empirical property of the contacting materials,

- F_n is the normal force exerted by each surface on the other, directed perpendicular (normal) to the surface.

The Coulomb friction F_f may take any value from zero up to μF_n , and the direction of the frictional force against a surface is opposite to the motion that surface would experience in the absence of friction. Thus, in the static case, the frictional force is exactly what it must be in order to prevent motion between the surfaces; it balances the net force tending to cause such motion. In this case, rather than providing an estimate of the actual frictional force, the Coulomb approximation provides a threshold value for this force, above which motion would commence. This maximum force is known as traction.

The force of friction is always exerted in a direction that opposes movement (for kinetic friction) or potential movement (for static friction) between the two surfaces. For example, a curling stone sliding along the ice experiences a kinetic force slowing it down. For an example of potential movement, the drive wheels of an accelerating car experience a frictional force pointing forward; if they did not, the wheels would spin, and the rubber would slide backwards along the pavement. Note that it is not the direction of movement of the vehicle they oppose, it is the direction of (potential) sliding between tire and road.

25.3.1   Normal force

Main article: Normal force

The normal force is defined as the net force compressing two parallel surfaces together; and its direction is perpendicular to the surfaces. In the simple case of a mass resting on a horizontal surface, the only component of the normal force is the force due to gravity, where $N = mg$. In this case, the magnitude of the friction force is the product of the mass of the object, the acceleration due to gravity, and

A block on a ramp

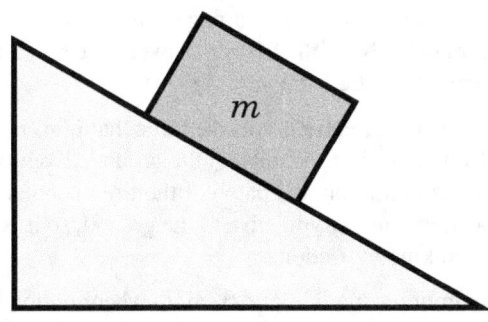

Free body diagram of just the block

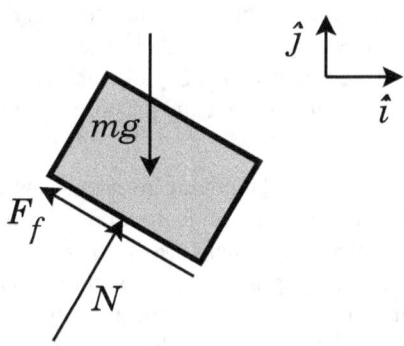

Free-body diagram for a block on a ramp. Arrows are vectors indicating directions and magnitudes of forces. N is the normal force, mg is the force of gravity, and F_f is the force of friction.

the coefficient of friction. However, the coefficient of friction is not a function of mass or volume; it depends only on the material. For instance, a large aluminum block has the same coefficient of friction as a small aluminum block. However, the magnitude of the friction force itself depends on the normal force, and hence on the mass of the block.

If an **object is on a level surface** and the force tending to cause it to slide is horizontal, the normal force N between the object and the surface is just its weight, which is equal to its mass multiplied by the acceleration due to earth's gravity, g. If the **object is on a tilted surface** such as an inclined plane, the normal force is less, because less of the force of

gravity is perpendicular to the face of the plane. Therefore, the normal force, and ultimately the frictional force, is determined using vector analysis, usually via a free body diagram. Depending on the situation, the calculation of the normal force may include forces other than gravity.

25.3.2   Coefficient of friction

The **coefficient of friction** (COF), often symbolized by the Greek letter μ, is a dimensionless scalar value which describes the ratio of the force of friction between two bodies and the force pressing them together. The coefficient of friction depends on the materials used; for example, ice on steel has a low coefficient of friction, while rubber on pavement has a high coefficient of friction. Coefficients of friction range from near zero to greater than one.

For surfaces at rest relative to each other $\mu = \mu_s$, where μ_s is the *coefficient of static friction*. This is usually larger than its kinetic counterpart.

For surfaces in relative motion $\mu = \mu_k$, where μ_k is the *coefficient of kinetic friction*. The Coulomb friction is equal to F_f , and the frictional force on each surface is exerted in the direction opposite to its motion relative to the other surface.

Arthur Morin introduced the term and demonstrated the utility of the coefficient of friction.[7] The coefficient of friction is an empirical measurement – it has to be measured experimentally, and cannot be found through calculations. Rougher surfaces tend to have higher effective values. Both static and kinetic coefficients of friction depend on the pair of surfaces in contact; for a given pair of surfaces, the coefficient of static friction is *usually* larger than that of kinetic friction; in some sets the two coefficients are equal, such as teflon-on-teflon.

Most dry materials in combination have friction coefficient values between 0.3 and 0.6. Values outside this range are rarer, but teflon, for example, can have a coefficient as low as 0.04. A value of zero would mean no friction at all, an elusive property. Rubber in contact with other surfaces can yield friction coefficients from 1 to 2. Occasionally it is maintained that μ is always < 1, but this is not true. While in most relevant applications μ < 1, a value above 1 merely implies that the force required to slide an object along the surface is greater than the normal force of the surface on the object. For example, silicone rubber or acrylic rubber-coated surfaces have a coefficient of friction that can be substantially larger than 1.

While it is often stated that the COF is a "material property," it is better categorized as a "system property." Unlike true material properties (such as conductivity, dielectric constant, yield strength), the COF for any two materials depends on system variables like temperature, velocity, atmosphere and also what are now popularly described as aging and deaging times; as well as on geometric properties of the interface between the materials. For example, a copper pin sliding against a thick copper plate can have a COF that varies from 0.6 at low speeds (metal sliding against metal) to below 0.2 at high speeds when the copper surface begins to melt due to frictional heating. The latter speed, of course, does not determine the COF uniquely; if the pin diameter is increased so that the frictional heating is removed rapidly, the temperature drops, the pin remains solid and the COF rises to that of a 'low speed' test.

Approximate coefficients of friction

Under certain conditions some materials have very low friction coefficients. An example is (highly ordered pyrolytic) graphite which can have a friction coefficient below 0.01.[24] This ultralow-friction regime is called superlubricity.

25.3.3 Static friction

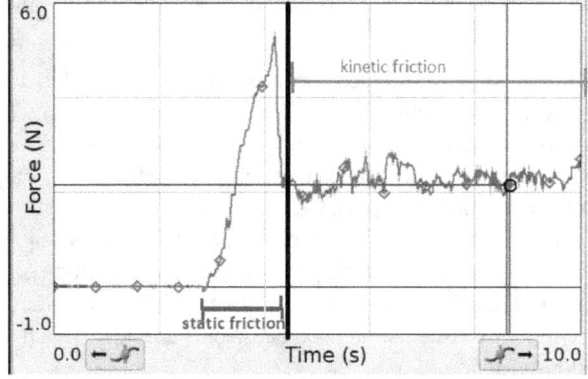

When the mass is not moving, the object experiences static friction. The friction increases as the applied force increases until the block moves. After the block moves, it experiences kinetic friction, which is less than the maximum static friction.

Static friction is friction between two or more solid objects that are not moving relative to each other. For example, static friction can prevent an object from sliding down a sloped surface. The coefficient of static friction, typically denoted as μ_s, is usually higher than the coefficient of kinetic friction.

The static friction force must be overcome by an applied force before an object can move. The maximum possible friction force between two surfaces before sliding begins is the product of the coefficient of static friction and the normal force: $F_{max} = \mu_s F_n$. When there is no sliding occurring, the friction force can have any value from zero up to

F_{max} . Any force smaller than F_{max} attempting to slide one surface over the other is opposed by a frictional force of equal magnitude and opposite direction. Any force larger than F_{max} overcomes the force of static friction and causes sliding to occur. The instant sliding occurs, static friction is no longer applicable—the friction between the two surfaces is then called kinetic friction.

An example of static friction is the force that prevents a car wheel from slipping as it rolls on the ground. Even though the wheel is in motion, the patch of the tire in contact with the ground is stationary relative to the ground, so it is static rather than kinetic friction.

The maximum value of static friction, when motion is impending, is sometimes referred to as **limiting friction**,[25] although this term is not used universally.[2]

25.3.4 Kinetic friction

Kinetic (or dynamic) friction occurs when two objects are moving relative to each other and rub together (like a sled on the ground). The coefficient of kinetic friction is typically denoted as μ_k, and is usually less than the coefficient of static friction for the same materials.[26][27] However, Richard Feynman comments that "with dry metals it is very hard to show any difference."[28] The friction force between two surfaces after sliding begins is the product of the coefficient of kinetic friction and the normal force: $F_k = \mu_k F_n$.

New models are beginning to show how kinetic friction can be greater than static friction.[29] Kinetic friction is now understood, in many cases, to be primarily caused by chemical bonding between the surfaces, rather than interlocking asperities;[30] however, in many other cases roughness effects are dominant, for example in rubber to road friction.[29] Surface roughness and contact area affect kinetic friction for micro- and nano-scale objects where surface area forces dominate inertial forces.[31]

The origin of kinetic friction at nanoscale can be explained by thermodynamics.[32] Upon sliding, new surface forms at the back of a sliding true contact, and existing surface disappears at the front of it. Since all surfaces involve the thermodynamic surface energy, work must be spent in creating the new surface, and energy is released as heat in removing the surface. Thus, a force is required to move the back of the contact, and frictional heat is released at the front.

25.3.5 Angle of friction

For the maximum angle of static friction between granular materials, see Angle of repose.

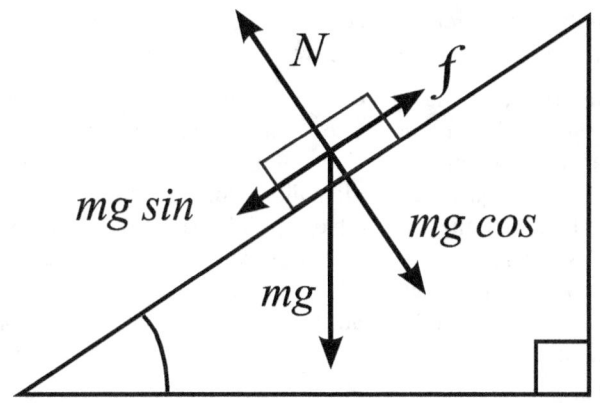

Angle of friction, θ, when block just starts to slide.

For certain applications it is more useful to define static friction in terms of the maximum angle before which one of the items will begin sliding. This is called the *angle of friction* or *friction angle*. It is defined as:

$$\tan \theta = \mu_s$$

where θ is the angle from horizontal and μs is the static coefficient of friction between the objects.[33] This formula can also be used to calculate μs from empirical measurements of the friction angle.

25.3.6 Friction at the atomic level

Determining the forces required to move atoms past each other is a challenge in designing nanomachines. In 2008 scientists for the first time were able to move a single atom across a surface, and measure the forces required. Using ultrahigh vacuum and nearly zero temperature (5 K), a modified atomic force microscope was used to drag a cobalt atom, and a carbon monoxide molecule, across surfaces of copper and platinum.[34]

25.3.7 Limitations of the Coulomb model

The Coulomb approximation mathematically follows from the assumptions that surfaces are in atomically close contact only over a small fraction of their overall area, that this contact area is proportional to the normal force (until saturation, which takes place when all area is in atomic contact), and that the frictional force is proportional to the applied normal force, independently of the contact area (you can see the experiments on friction from Leonardo da Vinci). Such reasoning aside, however, the approximation is fundamentally an empirical construct. It is a rule

of thumb describing the approximate outcome of an extremely complicated physical interaction. The strength of the approximation is its simplicity and versatility. Though in general the relationship between normal force and frictional force is not exactly linear (and so the frictional force is not entirely independent of the contact area of the surfaces), the Coulomb approximation is an adequate representation of friction for the analysis of many physical systems.

When the surfaces are conjoined, Coulomb friction becomes a very poor approximation (for example, adhesive tape resists sliding even when there is no normal force, or a negative normal force). In this case, the frictional force may depend strongly on the area of contact. Some drag racing tires are adhesive for this reason. However, despite the complexity of the fundamental physics behind friction, the relationships are accurate enough to be useful in many applications.

"Negative" coefficient of friction

As of 2012, a single study has demonstrated the potential for an *effectively negative coefficient of friction in the low-load regime*, meaning that a decrease in normal force leads to an increase in friction. This contradicts everyday experience in which an increase in normal force leads to an increase in friction.[35] This was reported in the journal *Nature* in October 2012 and involved the friction encountered by an atomic force microscope stylus when dragged across a graphene sheet in the presence of graphene-adsorbed oxygen.[35]

25.3.8 Numerical simulation of the Coulomb model

Despite being a simplified model of friction, the Coulomb model is useful in many numerical simulation applications such as multibody systems and granular material. Even its most simple expression encapsulates the fundamental effects of sticking and sliding which are required in many applied cases, although specific algorithms have to be designed in order to efficiently numerically integrate mechanical systems with Coulomb friction and bilateral and/or unilateral contact.[36][37][38][39][40] Some quite nonlinear effects, such as the so-called Painlevé paradoxes, may be encountered with Coulomb friction.[41]

25.3.9 Dry friction and instabilities

Dry friction can induce several types of instabilities in mechanical systems which display a stable behaviour in the absence of friction.[42] These instabilities may be caused by the decrease of the friction force with an increasing velocity of sliding, by material expansion due to heat generation during friction (the thermo-elastic instabilities), or by pure dynamic effects of sliding of two elastic materials (the Adams-Martins instabilities). The latter were originally discovered in 1995 by George G. Adams and João Arménio Correia Martins for smooth surfaces [43][44] and were later found in periodic rough surfaces.[45] In particular, friction-related dynamical instabilities are thought to be responsible for brake squeal and the 'song' of a glass harp,[46][47] phenomena which involve stick and slip, modelled as a drop of friction coefficient with velocity.[48]

A practically important case is the self-oscillation of the strings of bowed instruments such as the violin, cello, hurdy-gurdy, erhu etc.

A connection between dry friction and flutter instability in a simple mechanical system has been discovered,[49] watch the movie for more details.

Frictional instabilities can lead to the formation of new self-organized patterns (or "secondary structures") at the sliding interface, such as in-situ formed tribofilms which are utilized for the reduction of friction and wear in so-called self-lubricating materials.[50]

25.4 Fluid friction

Main article: Viscosity

Fluid friction occurs between fluid layers that are moving relative to each other. This internal resistance to flow is named *viscosity*. In everyday terms, the viscosity of a fluid is described as its "thickness". Thus, water is "thin", having a lower viscosity, while honey is "thick", having a higher viscosity. The less viscous the fluid, the greater its ease of deformation or movement.

All real fluids (except superfluids) offer some resistance to shearing and therefore are viscous. For teaching and explanatory purposes it is helpful to use the concept of an inviscid fluid or an ideal fluid which offers no resistance to shearing and so is not viscous.

25.5 Lubricated friction

Main article: Lubrication

Lubricated friction is a case of fluid friction where a fluid separates two solid surfaces. Lubrication is a technique employed to reduce wear of one or both surfaces in close prox-

imity moving relative to each another by interposing a substance called a lubricant between the surfaces.

In most cases the applied load is carried by pressure generated within the fluid due to the frictional viscous resistance to motion of the lubricating fluid between the surfaces. Adequate lubrication allows smooth continuous operation of equipment, with only mild wear, and without excessive stresses or seizures at bearings. When lubrication breaks down, metal or other components can rub destructively over each other, causing heat and possibly damage or failure.

25.6 Skin friction

Main article: Parasitic drag

Skin friction arises from the interaction between the fluid and the skin of the body, and is directly related to the area of the surface of the body that is in contact with the fluid. Skin friction follows the drag equation and rises with the square of the velocity.

Skin friction is caused by viscous drag in the boundary layer around the object. There are two ways to decrease skin friction: the first is to shape the moving body so that smooth flow is possible, like an airfoil. The second method is to decrease the length and cross-section of the moving object as much as is practicable.

25.7 Internal friction

Main article: Plastic deformation of solids
See also: Deformation (engineering)

Internal friction is the force resisting motion between the elements making up a solid material while it undergoes deformation.

Plastic deformation in solids is an irreversible change in the internal molecular structure of an object. This change may be due to either (or both) an applied force or a change in temperature. The change of an object's shape is called strain. The force causing it is called stress.

Elastic deformation in solids is reversible change in the internal molecular structure of an object. Stress does not necessarily cause permanent change. As deformation occurs, internal forces oppose the applied force. If the applied stress is not too large these opposing forces may completely resist the applied force, allowing the object to assume a new equilibrium state and to return to its original shape when the force is removed. This is known as elastic deformation or elasticity.

25.8 Other types of friction

25.8.1 Rolling resistance

Main article: Rolling resistance

Rolling resistance is the force that resists the rolling of a wheel or other circular object along a surface caused by deformations in the object and/or surface. Generally the force of rolling resistance is less than that associated with kinetic friction.[51] Typical values for the coefficient of rolling resistance are 0.001.[52] One of the most common examples of rolling resistance is the movement of motor vehicle tires on a road, a process which generates heat and sound as by-products.[53]

25.8.2 Braking friction

Any wheel equipped with a brake is capable of generating a large retarding force, usually for the purpose of slowing and stopping a vehicle or piece of rotating machinery. Braking friction differs from rolling friction because the coefficient of friction for rolling friction is small whereas the coefficient of friction for braking friction is designed to be large by choice of materials for brake pads.

25.8.3 Triboelectric effect

Main article: Triboelectric effect

Rubbing dissimilar materials against one another can cause a build-up of electrostatic charge, which can be hazardous if flammable gases or vapours are present. When the static build-up discharges, explosions can be caused by ignition of the flammable mixture.

25.8.4 Belt friction

Main article: Belt friction

Belt friction is a physical property observed from the forces acting on a belt wrapped around a pulley, when one end is being pulled. The resulting tension, which acts on both ends of the belt, can be modeled by the belt friction equation.

In practice, the theoretical tension acting on the belt or rope calculated by the belt friction equation can be compared to

the maximum tension the belt can support. This helps a designer of such a rig to know how many times the belt or rope must be wrapped around the pulley to prevent it from slipping. Mountain climbers and sailing crews demonstrate a standard knowledge of belt friction when accomplishing basic tasks.

25.9 Reducing friction

25.9.1 Devices

Devices such as wheels, ball bearings, roller bearings, and air cushion or other types of fluid bearings can change sliding friction into a much smaller type of rolling friction.

Many thermoplastic materials such as nylon, HDPE and PTFE are commonly used in low friction bearings. They are especially useful because the coefficient of friction falls with increasing imposed load. For improved wear resistance, very high molecular weight grades are usually specified for heavy duty or critical bearings.

25.9.2 Lubricants

A common way to reduce friction is by using a lubricant, such as oil, water, or grease, which is placed between the two surfaces, often dramatically lessening the coefficient of friction. The science of friction and lubrication is called tribology. Lubricant technology is when lubricants are mixed with the application of science, especially to industrial or commercial objectives.

Superlubricity, a recently discovered effect, has been observed in graphite: it is the substantial decrease of friction between two sliding objects, approaching zero levels. A very small amount of frictional energy would still be dissipated.

Lubricants to overcome friction need not always be thin, turbulent fluids or powdery solids such as graphite and talc; acoustic lubrication actually uses sound as a lubricant.

Another way to reduce friction between two parts is to superimpose micro-scale vibration to one of the parts. This can be sinusoidal vibration as used in ultrasound-assisted cutting or vibration noise, known as dither.

25.10 Energy of friction

According to the law of conservation of energy, no energy is destroyed due to friction, though it may be lost to the system of concern. Energy is transformed from other forms into thermal energy. A sliding hockey puck comes to rest because friction converts its kinetic energy into heat which raises the thermal energy of the puck and the ice surface. Since heat quickly dissipates, many early philosophers, including Aristotle, wrongly concluded that moving objects lose energy without a driving force.

When an object is pushed along a surface along a path C, the energy converted to heat is given by a line integral, in accordance with the definition of work

$$E_{th} = \int_C \mathbf{F}_{\text{fric}}(\mathbf{x}) \cdot d\mathbf{x} = \int_C \mu_k \, \mathbf{F}_n(\mathbf{x}) \cdot d\mathbf{x},$$

where

> $\mathbf{F}_{\text{fric}}$ is the friction force,
>
> $\mathbf{F}_n$ is the vector obtained by multiplying the magnitude of the normal force by a unit vector pointing *against* the object's motion,
>
> μ_k is the coefficient of kinetic friction, which is inside the integral because it may vary from location to location (e.g. if the material changes along the path),
>
> $\mathbf{x}$ is the position of the object.

Energy lost to a system as a result of friction is a classic example of thermodynamic irreversibility.

25.10.1 Work of friction

In the reference frame of the interface between two surfaces, static friction does *no* work, because there is never displacement between the surfaces. In the same reference frame, kinetic friction is always in the direction opposite the motion, and does *negative* work.[54] However, friction can do *positive* work in certain frames of reference. One can see this by placing a heavy box on a rug, then pulling on the rug quickly. In this case, the box slides backwards relative to the rug, but moves forward relative to the frame of reference in which the floor is stationary. Thus, the kinetic friction between the box and rug accelerates the box in the same direction that the box moves, doing *positive* work.[55]

The work done by friction can translate into deformation, wear, and heat that can affect the contact surface properties (even the coefficient of friction between the surfaces). This can be beneficial as in polishing. The work of friction is used to mix and join materials such as in the process of friction welding. Excessive erosion or wear of mating sliding surfaces occurs when work due to frictional forces rise to unacceptable levels. Harder corrosion particles caught between mating surfaces in relative motion (fretting) exacerbates wear of frictional forces. Bearing seizure or failure may result from excessive wear due to work of friction. As surfaces are worn by work due to friction, fit and surface finish of an object may degrade until it no longer functions properly.[56]

25.11 Applications

Friction is an important factor in many engineering disciplines.

25.11.1 Transportation

- Automobile brakes inherently rely on friction, slowing a vehicle by converting its kinetic energy into heat. Incidentally, dispersing this large amount of heat safely is one technical challenge in designing brake systems.

- Rail adhesion refers to the grip wheels of a train have on the rails, see Frictional contact mechanics.

- Road slipperiness is an important design and safety factor for automobiles

  - Split friction is a particularly dangerous condition arising due to varying friction on either side of a car.

  - Road texture affects the interaction of tires and the driving surface.

25.11.2 Measurement

- A tribometer is an instrument that measures friction on a surface.

- A profilograph is a device used to measure pavement surface roughness.

25.11.3 Household usage

- Friction is used to heat and ignite matchsticks (friction between the head of a matchstick and the rubbing surface of the match box).

25.12 See also

- Contact dynamics
- Contact mechanics

- Factor of adhesion

- Frictionless plane

- Galling

- Non-smooth mechanics

- Stick-slip phenomenon

- Transient friction loading

- Triboelectric effect

- Unilateral contact

- Friction torque

25.13   References

[1] http://www.merriam-webster.com/dictionary/friction

[2] Beer, Ferdinand P.; E. Russel Johnston, Jr. (1996). *Vector Mechanics for Engineers* (Sixth ed.). McGraw-Hill. p. 397. ISBN 0-07-297688-8.

[3] Meriam, J. L.; L. G. Kraige (2002). *Engineering Mechanics* (fifth ed.). John Wiley & Sons. p. 328. ISBN 0-471-60293-0.

[4] Ruina, Andy; Rudra Pratap (2002). *Introduction to Statics and Dynamics* (PDF). Oxford University Press. p. 713.

[5] Hibbeler, R. C. (2007). *Engineering Mechanics* (Eleventh ed.). Pearson, Prentice Hall. p. 393. ISBN 0-13-127146-6.

[6] Soutas-Little, Robert W.; Inman, Balint (2008). *Engineering Mechanics*. Thomson. p. 329. ISBN 0-495-29610-4.

[7] Dowson, Duncan (1997). *History of Tribology, 2nd Edition*. Professional Engineering Publishing. ISBN 1-86058-070-X.

[8] Armstrong-Hélouvry, Brian (1991). *Control of machines with friction*. USA: Springer. p. 10. ISBN 0-7923-9133-0.

[9] van Beek, Anton. "History of Science Friction". tribology-abc.com. Retrieved 2011-03-24.

[10] "Leonardo da Vinci's studies of friction" (PDF).

[11] Hutchings, Ian M. (2016-08-15). "Leonardo da Vinci´s studies of friction". *Wear*. 360–361: 51–66. doi:10.1016/j.wear.2016.04.019.

[12] "Study reveals Leonardo da Vinci's 'irrelevant' scribbles mark the spot where he first recorded the laws of friction". Retrieved 2016-07-26.

[13] Forest de Bélidor, Bernard. "Richtige Grund-Sätze der Friction-Berechnung" ("Correct Basics of Friction Calculation"), 1737, (in German)

[14] "Leonhard Euler". *Friction Module*. Nano World. 2002. Retrieved 2011-03-25.

[15] Fleeming Jenkin & James Alfred Ewing (1877) "On Friction between Surfaces moving at Low Speeds", *Philosophical Magazine* Series 5, volume 4, pp 308–10; link from Biodiversity Heritage Library

[16] "Friction Factors - Coefficients of Friction". Retrieved 2015-04-27.

[17] "Ultra-low friction coefficient in alumina–silicon nitride pair lubricated with water". *Wear*. **296**: 656–659. doi:10.1016/j.wear.2012.07.030. Retrieved 2015-04-27.

[18] Tian, Y.; Bastawros, A. F.; Lo, C. C. H.; Constant, A. P.; Russell, A.M.; Cook, B. A. (2003). "Superhard self-lubricating AlMgB[sub 14] films for microelectromechanical devices". *Applied Physics Letters*. **83** (14): 2781. Bibcode:2003ApPhL..83.2781T. doi:10.1063/1.1615677.

[19] Kleiner, Kurt (2008-11-21). "Material slicker than Teflon discovered by accident". Retrieved 2008-12-25.

[20] Higdon, C.; Cook, B.; Harringa, J.; Russell, A.; Goldsmith, J.; Qu, J.; Blau, P. (2011). "Friction and wear mechanisms in AlMgB14-TiB2 nanocoatings". *Wear*. **271** (9–10): 2111–2115. doi:10.1016/j.wear.2010.11.044.

[21] Coefficient of Friction. EngineersHandbook.com

[22] "Coefficients of Friction of Human Joints". Retrieved 2015-04-27.

[23] "The Engineering Toolbox: Friction and Coefficients of Friction". Retrieved 2008-11-23.

[24] Dienwiebel, Martin; et al. (2004). "Superlubricity of Graphite" (PDF). *Phys. Rev. Lett.* **92** (12): 126101. Bibcode:2004PhRvL..92l6101D. doi:10.1103/PhysRevLett.92.126101.

[25] Bhavikatti, S. S.; K. G. Rajashekarappa (1994). *Engineering Mechanics*. New Age International. p. 112. ISBN 978-81-224-0617-7. Retrieved 2007-10-21.

[26] Sheppard, Sheri; Tongue, Benson H.; Anagnos, Thalia (2005). *Statics: Analysis and Design of Systems in Equilibrium*. Wiley and Sons. p. 618. ISBN 0-471-37299-4. In general, for given contacting surfaces, $\mu_k < \mu_s$

[27] Meriam, James L.; Kraige, L. Glenn; Palm, William John (2002). *Engineering Mechanics: Statics*. Wiley and Sons. p. 330. ISBN 0-471-40646-5. Kinetic friction force is usually somewhat less than the maximum static friction force.

[28] Feynman, Richard P.; Leighton, Robert B.; Sands, Matthew (1964). "The Feynman Lectures on Physics, Vol. I, p. 12-5". Addison-Wesley. Retrieved 2009-10-16.

[29] Persson, B. N.; Volokitin, A. I (2002). "Theory of rubber friction: Nonstationary sliding". *Physical Review B*. **65** (13): 134106. Bibcode:2002PhRvB..65m4106P. doi:10.1103/PhysRevB.65.134106.

[30] Beatty, William J. "Recurring science misconceptions in K-6 textbooks". Retrieved 2007-06-08.

[31] Persson, B. N. J. (2000). *Sliding friction: physical principles and applications*. Springer. ISBN 978-3-540-67192-3. Retrieved 2016-01-23.

[32] Makkonen, L (2012). "A thermodynamic model of sliding friction". *AIP Advances*. **2**: 012179. doi:10.1063/1.3699027.

[33] Nichols, Edward Leamington; Franklin, William Suddards (1898). *The Elements of Physics*. **1**. Macmillan. p. 101.

[34] Ternes, Markus; Lutz, Christopher P.; Hirjibehedin, Cyrus F.; Giessibl, Franz J.; Heinrich, Andreas J. (2008-02-22). "The Force Needed to Move an Atom on a Surface". *Science*. **319** (5866): 1066–1069. Bibcode:2008Sci...319.1066T. doi:10.1126/science.1150288. PMID 18292336.

[35] Deng, Zhao; et al. (October 14, 2012), "Adhesion-dependent negative friction coefficient on chemically modified graphite at the nanoscale", *Nature*, Bibcode:2012NatMa..11.1032D, doi:10.1038/nmat3452, retrieved November 18, 2012, lay summary – *R&D Magazine* (October 17, 2012)

[36] Haslinger, J.; Nedlec, J.C. (1983). "Approximation of the Signorini problem with friction, obeying the Coulomb law". *Mathematical Methods in the Applied Sciences*. **5**: 422–437. Bibcode:1983MMAS....5..422H. doi:10.1002/mma.1670050127.

[37] Alart, P.; Curnier, A. (1991). "A mixed formulation for frictional contact problems prone to Newton like solution method". *Computer Methods in Applied Mechanics and Engineering*. **92** (3): 353–375. Bibcode:1991CMAME..92..353A. doi:10.1016/0045-7825(91)90022-X.

[38] Acary, V.; Cadoux, F.; Lemaréchal, C.; Malick, J. (2011). "A formulation of the linear discrete Coulomb friction problem via convex optimization". *Journal of Applied Mathematics and Mechanics / Zeitschrift für Angewandte Mathematik und Mechanik*. **91** (2): 155–175. Bibcode:2011ZaMM...91..155A. doi:10.1002/zamm.201000073.

[39] De Saxcé, G.; Feng, Z.-Q. (1998). "The bipotential method: A constructive approach to design the complete contact law with friction and improved numerical algorithms". *Mathematical and Computer Modelling*. **28** (4): 225–245. doi:10.1016/S0895-7177(98)00119-8.

[40] Simo, J.C.; Laursen, T.A. (1992). "An augmented lagrangian treatment of contact problems involving friction". *Computers and Structures*. **42** (2): 97–116. doi:10.1016/0045-7949(92)90540-G.

[41] Acary, V.; Brogliato, B. (2008). *Numerical Methods for Nonsmooth Dynamical Systems. Applications in Mechanics and Electronics*. **35**. Springer Verlag Heidelberg.

[42] Bigoni, D. *Nonlinear Solid Mechanics: Bifurcation Theory and Material Instability*. Cambridge University Press, 2012. ISBN 9781107025417.

[43] Adams, G. G. (1995). "Self-excited oscillations of two elastic half-spaces sliding with a constant coefficient of friction". *Journal of Applied Mechanics*. **62**: 867–872. Bibcode:1995JAM....62..867A. doi:10.1115/1.2896013.

[44] Martins, J.A., Faria, L.O. & Guimarães, J. (1995). "Dynamic surface solutions in linear elasticity and viscoelasticity with frictional boundary conditions". *Journal of Vibration and Acoustics*. **117**: 445–451. doi:10.1115/1.2874477.

[45] M, Nosonovsky,; G., Adams G. (2004). "Vibration and stability of frictional sliding of two elastic bodies with a wavy contact interface". *Journal of Applied Mechanics*. **71**: 154–161. Bibcode:2004JAM....71..154N. doi:10.1115/1.1653684.

[46] J., Flint,; J., Hultén, (2002). "Lining-deformation-induced modal coupling as squeal generator in a distributed parameter disk brake model". *J. Sound and Vibration*. **254**: 1–21. Bibcode:2002JSV...254....1F. doi:10.1006/jsvi.2001.4052.

[47] M., Kröger,; M., Neubauer,; K., Popp, (2008). "Experimental investigation on the avoidance of self-excited vibrations". *Phil. Trans. R. Soc. A*. **366** (1866): 785–810. Bibcode:2008RSPTA.366..785K. doi:10.1098/rsta.2007.2127. PMID 17947204.

[48] R., Rice, J.; L., Ruina, A. (1983). "Stability of Steady Frictional Slipping" (PDF). *Journal of Applied Mechanics*. **50** (2): 343–349. Bibcode:1983JAM....50..343R. doi:10.1115/1.3167042.

[49] Bigoni, D.; Noselli, G. (2011). "Experimental evidence of flutter and divergence instabilities induced by dry friction". *Journal of the Mechanics and Physics of Solids*. **59** (10): 2208–2226. Bibcode:2011JMPSo..59.2208B. doi:10.1016/j.jmps.2011.05.007.

[50] Nosonovsky, Michael (2013). *Friction-Induced Vibrations and Self-Organization: Mechanics and Non-Equilibrium Thermodynamics of Sliding Contact*. CRC Press. p. 333. ISBN 978-1466504011.

[51] Silliman, Benjamin (1871) *Principles of Physics, Or Natural Philosophy*, Ivison, Blakeman, Taylor & company publishers

[52] Butt, Hans-Jürgen; Graf, Karlheinz and Kappl, Michael (2006) *Physics and Chemistry of Interfaces*, Wiley, ISBN 3-527-40413-9

[53] Hogan, C. Michael (1973). "Analysis of highway noise". *Water, Air, & Soil Pollution*. **2** (3): 387–392. doi:10.1007/BF00159677.

[54] Den Hartog, J. P. (1961). *Mechanics*. Courier Dover Publications. p. 142. ISBN 0-486-60754-2.

[55] Leonard, William J (2000). *Minds-on Physics*. Kendall/Hunt. p. 603. ISBN 0-7872-3932-1.

[56] Bayer, Raymond George (2004). *Mechanical wear*. CRC Press. pp. 1, 2. ISBN 0-8247-4620-1. Retrieved 2008-07-07.

25.14 External links

- "Friction". *Encyclopædia Britannica*. **11** (11th ed.). 1911.

- Coefficients of Friction – tables of coefficients, plus many links

- Physclips: Mechanics with animations and video clips from the University of New South Wales

- CRC Handbook of Chemistry & Physics – Values for Coefficient of Friction

- Characteristic Phenomena in Conveyor Chain

- Atomic-scale Friction Research and Education Synergy Hub (AFRESH) an Engineering Virtual Organization for the atomic-scale friction community to share, archive, link, and discuss data, knowledge and tools related to atomic-scale friction.

Chapter 26

Shear force

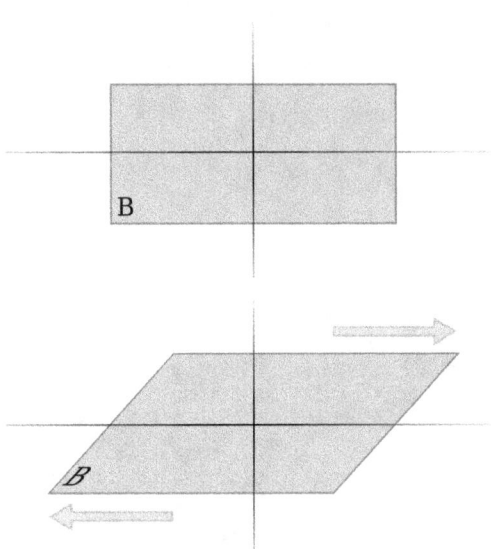

Shearing forces push in one direction at the top, and the opposite direction at the bottom, causing shearing deformation.

Further information: Shear stress

Shearing forces are unaligned forces pushing one part of a body in one direction, and another part of the body in the opposite direction. When the forces are aligned into each other, they are called compression forces. An example is a deck of cards being pushed one way on the top, and the other at the bottom, causing the cards to slide. Another example is when wind blows at the side of a peaked roof of a home - the side walls experience a force at their top pushing in the direction of the wind, and their bottom in the opposite direction, from the ground or foundation. William A. Nash defines shear force in terms of planes: "If a plane is passed through a body, a force acting along this plane is called a *shear force* or *shearing force*."[1]

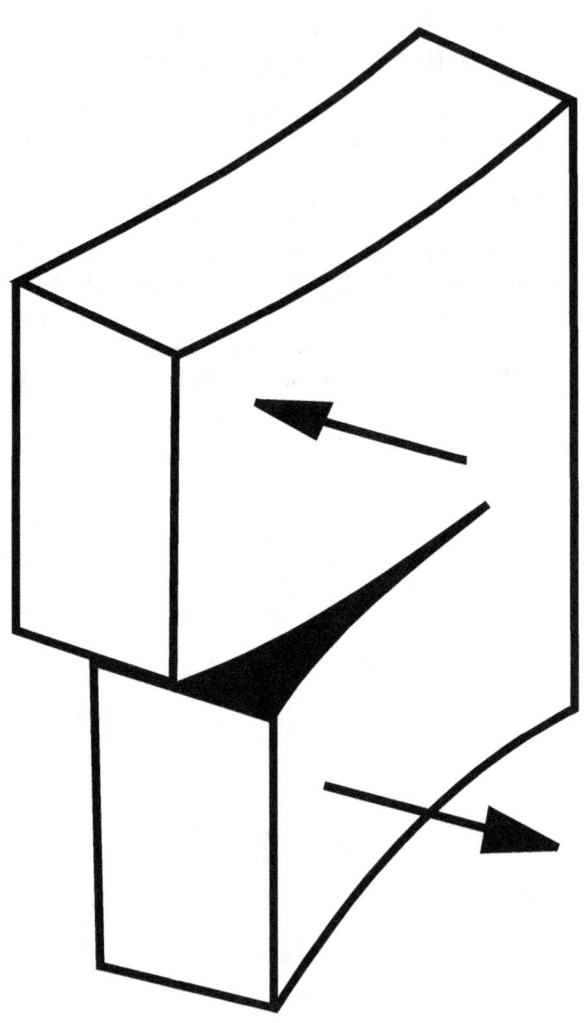

A crack or tear may develop in a body from parallel shearing forces pushing in opposite directions at different points of the body. If the forces were aligned and aimed straight into each other, they would pinch or compress the body, rather than tear or crack it.

26.1 Shear force of steel and bolts

Here follows a short example of how to work out the shear force of a piece of steel. The factor of 0.6 used to change from tensile to shear force could vary from 0.58 - 0.62 and will depend on application.

Steel called EN8 bright has a tensile strength of 800 MPa and Mild steel has a tensile strength of 400 MPa.

To work out the force to shear a 25 mm diameter round steel EN8 bright;

Area of the 25 mm round steel in mm^2 = $(12.5^2)(\pi) = 490.8\ mm^2$

$0.8\ kN/mm^2 \times 490.8\ mm^2 = 392.64\ kN = 40\ ton$ $\times\ 0.6$ (to change force from tensile to shear) = 24 ton

When working with a bolted joint, the strength comes from friction between the materials bolted together. Bolts are correctly torqued to maintain the friction. The shear force only becomes relevant when the bolts are not torqued.

A bolt with property class 12.9 has a tensile strength of 1200 MPa (1 MPa = 1 N/mm^2) or 1.2 kN/mm^2 and the yield strength is 0.90 times tensile strength, 1080 MPa in this case.

A bolt with property class 4.6 has a tensile strength of 400 MPa (1 MPa = 1 N/mm^2) or 0.4 kN/mm^2 and yield strength is 0.60 times tensile strength, 240 MPa in this case.

In case of fastners, proof load is specified as it gives a real life picture about the characteristics of the bolt.

26.2 See also

- ASTM F568M, shear force resistance
- Cantilever method

26.3 References

[1] William A. Nash (1 July 1998). *Schaum's Outline of Theory and Problems of Strength of Materials*. McGraw-Hill Professional. p. 82. ISBN 978-0-07-046617-3. Retrieved 20 May 2012.

Chapter 27

Tidal force

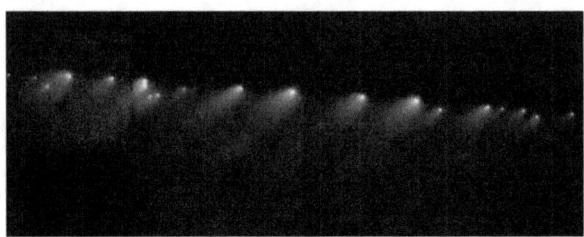

Figure 1: Comet Shoemaker-Levy 9 in 1994 after breaking up under the influence of Jupiter's tidal forces during a previous pass in 1992.

The **tidal force** is a secondary effect of the force of gravity and is responsible for the tides. It arises because the gravitational force exerted by one body on another is not constant across it; the nearest side is attracted more strongly than the farthest side. Thus, the tidal force is differential. Consider the gravitational attraction of the moon on the oceans nearest to the moon, the solid Earth and the oceans farthest from the moon. There is a mutual attraction between the moon and the solid earth which can be considered to act on its centre of mass. However, the near oceans are more strongly attracted and, since they are fluid, they approach the moon slightly, causing a high tide. The far oceans are attracted less. The attraction on the far-side oceans could be expected to cause a low tide but since the solid earth is attracted (accelerated) more strongly towards the moon, there is a *relative* acceleration of those waters in the outwards direction. Viewing the Earth as a whole, we see that all its mass experiences a mutual attraction with that of the moon but the near oceans more so than the far oceans, leading to a separation of the two.

In a more general usage in celestial mechanics, the expression 'tidal force' can refer to a situation in which a body or material (for example, tidal water) is mainly under the gravitational influence of a second body (for example, the Earth), but is also perturbed by the gravitational effects of a third body (for example, the Moon). The perturbing force is sometimes in such cases called a tidal force[1] (for example, the perturbing force on the Moon): it is the difference between the force exerted by the third body on the second and the force exerted by the third body on the first.[2]

27.1 Explanation

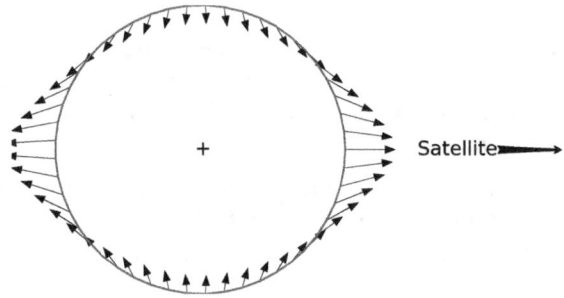

Figure 2: The Moon's gravity differential field at the surface of the Earth is known (along with another and weaker differential effect due to the Sun) as the Tide Generating Force. This is the primary mechanism driving tidal action, explaining two tidal equipotential bulges, and accounting for two high tides per day. In this figure, the Earth is the central blue circle while the Moon is far off to the right. The **outward** direction of the arrows on the right and left indicates that where the Moon is overhead (or at the nadir) its perturbing force opposes that between the earth and ocean.

When a body (body 1) is acted on by the gravity of another body (body 2), the field can vary significantly on body 1 between the side of the body facing body 2 and the side facing away from body 2. Figure 2 shows the differential force of gravity on a spherical body (body 1) exerted by another body (body 2). These so-called *tidal forces* cause strains on both bodies and may distort them or even, in extreme cases, break one or the other apart.[3] The Roche limit is the distance from a planet at which tidal effects would cause an object to disintegrate because the differential force of gravity from the planet overcomes the attraction of the parts of the object for one another.[4] These strains would not occur if the gravitational field were uniform, because a uniform field only causes the entire body to accelerate together in the same direction and at the same rate.

27.2   Effects of tidal forces

Figure 3: Saturn's rings are inside the orbits of its principal moons. Tidal forces oppose gravitational coalescence of the material in the rings to form moons.[5]

In the case of an infinitesimally small elastic sphere, the effect of a tidal force is to distort the shape of the body without any change in volume. The sphere becomes an ellipsoid with two bulges, pointing towards and away from the other body. Larger objects distort into an ovoid, and are slightly compressed, which is what happens to the Earth's oceans under the action of the Moon. The Earth and Moon rotate about their common center of mass or barycenter, and their gravitational attraction provides the centripetal force necessary to maintain this motion. To an observer on the Earth, very close to this barycenter, the situation is one of the Earth as body 1 acted upon by the gravity of the Moon as body 2. All parts of the Earth are subject to the Moon's gravitational forces, causing the water in the oceans to redistribute, forming bulges on the sides near the Moon and far from the Moon.[6]

When a body rotates while subject to tidal forces, internal friction results in the gradual dissipation of its rotational kinetic energy as heat. If the body is close enough to its primary, this can result in a rotation which is tidally locked to the orbital motion, as in the case of the Earth's moon. Tidal heating produces dramatic volcanic effects on Jupiter's moon Io. Stresses caused by tidal forces also cause a regular monthly pattern of moonquakes on Earth's Moon.

Tidal forces contribute to ocean currents, which moderate global temperatures by transporting heat energy toward the poles. It has been suggested that in addition to other factors, harmonic beat variations in tidal forcing may contribute to climate changes. However, no strong link has been found to date.[7]

Tidal effects become particularly pronounced near small bodies of high mass, such as neutron stars or black holes,

where they are responsible for the "spaghettification" of infalling matter. Tidal forces create the oceanic tide of Earth's oceans, where the attracting bodies are the Moon and, to a lesser extent, the Sun. Tidal forces are also responsible for tidal locking, tidal acceleration, and tidal heating.

By generating conducting fluids within the interior of the Earth, tidal forces also affect the Earth's magnetic field.[8]

27.3   Mathematical treatment

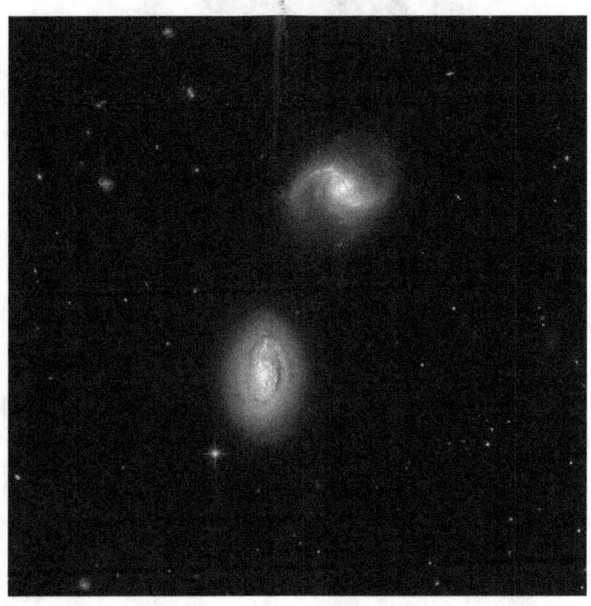

Tidal force is responsible for the merge of galactic pair MRK 1034.[9]

For a given (externally generated) gravitational field, the **tidal acceleration** at a point with respect to a body is obtained by vectorially subtracting the gravitational acceleration at the center of the body (due to the given externally generated field) from the gravitational acceleration (due to the same field) at the given point. Correspondingly, the term *tidal force* is used to describe the forces due to tidal acceleration. Note that for these purposes the only gravitational field considered is the external one; the gravitational field of the body (as shown in the graphic) is not relevant. (In other words, the comparison is with the conditions at the given point as they would be if there were no externally generated field acting unequally at the given point and at the center of the reference body. The externally generated field is usually that produced by a perturbing third body, often the Sun or the Moon in the frequent example-cases of points on or above the Earth's surface in a geocentric reference frame.)

Tidal acceleration does not require rotation or orbiting bodies; for example, the body may be freefalling in a straight

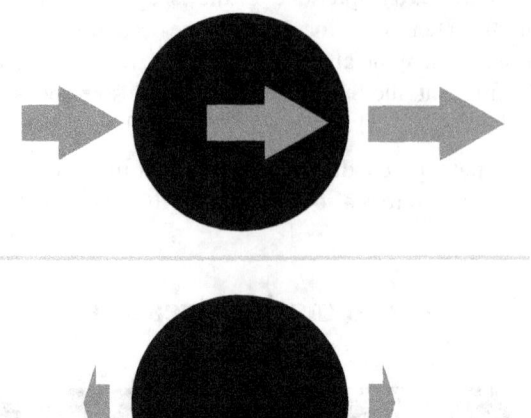

Figure 4: Graphic of tidal forces. The top picture shows the gravity field of a body to the right, the lower shows their residual once the field at the centre of the sphere is subtracted; this is the tidal force. See Figure 2 for a more detailed version

line under the influence of a gravitational field while still being influenced by (changing) tidal acceleration.

By Newton's law of universal gravitation and laws of motion, a body of mass m at distance R from the center of a sphere of mass M feels a force $\vec{F}_g$,

$$\vec{F}_g = -\hat{r}\, G\, \frac{Mm}{R^2}$$

equivalent to an acceleration $\vec{a}_g$,

$$\vec{a}_g = -\hat{r}\, G\, \frac{M}{R^2}$$

where $\hat{r}$ is a unit vector pointing from the body M to the body m (here, acceleration from m towards M has negative sign).

Consider now the acceleration due to the sphere of mass M experienced by a particle in the vicinity of the body of mass m. With R as the distance from the center of M to the center of m, let Δr be the (relatively small) distance of the particle from the center of the body of mass m. For simplicity, distances are first considered only in the direction pointing towards or away from the sphere of mass M. If the body of mass m is itself a sphere of radius Δr, then the new particle considered may be located on its surface, at a distance $(R \pm \Delta r)$ from the centre of the sphere of mass M, and Δr may be taken as positive where the particle's distance from

M is greater than R. Leaving aside whatever gravitational acceleration may be experienced by the particle towards m on account of m's own mass, we have the acceleration on the particle due to gravitational force towards M as:

$$\vec{a}_g = -\hat{r}\, G\, \frac{M}{(R \pm \Delta r)^2}$$

Pulling out the R^2 term from the denominator gives:

$$\vec{a}_g = -\hat{r}\, G\, \frac{M}{R^2}\, \frac{1}{(1 \pm \Delta r/R)^2}$$

The Maclaurin series of $1/(1 \pm x)^2$ is $1 \mp 2x + 3x^2 \mp \cdots$ which gives a series expansion of:

$$\vec{a}_g = -\hat{r}\, G\, \frac{M}{R^2} \pm \hat{r}\, G\, \frac{2M}{R^2}\, \frac{\Delta r}{R} + \cdots$$

The first term is the gravitational acceleration due to M at the center of the reference body m, i.e., at the point where Δr is zero. This term does not affect the observed acceleration of particles on the surface of m because with respect to M, m (and everything on its surface) is in free fall. When the force on the far particle is subtracted from the force on the near particle, this first term cancels, as do all other even-order terms. The remaining (residual) terms represent the difference mentioned above and are tidal force (acceleration) terms. When Δr is small compared to R, the terms after the first residual term are very small and can be neglected, giving the approximate tidal acceleration $\vec{a}_t$ (axial) for the distances Δr considered, along the axis joining the centers of m and M:

$$\vec{a}_t \text{ (axial)} \quad \approx \quad \pm\, \hat{r}\, 2\Delta r\, G\, \frac{M}{R^3}$$

When calculated in this way for the case where Δr is a distance along the axis joining the centers of m and M, $\vec{a}_t$ is directed outwards from to the center of m (where Δr is zero).

Tidal accelerations can also be calculated away from the axis connecting the bodies m and M, requiring a vector calculation. In the plane perpendicular to that axis, the tidal acceleration is directed inwards (towards the center where Δr is zero), and its magnitude is $|\vec{a}_t \text{ (axial)}|/2$ in linear approximation as in Figure 2.

The tidal accelerations at the surfaces of planets in the Solar System are generally very small. For example, the lunar tidal acceleration at the Earth's surface along the Moon-Earth axis is about 1.1×10^{-7} g, while the solar tidal acceleration at the Earth's surface along the Sun-Earth axis is about 0.52×10^{-7} g, where g is the gravitational acceleration at the Earth's surface. Hence the tide-raising force (acceleration) due to the Sun is about 45% of that due to the Moon.[10] The solar tidal acceleration at the Earth's surface was first given by Newton in the *Principia*.[11]

27.4   See also

- Amphidromic point

- Galactic tide

- Tidal resonance

- Spacetime curvature

27.5   References

[1] "On the tidal force", I N Avsiuk, in "Soviet Astronomy Letters", vol.3 (1977), pp. 96–99

[2] See p.509 in "Astronomy: a physical perspective", M L Kutner (2003).

[3] R Penrose (1999). *The Emperor's New Mind: Concerning Computers, Minds, and the Laws of Physics*. Oxford University Press. p. 264. ISBN 0-19-286198-0.

[4] Thérèse Encrenaz; J -P Bibring; M Blanc (2003). *The Solar System*. Springer. p. 16. ISBN 3-540-00241-3.

[5] R. S. MacKay; J. D. Meiss (1987). *Hamiltonian Dynamical Systems: A Reprint Selection*. CRC Press. p. 36. ISBN 0-85274-205-3.

[6] Rollin A Harris (1920). *The Encyclopedia Americana: A Library of Universal Knowledge*. **26**. Encyclopedia Americana Corp. pp. 611–617.

[7] "Millennial Climate Variability: Is There a Tidal Connection?".

[8] "Hungry for Power in Space". *New Scienctist*. New Science Pub. **123**: 52. 23 September 1989. Retrieved 14 March 2016.

[9] "Inseparable galactic twins". *ESA/Hubble Picture of the Week*. Retrieved 12 July 2013.

[10] The Admiralty (1987). *Admiralty manual of navigation*. **1**. The Stationery Office. p. 277. ISBN 0-11-772880-2., Chapter 11, p. 277

[11] Newton, Isaac (1729). *The mathematical principles of natural philosophy*. **2**. p. 307. ISBN 0-11-772880-2., Book 3, Proposition 36, Page 307 Newton put the force to depress the sea at places 90 degrees distant from the Sun at "1 to 38604600" (in terms of g), and wrote that the force to raise the sea along the Sun-Earth axis is "twice as great", i.e. 2 to 38604600, which comes to about 0.52×10^{-7} g as expressed in the text.

27.6   External links

- Gravitational Tides by J. Christopher Mihos of Case Western Reserve University

- Audio: Cain/Gay – Astronomy Cast Tidal Forces – July 2007.

- Gray, Meghan; Merrifield, Michael. "Tidal Forces". *Sixty Symbols*. Brady Haran for the University of Nottingham.

- "Pau Amaro Seoane MODEST working group 4 "Tidal disruption of a star by a massive black hole"". Retrieved 2013-05-30.

- Myths about Gravity and Tides by Mikolaj Sawicki of John A. Logan College and the University of Colorado.

- Tidal Misconceptions by Donald E. Simanek

27.7 Text and image sources, contributors, and licenses

27.7.1 Text

- **Force** *Source:* https://en.wikipedia.org/wiki/Force?oldid=724423501 *Contributors:* AxelBoldt, Eloquence, Bryan Derksen, Tarquin, AstroNomer, Seb, Andre Engels, XJaM, William Avery, Heron, Camembert, Ryguasu, Isis~enwiki, Stevertigo, Patrick, Michael Hardy, Tim Starling, Delirium, Ellywa, Cyp, Darkwind, Glenn, Netsnipe, Bassington, Charles Matthews, Jitse Niesen, Saltine, Phys, Thue, Bevo, Mackensen, Jusjih, Finlay McWalter, Phil Boswell, Robbot, Altenmann, F3meyer, Modulatum, Ashley Y, Merovingian, Academic Challenger, Hadal, Rho~enwiki, Tea2min, Alan Liefting, Matt Gies, Giftlite, YanA, Philwelch, Wolfkeeper, Tom harrison, Herbee, Jason Le Vaillant, Michael Devore, Bensaccount, Yekrats, Jorge Stolfi, Adam McMaster, Alan Au, Mooquackwooftweetmeow, Gyrofrog, Woggly, Utcursch, Andycjp, Pcarbonn, Antandrus, Aulis Eskola, Kaldari, Jossi, Karol Langner, Johnflux, RetiredUser2, Icairns, Tail, Jimwilliams57, Sam Hocevar, NoPetrol, Sonett72, Zondor, Mike Rosoft, Mr Bound, ThreeE, CALR, RossPatterson, Discospinster, Guanabot, Pak21, Hidaspal, Pjacobi, Vsmith, Autiger, Wadewitz, Bender235, ESkog, Kbh3rd, Nabla, Pmetzger, Miraceti, Joanjoc~enwiki, Laurascudder, RoyBoy, Femto, CDN99, Bobo192, Army1987, Harley peters, Whosyourjudas, John Vandenberg, Angie Y., Maurreen, I9Q79oL78KiL0QTFHgyc, Kevin Lamoreau, SpeedyGonsales, Kjkolb, Nk, Slipperyweasel, Clyde frogg, Sam Korn, Pearle, Nsaa, Grutness, Alansohn, MrB, Arthena, Mac Davis, Rudchenko, Ynhockey, Velella, ReyBrujo, Count Iblis, Dzhim, W7KyzmJt, Gene Nygaard, HenryLi, Ceyockey, Jakes18, Oleg Alexandrov, Natalya, Nuno Tavares, Velho, Orchew, StradivariusTV, Uncle G, BillC, Rtdrury, Mreult~enwiki, Sholtar, Bbatsell, Zzyzx11, Christopher Thomas, Pfalstad, Dysepsion, Graham87, Zeroparallax, Chun-hian, FreplySpang, RadioActive~enwiki, Jclemens, Enzo Aquarius, Josh Parris, Ketiltrout, Sjö, Sjakkalle, Rjwilmsi, Koavf, Саша Стефановић, Zbxgscqf, Pleiotrop3, Vary, Strait, Lordkinbote, Crazynas, Everton137, SeanMack, Ems57fcva, DoubleBlue, Volfy, Maxim Razin, Sango123, Meok, Ravidreams, FlaBot, Windchaser, Alfred Centauri, RexNL, Ayla, Jrtayloriv, Alphachimp, Sodin, Supersaiyanplough, Srleffler, Kri, Physchim62, Imnotminkus, Chobot, Jaraalbe, DVdm, Ariele, Sanpaz, Gwernol, Roboto de Ajvol, YurikBot, Wavelength, Freedom gundam200, Eraserhead1, Jimp, Bobby1011, RussBot, Crazytales, WAvegetarian, Splash, Jabber-Wok, Yosef1987, Stephenb, Gaius Cornelius, Wimt, Muijzo, NawlinWiki, Kiaparowits, Vanished user 1029384756, Irishguy, Ravedave, Seltar, Vb, CptnMisc, Roy Brumback, Rwalker, DeadEyeArrow, Bota47, Wknight94, Ms2ger, Light current, Enormousdude, Tanjir, Covington, Jwissick, Spondoolicks, Arthur Rubin, Sean Whitton, JoanneB, Urocyon, NielsenGW, Ilmari Karonen, Katieh5584, RG2, Cmglee, Sbyrnes321, Janek Kozicki, Luk, Treesmill, SmackBot, RDBury, KnowledgeOfSelf, Jagged 85, Delldot, Yamaguchi⬚⬚, Gilliam, Kmarinas86, Valley2city, Joefaust, MK8, MalafayaBot, Silly rabbit, Complexica, DHN-bot~enwiki, Sbharris, Colonies Chris, Darth Panda, Bil1, Scott3, Yidisheryid, Pevarnj, Krich, Fuhghettaboutit, Theonlyedge, Savidan, T-borg, MichaelBillington, Dreadstar, Akshaysrinivasan, Rpf, BryanG, DMacks, Andeggs, Synth3t1c, Stefano85, Bidabadi~enwiki, Sadi Carnot, Yevgeny Kats, Srikeit, Sure kr06~enwiki, Zahid Abdassabur, Aïki, John, AmiDaniel, Killahrich, UberCryxic, Mike1901, Loodog, Gobonobo, Warniats, Gang65, Hemmingsen, Nonsuch, Ckatz, Slakr, Timmeh, Beetstra, Peter Horn, Zapvet, KJS77, Dan Gluck, Wizard191, Iridescent, Clarityfiend, Theone00, JoeBot, Shoeofdeath, Casull, JulianDalloway, Courcelles, Chovain, Tawkerbot2, Filelakeshoe, Alexbrewer, Calumm, The Haunted Angel, JForget, CmdrObot, Zakian49, Judyll, Dycedarg, Fieldmarshal Miyagi, RedRollerskate, JohnCD, Usgnus, Kineticman, NickW557, OMGsplosion, ShelfSkewed, Outriggr (2006-2009), MarsRover, Leujohn, Gregbard, AndrewHowse, Badseed, Cydebot, Grahamec, Bvcrist, Xaariz, Dpjoh32, Gogo Dodo, Travelbird, ST47, Rracecarr, Alucard (Dr.), DumbBOT, Inkington, SpK, Zalgo, Gimmetrow, GeneChase, Malleus Fatuorum, Epbr123, Bot-maru, D4g0thur, Pajz, Qwyrxian, LeBofSportif, AndrewDressel, Kahriman~enwiki, N5iln, Headbomb, Marek69, Thljcl, Philippe, D.H, Dawnseeker2000, CTZMSC3, Futurebird, Hmrox, Cyclonenim, AntiVandalBot, Majorly, Luna Santin, Matthew1992, Quintote, Prolog, Gnixon, Paste, Jj137, Postlewaight, Tomasr, Farosdaughter, AubreyEllenShomo, Figma, Canadian-Bacon, JAnDbot, Husond, MER-C, Kprateek88, Fetchcomms, PhilKnight, Savant13, Acroterion, Penubag, Bongwarrior, VoABot II, Wikidudeman, JamesBWatson, Nitpicker06, BigChicken, Cardamon, Cgingold, Allstarecho, David Eppstein, Seckelberry1, MarcusMaximus, Vssun, DerHexer, JaGa, Cyberphysics, Pax:Vobiscum, Charitwo, XandroZ, Wassupwestcoast, MartinBot, Mårten Berglund, Arjun01, Rupesh.ravi, R'n'B, Leyo, Igit~enwiki, LittleOldMe old, Smokizzy, Pomte, J.delanoy, Trusilver, Rlsheehan, Uncle Dick, Ginsengbomb, Ball3r90, G. Campbell, Tdadamemd, Rod57, Dispenser, Mahewa, Parallel Child, Sman789, WaiteDavid137, AntiSpamBot, M-le-mot-dit, WHeimbigner, SJP, SriMesh, JohnnyRush10, Fairz, Ionescuac, 2help, Atropos235, Juliancolton, Bonadea, Lseixas, Squids and Chips, CardinalDan, Idioma-bot, Dancergal, Signalhead, X!, VolkovBot, CWii, Jeff G., Soliloquial, TheOtherJesse, Ryan032, Philip Trueman, TXiKiBoT, Oshwah, Jacob Lundberg, Davehi1, A4bot, Caster23, Qxz, Anna Lincoln, Martin451, LeaveSleaves, Andrewrost3241981, Skslater, Master Bigode, Kaenneth, Pishogue, Davwillev, Saturn star, ARUNKUMAR P.R, Venny85, RadiantRay, Greswik, Meters, Falcon8765, Musicaddict36, Yankeesfan7602, Shockkorea, Brianga, D. Recorder, Centerfold07, Crouchingtigershiningdragon, PaddyLeahy, Newbyguesses, Mr. dick 008, Frxstrem, SieBot, Coffee, Dusti, Gamesguru2, YonaBot, Tiddly Tom, Graham Beards, BotMultichill, ToePeu.bot, Dawn Bard, Jbmurray, Caltas, Matthew Yeager, Smenge32, The way, the truth, and the light, Bentogoa, JD554, Hxhbot, Eakaphray Udeday, Paolo.dL, Beast of traal, Steven Crossin, Lightmouse, Tombomp, Kevinbboobb, Alex.muller, OKBot, Guiltycivilian, Ward20, Randomblue, Hamiltondaniel, WikiLaurent, PerryTachett, Dolphin51, Denisarona, Nondistinguished, Loren.wilton, ClueBot, LAX, NickCT, Leasky24, PipepBot, Snigbrook, The Thing That Should Not Be, Helenabella, Ariadacapo, Jeff Maniac, Wysprgr2005, Arakunem, RYNORT, Ann dan88, F-j123, J8079s, SuperHamster, CounterVandalismBot, LizardJr8, Agge1000, Starelda, Carlmen2, DragonBot, Blackjackdavie1992, Frau369, Excirial, Vivio Testarossa, Brews ohare, Cenarium, Jotterbot, PhySusie, Praveen java, Morel, Elizium23, El bot de la dieta, La Pianista, Unmerklich, Aitias, Versus22, Mythdon, Crowsnest, Indopug, ClanCC, Cowardly Lion, Thisismymsnaddie, XLinkBot, Stickee, Hibernate SWE, Little Mountain 5, Mitch Ames, WikHead, SilvonenBot, Vianello, Truthnlove, Airplaneman, CàlculIntegral, HexaChord, Addbot, Thright, Substar, DOI bot, Fyrael, Landon1980, DaughterofSun, Gabrielmulenga, Fgnievinski, Pkkphysicist, Fieldday-sunday, KorinoChikara, CanadianLinuxUser, Leszek Jańczuk, MrOllie, Thom443, Chzz, Doniago, 5 albert square, Numbo3-bot, Wolfeye90, VASANTH S.N., Tide rolls, Teles, Superboy112233, MissAlyx, Ttasterul, Luckas-bot, Yobot, 2D, Tohd8BohaithuGh1, Ptbotgourou, Snydale, II MusLiM HyBRiD II, ArchonMagnus, Mmxx, THEN WHO WAS PHONE?, Kåre-Olav, KamikazeBot, Qui-Gon Jinn, IW.HG, Tempodivalse, Szajci, AnomieBOT, Jim1138, IRP, Yachtsman1, VCleemput, Materialscientist, Chaitanyatikule, Citation bot, Another Stickler, Maxis ftw, GB fan, ArthurBot, LilHelpa, Xqbot, S h i v a (Visnu), Sionus, Capricorn42, Karinsteinbock, Nickkid5, Tad Lincoln, Shadowdragonofdoom, Livrocaneca, Maddie!, Azimsultan, Pmlineditor, GrouchoBot, Nayvik, Omnipaedista, RibotBOT, Mars786420, Doulos Christos, Twin Hills student, N419BH, Trogdorcronus27, Srr712, WaysToEscape, Shadowfireinside, Aaron Kauppi, Miz213, Samwb123, Griffinofwales, ⬚⬚, CES1596, FrescoBot, Wppo, Kaneman14, LucienBOT, Kloddant, Cargoking, Dger, Machine Elf 1735, Haein45, Cannolis, Citation bot 1, Javert, AstaBOTh15, Gautier lebon, Pinethicket, I dream of horses, Boulaur, MJ94, Rushbugled13, A8UDI, RedBot, Ongar the World-Weary, Serols, SpaceFlight89, Marcmarroquin, Pbsouthwood, Beao, Full-date unlinking bot, Dude1818, What do i want to..., Jauhienij, Meier99, FoxBot, Thrissel, TobeBot, Jordgette, Womenwithhorses, Lotje, Dinamik-bot, Vrenator, Alex2009258, ..Playa187.., Theo10011, Reaper Eternal, Physicstutorials, Peacedance, Jwilso911, Shanker Pur, Reach Out to the Truth, DARTH SIDIOUS 2, Jfmantis, Nacho999, Makrtbear, WildBot, SM262009, LibertyDodzo, EmausBot, John of Reading, Acather96, Milkunderwood, Funkysam123, Hula Hup, Syncategoremata, RA0808, Philipp Wetzlar, Dddin12345, Solarra, Jmencisom, Tommy2010, TuHan-Bot, Wikipelli, Dcirovic, K6ka, Soup with noodles, Hhhippo, JSquish, ZéroBot, Traxs7, Taymonkey, Hazard-SJ, H3llBot, Cymru.lass, Wayne Slam, Ocaasi, IGeMiNix, Gray eyes, L Kensington, Larryg510, Zayzya, Donner60, Zueignung, Puffin, Status, Ilevanat, Akash Attri, RockMagnetist, EdoBot, ClueBot NG, Accelerometer, This lousy T-shirt, Kikichugirl, Iiteehee, Gggbgggb, Enopet, Cntras, OLI Statics, Masssly, Widr, Baechljm, Shovan Luessi, Jorgenev, FluorineGas, Helpful Pixie Bot, Johnny C. Morse, Musclemane, Bibcode Bot, Lowercase sigmabot, Vagobot, Wiki13, Athanasios Kala, AvocatoBot, Allecher, Supremeaim, Lynskyder, Jedimindtrick56, Cky2250, Glacialfox, Achowat, EpicRevolution, Anbu121, Jonadin93, Vishakh24, Melanee55, KidA424, Mrt3366, Ray98872, ChrisGualtieri, Jeeef100, Dexbot, Feynrichard, ScitDei, Muhammadsikander97, TwoTwoHello, Jetbl44, 55452OH, Emin.gabrielyan, Wywin, Kevin12xd, Cadillac000, Samoht1236831, Mark viking, Jizzy420, Ruby Murray, AmericanLemming, Paikrishnan, Wuzh, Nanotribologist, Svjo-2, Sol1, Buffbills7701, Bleckneuhaus, Montyv, Newestcastleman, Monkbot, Yikkayaya, Qwertyxp2000, 115.241.241.2d, Mohamed sameh sayed rezk, Antonio heeralal, Asdklf;, Terraneuve, Sidsaurb, Intellectual Bookworm, Jesus the Light of my life, Arvind asia, 18jvilches, Saisirajahmed, ChemicalyImbalanced, Tetra quark, MangoKDogII, Aakamal, Bob46738, GeneralizationsAreBad, Amir7561, KasparBot, Mracidglee, Bishnu neupane602, Rithin Babu, Iwillchangeinfos, Pabebe09, Bulenica, DanLucey123, Daredevil1414, Roasteed, Gagz7, Robot psychiatrist and Anonymous: 1179

- **Classical mechanics** *Source:* https://en.wikipedia.org/wiki/Classical_mechanics?oldid=735531830 *Contributors:* AxelBoldt, CYD, Mav, Tarquin, AstroNomer, Ap, Josh Grosse, XJaM, William Avery, Roadrunner, Peterlin~enwiki, Maury Markowitz, FlorianMarquardt, Camembert, Isis~enwiki, Lir, Patrick, Michael Hardy, Tim Starling, Grahamp, Bcrowell, TakuyaMurata, Looxix~enwiki, Stevenj, Lupinoid, Glenn, Bogdangiusca, Rossami, Denny, Pizza Puzzle, Charles Matthews, Aravindet, Reddi, Dandrake, The Anomebot, Jeepien, Furrykef, Phys, Raul654, BenRG, RadicalBender, Phil Boswell, Robbot, F3meyer, Mayooranathan, Moink, Hadal, Papadopc, Fuelbottle, Anthony, Tea2min, Giftlite, Wolfkeeper, Tom harrison, Wwoods, Wgmccallum, Jorge Stolfi, Dan Gardner, PlatinumX, Mobius, Quadell, Antandrus, Beland, Karol Langner, APH, Gauss, Icairns, Zfr, Muijz, Guanabot, FT2, Dave souza, Paul August, SpookyMulder, Bender235, JoeSmack, Brian0918, MBisanz, Surachit, Bobo192, Nigelj, John Vandenberg, BrokenSegue, Haham hanuka, LucaB~enwiki, Mlessard, Sun King, Batmanand, Orionix, Velella, Evil Monkey, Dirac1933, Woodstone, Gene Nygaard, RandomWalk, Oleg Alexandrov, Nuno Tavares, Linas, StradivariusTV, Drostie, Ruud Koot, Dodiad, Jeff3000, Ulcph, Mayz, XaosBits, Phlebas, Leapfrog314, Graham87, Magister Mathematicae, Qwertyus, FreplySpang, Yurik, Seidenstud, Kinu, MarSch, Thechamelon, RE, Bhadani, Cethegus, DirkvdM, FlaBot, Mathbot, RexNL, Srleffler, Chobot, Krishnavedala, Sharkface217, DVdm, Sanpaz, Gwernol, Wavelength, Hairy Dude, Deeptrivia, Retodon8, RussBot, Carl T, JabberWok, David R. Ingham, Johann Wolfgang, Ragesoss, Chichui, Enormousdude, Covington, Thou shalt not have any gods before Willy on Wheels, RG2, Timothyarnold85, Sbyrnes321, SmackBot, Tom Lougheed, Hydrogen Iodide, Jagged 85, Ptpare, Harald88, Squiddy, Frédérick Lacasse, Saros136, Bluebot, TimBentley, SMP, Pieter Kuiper, Silly rabbit, Complexica, DHN-bot~enwiki, Salmar, Foxjwill, Berland, Rsm99833, Cybercobra, Chrylis, Dr. Sunglasses, Sure kr06~enwiki, Vgy7ujm, Loodog, Farid2053, Phancy Physicist, Xunex, SirFozzie, Mets501, Ssiruuk25, Anjor, Tawkerbot2, RSido, Sketch051, BeenAroundAWhile, Matthew Auger, Gregbard, Logicus, Cydebot, Rushbie, Rracecarr, Thijs!bot, Barticus88, TonyTheTiger, AndrewDressel, Kahriman~enwiki, MrXow, Imusade, Headbomb, James086, Memayer, Austin Maxwell, Seaphoto, Storkk, JAnDbot, CosineKitty, Db099221, Yill577, Magioladitis, VoABot II, Ling.Nut, Dfalcantara, Ryeterrell, David Eppstein, User A1, MarcusMaximus, JaGa, Ekotkie, Euneirophrenia, Rohan Ghatak, Nigholith, AtholM, Bcartolo, C quest000, CompuChip, Juliancolton, Treisijs, Useight, Idioma-bot, Pafcu, VolkovBot, JohnBlackburne, TXiKiBoT, The Original Wildbeast, BertSen, GroveGuy, Hqb, Sankalpdravid, Anna Lincoln, Costela, Windrixx, BotKung, Amd628, Gnf1, Tom Atwood, Synthebot, AlleborgoBot, Neparis, SieBot, ToePeu.bot, JerrySteal, Paolo.dL, Lisatwo, Duae Quartunciae, Tomasz Prochownik, ClueBot, WurmWoode, DeepBlueDiamond, Luke490, CyrilThePig4, Razimantv, Mild Bill Hiccup, Niceguyedc, Djr32, Excirial, Jomsborg, Gulmammad, Brews ohare, Arjayay, PhySusie, BOTarate, Crowsnest, XLinkBot, Rror, Saeed.Veradi, Andeasling, Truthnlove, Cholewa, Addbot, Willking1979, Atethnekos, Dgroseth, Njaelkies Lea, Fluffernutter, SpillingBot, Cst17, EconoPhysicist, Bassbonerocks, CUSENZA Mario, LinkFA-Bot, Tassedethe, Tide rolls, Lightbot, Lrrasd, Luckas-bot, Bunnyhop11, Tannkrem, AnomieBOT, Rubinbot, Keithbob, Jpc4031, Citation bot, Xqbot, Tripodian, Amareto2, Charvest, Aaron Kauppi, Thehelpfulbot, Dan6hell66, LucienBOT, Tobby72, Steve Quinn, Machine Elf 1735, Pinethicket, Codwiki, SpaceFlight89, Corinne68, TobeBot, Trappist the monk, Wdanbae, Lotje, Dinamik-bot, JLincoln, Diannaa, Onel5969, RjwilmsiBot, EmausBot, Syncategoremata, Elementaro, Amrator, Wikipelli, Dcirovic, JSquish, Cogiati, Knight1993, Stanford96, Empty Buffer, Vramasub, Maschen, ChuispastonBot, RockMagnetist, Wakebrdkid, ClueBot NG, Satellizer, SusikMkr, Enopet, Frietjes, Braincricket, Widr, ساجد امجد ساجد, Lincoln Josh, Helpful Pixie Bot, ଝିଙ୍କ କାପିଲେଶ୍, IzackN, Prof McCarthy, Brian Tomasik, Blue Mist 1, Sparkie82, Snow Blizzard, Zedshort, Miszatomic, StopTheCrackpots, YFdyh-bot, Khazar2, Dexbot, Jamgoodman, Fragapanagos, Thatguy1234352, Rahulsehwag, Reatlas, Devinray1991, Howicus, Duckduckstop, Fidasty, Jburnett63, JaconaFrere, Arachmen, Kiranbg12, Trackteur, ElectronicKing888, Peterzipfel37, Crystallizedcarbon, Mars wanderer, Jarjarbinks123455555, FeatheredOrcian, KasparBot, Kafishabbir, Qazplmqwertybchefbiusdb wieuhdcbiqhedcbwhuid, DirtyRotten, Gagz7 and Anonymous: 240

- **Newton's laws of motion** *Source:* https://en.wikipedia.org/wiki/Newton'{}s_laws_of_motion?oldid=725381690 *Contributors:* AxelBoldt, Sodium, AstroNomer, Matusz, SimonP, Peterlin~enwiki, Heron, Bth, Isis~enwiki, Sunilbajpai, Stevertigo, Bdesham, Marvinfreeman, Lir, Patrick, Nommonomanac, D, Michael Hardy, Ixfd64, Bcrowell, Karada, Eric119, Looxix~enwiki, Ahoerstemeier, Docu, Salsa Shark, Andres, Mxn, Smack, Raven in Orbit, Pizza Puzzle, Andrevan, Magnus.de, Doradus, DJ Clayworth, Xaven, Shizhao, Betterworld, Fvw, Sokane, Rogper~enwiki, Donarreiskoffer, Robbot, Fredrik, Moondyne, Gandalf61, Stewartadcock, Pingveno, Sverdrup, AceMyth, Hadal, Wikibot, Robinh, Dave6, Ancheta Wis, Matthew Stannard, Decumanus, Connelly, Giftlite, Smjg, DavidCary, Sj, Robin Patterson, Wolfkeeper, BenFrantzDale, Lee J Haywood, MSGJ, Obli, Wwoods, Everyking, No Guru, Curps, Ehabkost, Jason Quinn, Mboverload, Tagishsimon, SoWhy, Knutux, LiDaobing, Antandrus, HorsePunchKid, MarkSweep, Karol Langner, Johnflux, Icairns, Urhixidur, Jp347, Sword~enwiki, M1ss1ontomars2k4, Trevor MacInnis, Brianjd, Solitude, Vsmith, Mecanismo, D-Notice, Vivers, Paul August, SpookyMulder, Bender235, Rubicon, Kipton, Kaisershatner, JoeSmack, Nabla, Pmetzger, Maclean25, El C, Rgdboer, Hayabusa future, Marx Gomes, Laurascudder, Shanes, RoyBoy, Adambro, Bobo192, Army1987, Robotje, Flxmghvgvk, I9Q79oL78KiL0QTFHgyc, SpeedyGonsales, Nicop (Usurp), Nk, Novakyu, Rje, Nhandler, Haham hanuka, Alansohn, Keenan Pepper, Ricky81682, Sl, AzaToth, Lectonar, Sowelilitokiemu, Redfarmer, WikiParker, Hwefhasvs, Idont Havaname, Sir Joseph, Jon Cates, RainbowOfLight, Sciurinæ, Mikeo, Gene Nygaard, Yurivict, Oleg Alexandrov, Joke dst, Feezo, Simetrical, Mindmatrix, Cimex, Daniel Case, StradivariusTV, Kzollman, Mazca, Pol098, Dodiad, Jok2000, Plrk, Zzyzx11, Brendanconway, Prashanthns, Palica, Pfalstad, Gerbrant, Mandarax, Mdale, MassGalactusUniversum, Rnt20, Graham87, Magister Mathematicae, BD2412, FreplySpang, Yurik, Josh Parris, Sjö, Rjwilmsi, MarSch, Nneonneo, HappyCamper, S Schaffter, The wub, DoubleBlue, Cethegus, Dougluce, Matt Deres, MapsMan, Yamamoto Ichiro, Leo44, KaiMartin, Wragge, Anskas, Spaceman85, Mathbot, Nihiltres, Crazycomputers, GnuDoyng, Jgmakin, B44H, Nickpowerz, Elmer Clark, RexNL, Gurch, Alexjohnc3, OrbitOne, SteveBaker, King of Hearts, Chobot, SirGrant, Celebere, Sanpaz, VolatileChemical, Cactus.man, The Rambling Man, Siddhant, YurikBot, Wavelength, Peregrine Fisher, X42bn6, Jimp, RussBot, Arado, Loom91, Splash, Mrboh, RadioFan, Stephenb, Polluxian, Ihope127, Cryptic, Wimt, NawlinWiki, Wiki alf, Grafen, Zephyr9, Długosz, Haranoh, Krea, Irishguy, Robdurbar, Cholmes75, RUL3R, Misza13, Zwobot, Biopresto, Dbfirs, T, Kyle Barbour, SFC9394, Mysid, BanditBubbles, Gadget850, Dead-EyeArrow, Botteville, FF2010, Jcrook1987, Light current, Enormousdude, 21655, Chase me ladies, I'm the Cavalry, Theda, Closedmouth, A-Hrafn, Ketsuekigata, Dspradau, BorgQueen, Mastercampbell, Aleksas, Fram, Emc2, Staxringold, Shastra, Stuhacking, RG2, Tyomitch, Dkasak, Cmglee, David.hillshafer, Phinnaeus, Crystallina, SmackBot, YellowMonkey, Tom Lougheed, InverseHypercube, KnowledgeOfSelf, Hydrogen Iodide, Shoy, Rokfaith, Bomac, Yuyudevil, Jagged 85, Arniep, Gabrielleitao, Edgar181, HalfShadow, Gary2863, Laky68, Cool3, Aksi great, Gilliam, Ohnoitsjamie, Isaac Dupree, ERcheck, Andy M. Wang, Kmarinas86, Fetofs, Saros136, Bh3u4m, Chris the speller, JCSantos, Simon123, Persian Poet Gal, ElTchanggo, NCurse, Ian13, Jayanta Sen, Mr Poo, Silly rabbit, SchfiftyThree, RayAYang, JoeBlogsDord, Kryic83, Baa, Yurigerhard, DHN-bot~enwiki, Hongooi, Rama's Arrow, Zsinj, Can't sleep, clown will eat me, MyNameIsVlad, Jzenksta, Chlewbot, Snowmanradio, Vgm3985, Voyajer, Pevarnj, Kool11, Addshore, Chcknwnm, SundarBot, Mrdempsey, Subheight640, Mayank Abhishek, Flyguy649, Cybercobra, Decltype, MichaelBillington, Dreadstar, J.Wolfe@unsw.edu.au, Salt Yeung, Kid A~enwiki, Bronzie, Mwtoews, DMacks, Kalathalan, Kukini, Yevgeny Kats, Ohconfucius, SashatoBot, Nishkid64, ArglebargleIV, Ser Amantio di Nicolao, Aïki, Kuru, Richard L. Peterson, Euchiasmus, Shaverc, Crypticfortune, Minish man, Cpastern, NongBot~enwiki, IronGargoyle, Axy, Slakr, Stwalkerster, Noah Salzman, Alana Shirley, Sethmiester, Fangfufu, Waggers, NJA, Ryulong, Citicat, MTSbot~enwiki, Laurapr, Caiaffa, Cisca Harrison, Squirepants101, Hu12, BranStark, Iridescent, Dilip rajeev, J Di, Toxic Spade, Richard75, Lilchicklet007, Courcelles, Dark Dragon Master1337, Tawkerbot2, JRSpriggs, Scigatt, Conrad.Irwin, Lahiru k, Zenzic, Durtysouthgurl, Xykon, SkyWalker, JForget, Dycedarg, NinjaKid, Ninetyone, JohnCD, McVities, Ambassador Quan, Korandder, Myasuda, Mct mht, Pewwer42, Nilfanion, Rudjek, Kribbeh, LaFoiblesse, Cydebot, Ubiq, MC10, VashiDonsk, Law200000, Gogo Dodo, Travelbird, Zginder, Flowerpotman, Xxanthippe, Sibusiso Mabaso, Rracecarr, Kotiwalo, DavidRF, Shirulashem, Christian75, DumbBOT, Sonia62585, Sharonlees, Taneli HUUSKONEN, JodyB, UberScienceNerd, HazeNZ, JamesAM, Epbr123, Knakts, TonyTheTiger, AndrewDressel, Headbomb, Marek69, John254, A3RO, Nicholeeeeo, RFerreira, Thljcl, Dfrg.msc, Orfen, Sean William, Natalie Erin, Valandil211, Nobar, Escarbot, Stui, Mentifisto, AntiVandalBot, BokicaK, Luna Santin, Ben pcc, Kliao93, Seaphoto, Opelio, QuiteUnusual, AstroLynx, Thranduil, Fashionslide, Jj137, Goku9821, Farosdaughter, Wienwei, Pasteman, Rico402, Carewolf, Dallas84, Kainino, Canadian-Bacon, 41523, Res2216firestar, Mikenorton, JAnDbot, Narssarssuaq, Leuko, MER-C, Yuksing, Bubla, Hamsterlopithecus, Andonic, Hut 8.5, TheEditrix2, Denimadept, Kerotan, Zadeez, Lmhjulian, Karlhahn, Pedro, Bongwarrior, VoABot II, Transcendence, Kq6up, AtticusX, JamesBWatson, SHCarter, Farquaadhnchmn, Ling.Nut, Aelyn, Cardamon, Animum, 28421u2232nfenfcenc, LookingGlass, SperryTS, Cpl Syx, MarcusMaximus, Avatar 06349, DerHexer, Coffeepusher, Hbent, Kayau, Krazyklink, S3000, Kornfan71, Perfgeek, Kabton14, Kevinsam, Gerhard.Brunthaler, Anaxial, R'n'B, Oysta, AlexiusHoratius, Smokizzy, Tblaxland, LedgendGamer, HEL, J.delanoy, Jorgenumata, Filll, Kimse, Ed-dg, R. Baley, Bogey97, Nbauman, Urmammasfat, Uncle Dick, Eliz81, Extransit, WarthogDemon, Teressa Keiner, Lordgilman, Ncmvocalist, Hakufu Sonsaku, Bdodo1992, Skier Dude, Djsolie, Gurchzilla, Armaetin, NewEnglandYankee, Half-Blood Auror, Wesino, KChiu7, Vanished user 47736712, Malerin, Neyshan, Hellojoshhowareyou, Chairmclee, BrettAllen, Babygene52, JamesM123, Sarregouset, Natl1, Cuckooman4, Bonadea, Tameradly, Hulten, Lseixas, SoCalSuperEagle, Specter01010, Puneetbahri 82, CardinalDan, Dvyjones, Idioma-bot, Sheliak, Funandtrvl, 1812ahill, Chaseyoung1500, Wikieditor06, Gogobera, X!, Deor, VolkovBot, CWii, Pasixxxx, Pleasantville, Science4sail, Jeff G., Imgoş, CART fan, Philip Trueman, Af648, TXiKiBoT, Nicholasnice, BertSen, Antoni Barau, Vipinhari, Hqb, Captain Wikify, Aymatth2, Anna Lincoln, CanOfWorms, LeaveSleaves, Guldenat, Ilyushka88, Maxim, Wenli, RandomXYZb, Proud Muslim, Synthebot, Falcon8765, @pple, Enviroboy, Vector Potential, TrippingTroubadour, Spinningspark, Al pope, Shockkorea, Dodo von den Bergen, PaddyLeahy, EJF, Digmaster, HaLoGuY007, Dogah, SieBot, Tiddly Tom, WereSpielChequers, RJaguar3, Susanwangrules, Cb77305, Idk my bff jill, Keilana, Lpug21, Oolongy, Tiptoety, Radon210, Oda Mari, Paolo.dL, Pele' boy, Oxymoron83, Mankar Camoran, Foxtrotman, Faradayplank, Burntmonkey5, AnonGuy, Lightmouse, Tombomp, MASQUERAID, KathrynLybarger, Mchhabria, Imlost20, Nancy, Homerguy, Silvergoat, C'est moi, StaticGull, Susan118, Nn123645, Dolphin51, Quinling, Alfredsimpson, Explicit, Muhends, FlamingSilmaril, ClueBot, LAX, Avenged Eightfold, Snigbrook, Foxj, Narom, The Thing That Should Not Be, Aaronsclee, Blackangel25, Meisterkoch, Venfranc, Pomona17, Pkbharti, Arakunem, Control-alt-delete, Frdayeen, Maaru, J8079s, Joao Xavier, SuperHamster, Matt 888, Rolo Tamasi, Make91, Heyyal77, LizardJr8, LonelyBeacon, Agge1000, Neverquick, Flame1009, Puchiko, PMDrive1061, Manishearth, Darkmaster333, DragonBot, Djr32, Abdullah Köroğlu~enwiki, Excirial, Alexbot, Jusdafax, Bender2k14, Eeekster, Graham77, Gtstricky, Notalex, Jhud89, Lartoven, Spaceneil8, Kman229, Brews ohare, Lunchscale, Jotterbot, PhySusie, Psinu, JamieS93, Tnxman307, DeltaQuad, SchreiberBike, La Pianista, Unmerklich, Thingg, Jtle515, Aitias, 7, Versus22, Anon126, Egmontaz, Crowsnest, Rishi.bedi, LightAnkh, XLinkBot, Candy12324, Terry0051, Stickee, Jovianeye, Dthomsen8, Avoided, Sergay, Firebat08, Clayboskie, Truthnlove, Antimatter33, Schipperke22, Totlmstr, 5IIHoova, Jshane04, Addbot, Travi1994, Proofreader77, Mortense, Guoguo12, Wsvlqc, Tcncv, Ezelenyv, Betterusername, Captain-tucker, Shearyears394, Berl95, Ronhjones, Hatashe, Mr. Wheely Guy, Jfkdklsjf, CanadianLinuxUser, SpillingBot, Yoyoyow, Dranorter, Trevor Marron, Kjkl, LaaknorBot, EconoPhysicist, LAAFan, Glane23, Ld100, Cress Arvein, Favonian, TStein, Aktsu, Scizor55, Numbo3-bot, 🇲🇦🇷🇸🇳🇧, Schlongboymega, Obvioustrollisobvious, Tide rolls, Smeagol 17, Krano, Zorrobot, MuZemike, Arbitrarily0, Quantumobserver, Bartledan, IsaacNewton17, LuK3, Grandpsykick, Raziaex, Legobot, Luckas-bot, Yobot, Tseulik, Saxum, Kartano, Tohd8BohaithuGh1, Yiya91, RHB100, Yngvadottir, Tannkrem, Creektheleftcheeksneak, South Bay, Synchronism, Passamaquoddy boi, CoyoteG, AnomieBOT, A More Perfect Onion, Fil21, ThaddeusB, Killiondude, Jim1138, Ocvailes, Danielt998, AdjustShift, Kingpin13, RandomAct, Flewis, Materialscientist, Limideen, Spirit469, Citation bot, E2eamon, Cbgermany, Elm-39, Frankenpuppy, Apollo, Neurolysis, Sionus, Addihockey10, Capricorn42, AJKING182, 963kickemall, Ashman512, Ter890, The Evil IP address, Makeswell, J04n, Pmlineditor, Omnipaedista, Slow Phil, Amaury, Shadowjams, A. di M., Erik9, Imveracious, 🇲🇦, James R. Ward, Angelus Delapsus, Jilkmarine, FrescoBot, Devnullnor, Pepper, Jack1002009, HJ Mitchell, Steve Quinn, Haein45, DivineAlpha, Citation bot 1, I dream of horses, Elockid, HRoestBot, Edderso, Fuzbaby, PrincessofLlyr, Tom.Reding, Mutinus, Gautampratapsingh1993, RedBot, IceBlade710, Phearson, Reconsider the static, Spinachwrangler, Abc518, Yzha519, TobeBot, Jordgette, Lotje, IAcre, Tofutwitch11, Vrenator, Hankston, Diannaa, Brian the Editor, Stroppolo, Minimac, DARTH SIDIOUS 2, Difu Wu, The Utahraptor, RjwilmsiBot, Xcxcxc5k, Becritical, Slon02, Nishadpotdar, EmausBot, Immunize, Kourosch44, Ajraddatz, Heracles31, Dewritech, RebornX, Syncategoremata, Cellorau.murthy, IncognitoErgoSum, RA0808, BEWBS, Mmeijeri, Wikipelli, Jedi062, Wilowisp, Pedro1557, Solomonfromfinland, Hhhippo, Akhil 0950, AvicBot, JSquish, ZéroBot, Fæ, Crua9, Derekleungtszhei, Azuris, Quantumavik, Wayne Slam, Joshkb01, Iiar, L Kensington, Quantumor, Deutschgirl, Donner60, Knackers1, Zueignung, Armaghanwahid, Carmichael, Manubaba11, Ilevanat, Tej karani, Matthewrbowker, Hector2138, EternamenteAprendiz, LGSlayer1127, Sven Manguard, DASHBotAV, Nacho123456789987654321, Qwertyqwertyqwertyy, April to August, ClueBot NG, Arod125, Newton890, Keithdizon, Alexandria37, Nipper211, Bardnick187, Lord Chamberlain, the Renowned, Nscozzaro, Helpful Pixie Bot, පසිඳු කාවින්ද, Bibcode Bot, Elauminri, Lowercase sigmabot, Turtlesilove93, Arjenvreugd, Glevum, Rrronny, BattyBot, Timothy Gu,

786b6364, Wrestling94, Weidolovesmusic, Dexbot, TheZelos, Albina-belenkaya, Jochen Burghardt, Kevin12xd, Reatlas, Mapsurfer49, Paikrishnan, Jodosma, Montyv, Yasir72.multan, Stamptrader, ARUNEEK, GravyKnives, Garfield Garfield, Loraof, Y-S.Ko, Gouravsxs, KasparBot, Wikidalien and Anonymous: 1572

- **Four-force** *Source:* https://en.wikipedia.org/wiki/Four-force?oldid=717609307 *Contributors:* Michael Hardy, Wapcaplet, Cyp, MathKnight, Lockeownzj00, Keenan Pepper, Lerdsuwa, Mpatel, Salix alba, Ligulem, KasugaHuang, That Guy, From That Show!, SmackBot, Dr Greg, ShelfSkewed, Cydebot, Raoul NK, WaiteDavid137, XCelam, Mario briony, VQuakr, Forbes72, Addbot, Yobot, Citation bot, GrouchoBot, Helpful Pixie Bot, Photon89, Waitedavid137, Tach123, Elenceq, Kasuga, Sir Cumference and Anonymous: 5

- **Net force** *Source:* https://en.wikipedia.org/wiki/Net_force?oldid=721843099 *Contributors:* Mrwojo, Patrick, Gbleem, Ronz, Darkwind, Saltine, Raul654, Chuunen Baka, Bearcat, Merovingian, Ssd, ShakataGaNai, Icairns, Sonett72, Discospinster, Paul August, Arthena, Snowolf, Ketiltrout, Sjakkalle, Nihiltres, Nivix, Kri, Bgwhite, YurikBot, Pip2andahalf, Malcolma, Wknight94, Pb30, Phil Holmes, SmackBot, Incnis Mrsi, Melchoir, Anastrophe, Vanished user 3dk2049pot4, BiT, Bluebot, Silly rabbit, Para, Alex Huck, Huon, Dreadstar, Saippuakauppias, Nishkid64, Kuru, Gobonobo, Werdan7, Myasuda, Yaris678, Cydebot, Pajz, Hazmat2, Marek69, Doyley, Bongwarrior, VoABot II, R'n'B, J.delanoy, Melamed katz, Nothingofwater, MilosIvanovic, Vanished user 39948282, Pdcook, Tkgd2007, Signalhead, VolkovBot, Thurth, TXiKiBoT, Lkopeter, PaulTanenbaum, Staka, SieBot, Tiddly Tom, Hxhbot, Paolo.dL, Oxymoron83, Dolphin51, ClueBot, The Thing That Should Not Be, SkeletorUK, WestwoodMatt, Pointillist, Brews ohare, SchreiberBike, Thingg, Versus22, Life of Riley, Hotcrocodile, Hm29168, Addbot, Grayfell, Download, 5 albert square, Zorrobot, Yobot, Ptbotgourou, Piano non troppo, Materialscientist, OllieFury, RibotBOT, Kyng, Uhggnnn, FrescoBot, Pinethicket, HRoestBot, Tom.Reding, RedBot, RandomStringOfCharacters, DReifGalaxyM31, Patriot8790, J36miles, Rami radwan, MikeSchmit, ZéroBot, L Kensington, Donner60, Usb10, Ilevanat, DASHBotAV, 28bot, Petrb, ClueBot NG, Kenoguy123, Piast93, Jwchong, Prof McCarthy, Klilidiplomus, Maddiegraham345, Jellyjellyjellybear, ChrisGualtieri, Mediran, Ducknish, Webclient101, Navyseals1209850, Astro123456, JaconaFrere, St170e and Anonymous: 143

- **Normal force** *Source:* https://en.wikipedia.org/wiki/Normal_force?oldid=724660306 *Contributors:* Patrick, Mxn, Charles Matthews, Wolfkeeper, BenFrantzDale, Jorge Stolfi, Lucky 6.9, Discospinster, Vsmith, Freestylefrappe, Bender235, El C, Zachlipton, Gary, QVanillaQ, Malo, Shoefly, Gene Nygaard, StuTheSheep, Adam Field, JP Godfrey, FlaBot, YurikBot, Kymacpherson, Zephyr9, Deeday-UK, SmackBot, BiT, Persian Poet Gal, Tilin, Daniel.Cardenas, Vina-iwbot~enwiki, Ray9x, Aïki, Loodog, IronGargoyle, Wizard191, George100, Sabate, Cydebot, XduOv3r, JacobBramley, AndrewDressel, Escarbot, Stannered, AntiVandalBot, Email4mobile, EagleFan, R'n'B, J.delanoy, TXiKiBoT, Anonymous Dissident, Monty845, SieBot, ClueBot, Ideal gas equation, Wikijens, Estirabot, Jsondow, Shoemaker's Holiday, Addbot, Yobot, AnomieBOT, Götz, Materialscientist, ArthurBot, Xqbot, Erik9bot, KuroiShiroi, Ammatsun, OgreBot, Deathbeast, Pinethicket, Xopar, WikitanvirBot, Hhhippo, JSquish, Carultch, Donner60, Awesomestpersonintheworld42, RockMagnetist, ClueBot NG, Snotbot, O.Koslowski, HMSSolent, AwamerT, Lynskyder, Pratyya Ghosh, Uxy and Anonymous: 103

- **Equilibrant Force** *Source:* https://en.wikipedia.org/wiki/Equilibrant_Force?oldid=705451861 *Contributors:* Topbanana, Pinethicket, Dineshkumar Ponnusamy, ClueBot NG, AntanO, Mogism, OccultZone, Antrocent, Aguilus, Pokéfan95 and Anonymous: 5

- **Line of action** *Source:* https://en.wikipedia.org/wiki/Line_of_action?oldid=725456994 *Contributors:* Jitse Niesen, Rossumcapek, Bearcat, Rich Farmbrough, Hoary, DVdm, Algebraist, Conscious, Malcolma, Momus, SmackBot, Melchoir, AndrewDressel, David Eppstein, Cjmclark, Philip Trueman, PaulTanenbaum, James McStub, Paolo.dL, Biggerj1, SchreiberBike, Addbot, Jack VerE, Erik9bot, Pinethicket, Pointe blank, Wcherowi, OLI Statics, Sheraz.ahmed6, CAPTAIN RAJU and Anonymous: 10

- **Resultant force** *Source:* https://en.wikipedia.org/wiki/Resultant_force?oldid=731696713 *Contributors:* Raul654, Bearcat, Cydebot, DH85868993, Download, Omnipaedista, ClueBot NG, Widr, CasualVisitor, Prof McCarthy, Mogism, Qwertyxp2000, Swagstica, Eurodyne, CAPTCHA007 and Anonymous: 12

- **Force field (physics)** *Source:* https://en.wikipedia.org/wiki/Force_field_(physics)?oldid=717868039 *Contributors:* Bearcat, Everyking, Abdull, Jiy, Yosofun, Rich Farmbrough, Rune.welsh, Bgwhite, Conscious, Malcolma, Sbyrnes321, SmackBot, MalafayaBot, Kashami, Oatmeal batman, Vina-iwbot~enwiki, JRSpriggs, Chetvorno, No1lakersfan, Trevyn, Icep, MartinBot, Mårten Berglund, Hodja Nasreddin, Mossd, Caltas, Yintan, Paolo.dL, ClueBot, PipepBot, LolZ0r-Richt, DumZiBoT, Addbot, Brentdeezee, Zorrobot, Suiseiseki, Yobot, Sionus, Miracleworker5263, NOrbeck, Erik9bot, I dream of horses, Brandonian9150, Jzana, Cowabunga87, Black Shadow, Mentibot, ClueBot NG, Shantnup, Spidey568, Opsafag, Ahmadalkhairy and Anonymous: 23

- **Fundamental interaction** *Source:* https://en.wikipedia.org/wiki/Fundamental_interaction?oldid=735206111 *Contributors:* AxelBoldt, Zundark, The Anome, Tarquin, AstroNomer, William Avery, Roadrunner, Ellmist, Robert Foley, Heron, Isis~enwiki, Stevertigo, Patrick, Michael Hardy, Gdarin, CesarB, Looxix~enwiki, Cyp, William M. Connolley, Theresa knott, Mxn, Bemoeial, Reddi, Zoicon5, Finlay McWalter, Robbot, Lowellian, Brjaga, Roscoe x, Seth Ilys, Ancheta Wis, Giftlite, Christopher Parham, Herbee, Monedula, Xerxes314, Alison, Pcarbonn, Beland, Melikamp, Karol Langner, AmarChandra, Mike Rosoft, Jørgen Friis Bak, JimJast, Discospinster, Guanabot, FT2, Harriv, Quietly, GoldenRing, Clement Cherlin, El C, Lycurgus, Joanjoc~enwiki, Alereon, Euyyn, Kanzure, Army1987, Rbj, Haham hanuka, Nsaa, Jumbuck, Foant, Dachannien, Kdau, ReyBrujo, Reaverdrop, BDD, Someoneinmyheadbutit'snotme, DV8 2XL, Kazvorpal, Woohookitty, Linas, Mindmatrix, Sabejias, StradivariusTV, Mpatel, Miss Madeline, Isnow, Elvey, Chun-hian, Koavf, Strait, Jmcc150, RE, Gadha, FlaBot, DClement, ZoneSeek, Alfred Centauri, Lmatt, Rell Canis, Mstroeck, Chobot, Subtractive, Visor, GangofOne, Mysekurity, YurikBot, Ashleyisachild, Bambaiah, Lucinos~enwiki, Wavesmikey, Chaos, FFLaguna, Dbfirs, Trigger hippie77, Enormousdude, Shimei, RG2, Bweenie, Phr en, GrinBot~enwiki, SmackBot, Unyoyega, Andy M. Wang, Vvarkey, Jjalexand, Mithaca, Acipsen, DHN-bot~enwiki, Colonies Chris, Andy120290, Addshore, SundarBot, Jgwacker, LeoNomis, Sadi Carnot, TTE, SashatoBot, Lambiam, FrozenMan, Philosophus, A. Parrot, Fangfufu, GDallimore, Avanishsharma, CRGreathouse, Green caterpillar, McVities, MaxEnt, A. Exeunt, Scott.medling, LouisBB, Thijs!bot, Martin Hogbin, Mojo Hand, Headbomb, Dfrg.msc, Dodecahedron~enwiki, HolyT, JAnDbot, The penfool, Fordskydog, MER-C, TheEditrix2, Fabrictramp, Leyo, Trusilver, Joshuaali, Idioma-bot, VolkovBot, TXiKiBoT, Anonymous Dissident, MackSalmon, Don4of4, Praveen pillay, BotKung, Gnomon13, Lamro, RMW42, EmxBot, Neparis, SieBot, WereSpielChequers, ToePeu.bot, Avargasm, RadicalOne, Dhatfield, SuperSpy00bob, Sbowers3, BartekChom, Beast of traal, Lightmouse, Nskillen, Sunrise, OKBot, Bpeps, C0nanPayne, StewartMH, Sfan00 IMG, ClueBot, MichaelVernonDavis, SuperHamster, Djr32, Sadiqsaleem09, PixelBot, Eeekster, Zamis45, Yonskii, 1ForTheMoney, Noctibus, Truthnlove, Addbot, Mabdul, LinkFA-Bot, F Notebook, Lightbot, Legobot, Clay Juicer, Luckas-bot, Yobot, II MusLiM HyBRiD II, Rifter0x0000, AnomieBOT, Glen Dillon, Cantanchorus, Girl Scout cookie, Cleroth, JackieBot, Piano non troppo, Flewis, AthenaO, Xqbot, Gap9551, Omnipaedista, RibotBOT, Alvin Seville, A. di M., Ironboy11, Goodbye Galaxy, Jmbenham, Unkownkid2400, Rameshngbot, Jschnur, RedBot, Σ, Frankjohnson123, IVAN3MAN, Right-wing

Macaddct1984, Eras-mus, Gimboid13, Rnt20, Graham87, Qwertyus, Ando228, Rjwilmsi, Collins.mc, Tangotango, Ligulem, SeanMack, Bhadani, Yamamoto Ichiro, Cjpuffin, FlaBot, Nihiltres, Nivix, RexNL, Ewlyahoocom, Gurch, Otets, Fresheneesz, TeaDrinker, Srleffler, Physchim62, Chobot, Roboto de Ajvol, YurikBot, Wavelength, Spacepotato, JabberWok, CambridgeBayWeather, Salsb, Wimt, David R. Ingham, NawlinWiki, Bachrach44, Buster79, Tearlach, Anetode, Scottfisher, Figaro, Bota47, Nick123, FF2010, Light current, Orioane, Enormousdude, 21655, 2over0, JoanneB, Phil Holmes, Willtron, Sizarieldor, AGToth, Katieh5584, RG2, GrinBot~enwiki, Sbyrnes321, DVD R W, Luk, SmackBot, Manu0x0~enwiki, PEHowland, Prodego, KnowledgeOfSelf, McGeddon, Jagged 85, Jrockley, Swerdnaneb, Rpmorrow, Skizzik, JAn Dudík, LinguistAtLarge, Master of Puppets, Complexica, CMacMillan, TheGerm, Frap, Ioscius, Avoidance, SundarBot, Stevenmitchell, Cybercobra, MichaelBillington, Blake-, Akriasas, Illnab1024, Daniel.Cardenas, LeoNomis, Sadi Carnot, Apoorvchebolu, Skinnyweed, TTE, WayKurat, Sarfa, DJIndica, Nmnogueira, Ozhiker, Ser Amantio di Nicolao, Wvbailey, Finejon, UberCryxic, Philosophus, Cronholm144, Ckatz, El Dahveed, Grapetonix, Alatius, Sinistrum, Dicklyon, Jon186, Waggers, Ryulong, Describer, KJS77, Newone, Adambiswanger1, Courcelles, Ziusudra, Tawkerbot2, Yanah, Xcentaur, Mosaffa, JForget, KyraVixen, Baiji, Vyznev Xnebara, Fjomeli, MarsRover, Musicalantonio, Marly88, Peripitus, Fifo, Ssilvers, UberScienceNerd, Thijs!bot, VoABot, Jb.schneider-electric, Headbomb, Marek69, Gerry Ashton, Leon7, D.H, MichaelMaggs, AntiVandalBot, Majorly, Seaphoto, Quintote, Gnixon, Jnyanydts, Tyco.skinner, Tlabshier, Eleos, Steelpillow, JAnDbot, Matthew Fennell, Mkch, Bongwarrior, VoABot II, Tails4, SHCarter, TxAlien, GBYork, WhatamIdoing, Vanished user ty12kl89jq10, Adrian J. Hunter, 28421u2232nfenfcenc, Coldwarrier, User A1, Khalid Mahmood, Ruhihumphries, PrattTA1, InvertRect, MartinBot, Jargon777, LedgendGamer, J.delanoy, Sasajid, Abecedare, Bogey97, JohnPritchard, Maurice Carbonaro, Extransit, Tarotcards, NewEnglandYankee, Wesino, Shoessss, Cometstyles, ACBest, DorganBot, Treisijs, Bently34, Lights, 28bytes, Part Deux, Thedjatclubrock, Constant314, Philip Trueman, TXiKiBoT, The Original Wildbear, Guillaume2303, Anonymous Dissident, Kevin Steinhardt, Monkey Bounce, CaptinJohn, Imasleepviking, Mbarrieau, DoktorDec, Atomicswoosh, TongueSpeaker, Andy Dingley, Dirkbb, Lova Falk, @pple, Sylent, Doc James, PGWG, FlyingLeopard2014, SieBot, Coffee, Tresiden, Graham Beards, Work permit, Hertz1888, Avargasm, Winchelsea, Dawn Bard, Caltas, Yintan, Zoragotcha, Keilana, Flyer22 Reborn, Qst, Csblack, Klmeze~enwiki, Joseph Banks, Oxymoron83, Fbarw, Maelgwnbot, StaticGull, Dolphin51, TreeSmiler, Kanonkas, Ainlina, ElectronicsEnthusiast, Llywelyn2000, WikipedianMarlith, Mr. Granger, Bschaeffer~enwiki, Atif.t2, Martarius, ClueBot, Stevekirst7, GorillaWarfare, The Thing That Should Not Be, Jan1nad, Uncle Milty, Boing! said Zebedee, Blanchardb, Twicemost, LizardJr8, Jackey0105, Electromagnetic, Seanwal111111, Otto Tanaka, Excirial, Kocher2006, BlueLikeYou, Jusdafax, Abrech, Plastic Fish, MacedonianBoy, PhySusie, Friedlibend und tapfer, Thingg, Jesus.murphy55567, Versus22, InternetMeme, XLinkBot, Emmette Hernandez Coleman, Jovianeye, Rror, Boratlike, HarlandQPitt, Rogimoto, Deineka, Addbot, Dustbin123, Allfor12008, CanadianLinuxUser, Download, EconoPhysicist, Redheylin, WikiDegausser, K Eliza Coyne, LinkFA-Bot, Kisbesbot, AgadaUrbanit, VASANTH S.N., Tide rolls, Lightbot, Gail, Kurtis, Will.M.Thompson, Luckas-bot, Yobot, Ptbotgourou, Lichen from Hell, ScienceMind, Tempodivalse, Orion11M87, AnomieBOT, Jim1138, Piano non troppo, AdjustShift, Penguinatortoo, Materialscientist, Citation bot, Vuerqex, Xqbot, Konor org, Lolman33, Plumpurple, Melmann, DSisyphBot, GrouchoBot, Nayvik, Omnipaedista, RibotBOT, Jsleaby, Maplestory101, Slowart, A. di M., GliderMaven, Tobby72, Wikipe-tan, Sum33, Ian88800, Hippycaller, Steve Quinn, HamburgerRadio, Citation bot 1, Чахович Уладзіслаў, Pinethicket, I dream of horses, Elockid, Hard Sin, LittleWink, Tom.Reding, PvsKllKsVp, Smuckola, Tinton5, Yahia.barie, Jschnur, Ezhuttukari, Corinne68, FoxBot, Рыцарь поля, Retired user 0001, SchreyP, Hickorybark, ItsZippy, Lotje, Cjlim, LilyKitty, Antipastor, Reaper Eternal, Tbhotch, DARTH SIDIOUS 2, Triden, RjwilmsiBot, Alph Bot, Agent Smith (The Matrix), WildBot, DASHBot, EmausBot, John of Reading, WikitanvirBot, Mnkyman, Never give in, ITshnik, IncognitoErgoSum, RA0808, Tommy2010, Amrator, Wikipelli, Dcirovic, Thecheesykid, JSquish, ZéroBot, Harddk, Leptonoggin, Fæ, Maypigeon of Liberty, WFarver, H3llBot, Quondum, Git2010, Makecat, Sonygal, Coasterlover1994, L Kensington, Rr2wiki, Maschen, Donner60, Rjowsey, RockMagnetist, Peter Karlsen, Sir sachin, Teapeat, Planetscared, Cgt, Xanchester, ClueBot NG, Jack Greenmaven, MelbourneStar, Hiperfelix, Hallaman3, Ant.acke, Enopet, Frietjes, Milikguay, Widr, Antiqueight, Helpful Pixie Bot, Eemcginnis, Vagobot, Bolatbek, Manuelfeliz, MusikAnimal, Kagundu, Rm1271, Thekillerpenguin, Cadiomals, Mariano Blasi, Whyamiadampandalol, Franz99, Zedshort, Physicsch, Brad7777, Glacialfox, Neumannjo, Acul132, ChrisGualtieri, Dexbot, Marlowfrontier, Lugia2453, Frosty, Josophie, Hamerbro, SubratamindPal, Trillig, Abhaikumar10, CsDix, Ihatedirac2k13, Hell to earth88, CROY123, Zenibus, Jwratner1, YiFeiBot, DrRC, SpecialPiggy, Kdmeaney, Yashshroff97, Csutric, Mahusha, TE5ITA, Venomous Cobra, Trackteur, Gronk Oz, Meski33, Hkeyser, Lolbob12345, Nanophysisct12345, Simonessnygg, Fufulord, Noobmagnet, Mediavalia, Slayeredwarrior, Tetra quark, Isambard Kingdom, SergioCruz2015, Aenfinger, KasparBot, Zeke Essiestudy, MusikBot, Kafishabbir, JJMC89, Datbubblegumdoe, Cesarnajera56, Schweikhardt, CAPTAIN RAJU, LIGHTSSWITCH, BananaPuppet, JGS952, Shane ducharme, Marvellous Spider-Man and Anonymous: 723

• **Gravity** *Source:* https://en.wikipedia.org/wiki/Gravity?oldid=735778277 *Contributors:* Bryan Derksen, Danny, Rmhermen, Caltrop, Heron, Montrealais, Stevertigo, Patrick, D, Michael Hardy, Ixfd64, TakuyaMurata, GTBacchus, Wintran, Snarfies, Ahoerstemeier, Mac, Dgaubin, William M. Connolley, Jeff Relf, TonyClarke, Smack, Ec5618, Tpbradbury, Phoebe, Fairandbalanced, BenRG, Pollinator, Lumos3, Jni, Northgrove, Donarreiskoffer, Robbot, Sander123, Jakohn, TimothyPilgrim, Academic Challenger, Caknuck, Bkell, Nerval, Aetheling, Ruakh, Dina, Alan Liefting, Cedars, Ancheta Wis, Giftlite, Mikez, Haeleth, Wolfkeeper, Inkling, Fropuff, Everyking, Frencheigh, Aechols, Bobblewik, Utcursch, Keith Edkins, Antandrus, Phe, Quarl, Elembis, Kiteinthewind, Jossi, Karol Langner, Lindberg G Williams Jr, Demiurge, Trevor MacInnis, Grstain, Mike Rosoft, &Delta, DanielCD, Shipmaster, JimJast, Discospinster, Zaheen, Supercoop, Vsmith, Smyth, Notinasnaid, Dbachmann, Bender235, Rubicon, Loren36, El C, Huntster, Kwamikagami, Aude, Diomidis Spinellis, RoyBoy, Nrbelex, Gershwinrb, Bobo192, Smalljim, Elipongo, I9Q79oL78KiL0QTFHgyc, Larryv, MPerel, Danski14, Alansohn, JYolkowski, Anthony Appleyard, Davetcoleman, Atlant, Maya Levy, Paleorthid, Andrew Gray, Cjthellama, Riana, Goldom, Mac Davis, Wdfarmer, Hdeasy, Snowolf, Mononoke~enwiki, BRW, Yuckfoo, RainbowOfLight, Jesvane, Bookandcoffee, Kazvorpal, Falcorian, Stephen, Feezo, Richard Arthur Norton (1958-), OwenX, Linas, Camw, LOL, Benhocking, JFG, MONGO, Mpatel, Tabletop, GregorB, Eaolson, Isnow, Scm83x, SDC, Philbarker, TheAlphaWolf, Brownsteve, Radiant!, Dysepsion, Sin-man, Ashmoo, Chrispasquale, Chun-hian, Drbogdan, Rjwilmsi, Tangotango, Nichiran, Fel64, Kazrak, Ligulem, Ems57fcva, LjL, Boccobrock, Bhadani, MarnetteD, Matt Deres, Sango123, Syced, Yamamoto Ichiro, Fish and karate, JanSuchy, Magmafox, Titoxd, RobertG, Old Moonraker, Nihiltres, Nivix, RexNL, Gurch, Fresheneesz, Alphachimp, Tardis, Srleffler, King of Hearts, DVdm, Guliolopez, Hall Monitor, Bomb319, Gwernol, EamonnPKeane, Raelx, The Rambling Man, Wavelength, Drdisque, Sceptre, Hairy Dude, Jimp, Hillman, Brandmeister (old), Ohwilleke, Red Slash, Musicpvm, Anonymous editor, Loom91, Bhny, Pigman, Epolk, Philip Hazelden, DanMS, Scott5834, CambridgeBayWeather, Wimt, Ccmccm, NawlinWiki, Nowa, Wiki alf, Ytcracker, SigPig, SCZenz, Apokryltaros, Nick, Brandon, Katrina Graziano, Semperf, Aaron Schulz, Roy Brumback, Addps4cat, Jessemerriman, Rayc, Mgnbar, Dna-webmaster, Eurosong, Heeroyuy135, Enormousdude, 2over0, SFGiants, Chase me ladies, I'm the Cavalry, Lappado, Endomion, E Wing, Fmyhr, Smurrayinchester, Kevin, Geoffrey.landis, Anclation~enwiki, Emc2, Willtron, Kungfuadam, RG2, JDspeeder1, Bo Jacoby, Draicone, FyzixFighter, Mejor Los Indios, Sbyrnes321, DVD R W, Hide&Reason, That Guy, From That Show!, Shenhemu, Luk, TravisTX, Sardanaphalus, SmackBot, Ulterior19802005, Incnis Mrsi, Reedy, KnowledgeOfSelf, Hydrogen Iodide, NaiPiak, David Shear, Kilo-Lima, Jagged 85, Thunderboltz, Delldot, Hardyplants, Cessator, Syckls, BiT,

Timotheus Canens, GraemeMcRae, HalfShadow, Typhoonchaser, Yamaguchi⬚⬚, Gilliam, Algont, Hmains, Ppntori, ERcheck, Andy M. Wang, Kmarinas86, Marc Kupper, The monkeyhate, Saros136, Bluebot, Cush, Keegan, Raymond arritt, Fplay, Silly rabbit, Lehkost, Complexica, Bbq332, Jeff5102, Sbharris, Hallenrm, CharonM72, Scwlong, Can't sleep, clown will eat me, Scott3, UNHchabo, MJCdetroit, Apostolos Margaritis, Lesnail, Cryocide, Rsm99833, Amazins490, Jmlk17, Cybercobra, Nakon, Steve Pucci, TedE, Red1~enwiki, Jiddisch~enwiki, Dreadstar, Dave-ros, Weregerbil, Cockneyite, Crd721, Bryanmcdonald, Jklin, DMacks, Wizardman, Where, LeoNomis, Risker, Sadi Carnot, Carlosp420, TTE, Yevgeny Kats, Will Beback, Jonpalmer, Lambiam, Nathanael Bar-Aur L., Bcasterline, Geoffrey Wickham, Rklawton, Djeneba, Sophia, Dsantesteban, Kuru, Thefro552, Titus III, Richard L. Peterson, John, Scientizzle, Stephane Yelle, DocRocks1, Jaganath, Thegathering, Skoobieschnax, JorisvS, LestatdeLioncourt, Coredesat, Accurizer, Minna Sora no Shita, Mgiganteus1, A5y, Nonsuch, Ridersbydelta, Mr. Lefty, AtD, Jess Mars, Ben Moore, JHunterJ, MarkSutton, Slakr, Special-T, Momolee, LuYiSi, Mr Stephen, Samaster1991, Spiel496, Buttle, Novangelis, PSUMark2006, Inquisitus, Dl2000, ShakingSpirit, Hgrobe, Ginkgo100, Vanished user, JMK, Craigboy, Lakers, Newone, MOBle, J Di, StephenBuxton, Matt Bernius, Igoldste, Taucetiman, Tofoo, Tawkerbot2, Lsskys, George100, Kurtan~enwiki, Lahiru k, CalebNoble, SkyWalker, JForget, CmdrObot, Tanthalas39, Gholson, Porterjoh, Ale jrb, Scohoust, Aherunar, Galo1969X, Picaroon, Shakespeare87, User92361, Zureks, Basawala, Ruslik0, GHe, Dgw, OMGsplosion, ShelfSkewed, WHATaintNOcountryIeverHEARDofDOtheySPEAKenglishINwhat, MarsRover, Hi There, Groosh, Myasuda, Anthony Bradbury, Gregbard, Logicus, Cydebot, Steel, Travelbird, Red Director, Jon Stockton, A Softer Answer, Adolphus79, Nicesai, Rracecarr, Codingmasters, Ch0rx, Tawkerbot4, Alexnye, Christian75, M a s, DumbBOT, DarkLink, Interwiki gl, FastLizard4, Optimist on the run, Jimip, Waxigloo, SteveMcCluskey, Omicronpersei8, Stoked, Gimmetrow, Sevenaces, Raoul NK, FrancoGG, Thijs!bot, Epbr123, DarlingFriend, Opabinia regalis, Pajz, Ishdarian, Jamesluster, Andyjsmith, 24fan24, Gamer007, ClosedEyesSeeing, Headbomb, John254, Bobblehead, Neil916, Pogogunner, Grayshi, EdJohnston, HistoryMaster 1, Zachary, The Hybrid, Nick Number, Lithpiperpilot, CarbonX, MichaelMaggs, Sam42, J.S.B.Anderson, Escarbot, Eleuther, Mentifisto, Hmrox, AntiVandalBot, Yonatan, Luna Santin, CodeWeasel, Themaxeditor, Prolog, Yay unto the Chicken, Dylan Lake, Gdo01, Myanw, Ioeth, JAnDbot, Leuko, Husond, Vorpal blade, Kaobear, ThomasO1989, Roman à clef, MER-C, Nevadacall, Andonic, Hut 8.5, 100110100, Acroterion, N shaji, Pablothegreat85, Magioladitis, Foobird107, Murgh, Bennybp, Bongwarrior, VoABot II, AuburnPilot, Xn4, Wikidudeman, Hendrixjoseph, Careless hx, Aerographer1981, Crazytonyi, 9holdss, ThomasThePolishMan, Bubba hotep, Mi6agent00g off, BatteryIncluded, Beetfarm Louie, Adrian J. Hunter, Alexei Kojenov, Kane1047, LorenzoB, Mollwollfumble, Scot.parker, Andykass, Talon Artaine, Chris G, DerHexer, MeEricYay, WLU, Seph Vellius, TheRanger, Patstuart, Seba5618, Oroso, NatureA16, FisherQueen, Hdt83, E.vondarkmoor, MartinBot, Sinfear, Shimwell, Flamingpanda, The Ubik, UnfriendlyFire, APT, Rettetast, Juansidious, Anaxial, Sm8900, David J Wilson, Mschel, R'n'B, GarrisonGreen, AlexiusHoratius, Pekaje, LittleOldMe old, PrestonH, J.delanoy, Filll, Trusilver, Tonmoy Chowdhury, Bloomingiris, 72Dino, MoogleEXE, Lhynard, Ginsengbomb, WarthogDemon, Willow123~enwiki, SubwayEater, Yeti Hunter, Hisagi, James Mead, M C Y 1008, Wandering Ghost, Redmotherfive, Rod57, Vertigo900, Mr Rookles, Samtheboy, Gurchzilla, Supuhstar, Pyrospirit, AntiSpamBot, GhostPirate, Belovedfreak, Raichu Trainer, Ohms law, Mitchell is hollywood, SJP, Policron, Touch Of Light, Pwnasaurusrex38, MKoltnow, Mufka, FJPB, Blckavnger, Cmichael, Mohrflies, Stoned Proffesor, Kenneth M Burke, Cosmictinker, RB972, Tiggerjay, U.S.A.U.S.A.U.S.A., Treisijs, Mike V, Redrocket, Gtg204y, MtyQuinn, Darkfrog24, Jxzj, Annax3, Smartman10, Ronbo76, Micmic28, Yecril, Missphysics, GoldenGolem, Xiahou, Lorax835, Steel1943, Washboard6, Sheliak, Funandtrvl, Gravityc, Tecsup, Black Kite, Chinneeb, Deor, VolkovBot, TreasuryTag, TJ Elliot Scott, Meaningful Username, Danwills, DSRH, Mtesm, Indubitably, Gseletko, Cullaloe, Veddan, Boaex, Dominics Fire, Heerojyuy, Philip Trueman, Childhoodsend, TXiKiBoT, Oshwah, GVIlleneuve27, Davecas0, Adamwang, BuickCenturyDriver, 99DBSIMLR, MeStevo, Lolerballer~enwiki, Brocq 18, Andrius.v, Z.E.R.O., Anonymous Dissident, Jonnymagic, Ask123, Trentc, MattCarterSurrealist, IPSOS, Sarthella, Seraphim, Fizzackerly, Wvogeler, The One Cause, Tallcreek, Markp93, Drappel, PDFbot, Freak104, Manticore55, Cremepuff222, Blackdragon 1002, Wiae, Vrsixfire, Liberal Classic, Witchzilla, Noel rebeira, Inductiveload, Gladiator2155, Sydweighz, Spiral5800, Larklight, RobertFritzius, Dirkbb, SQL, SallyBoseman, Frag1983, Synthebot, ChillDeity, Speria, Heroesrule17, Enviroboy, Hollop09, Sylent, Kaseman519, Ballsucker22, Sesshomaru, Brianga, Skins88, Chickyfuzz14, AlleborgoBot, GavinTing, Shaidar cuebiyar, Happyhacker101, FlyingLeopard2014, Steven Weston, D. Recorder, Kastrel, Al.Glitch, Xarr, The Random Editor, Mr. dick 008, SieBot, Tosun, Crunchedfor6, High2lowo, K. Annoyomous, Thong123456789, Work permit, Scarian, Invmog, Gerakibot, Mazza uk04, Mickeyd24, LealandA, Caltas, BreakfastTom, Beethoven314, Pieman123456789, RJaguar3, Triwbe, Rangutan, Chs³, Tubular bells83, The way, the truth, and the light, Garrett gagne, Keilana, Elliott Fontain, CaptainIraq, RadicalOne, Flyer22 Reborn, Tiptoety, CombatCraig, Bhahn0125, Oda Mari, Sunayanaa, Cnormansen, Thin joe, M keshe, Jirachipokemon, Kcin213, Chives4life, Luskj, Oxymoron83, Faradayplank, Smilesfozwood, AngelOfSadness, Nuttycoconut, Edwardwittenstein, Zharradan.angelfire, John fromer, Lightmouse, AMCKen, Iain99, Techman224, Skateboards fly, Mr mimises, Blobbucket, BenoniBot~enwiki, Dillard421, Panicum, Fedosin, C'est moi, Hamiltondaniel, ShexRox, Dolphin51, M2Ys4U, Into The Fray, Canglesea, C0nanPayne, Albin&dani, Quinling, Muhends, Moony1000, Atif.t2, Tomasz Prochownik, ClueBot, Thejman123456789, Artichoker, PipepBot, The Thing That Should Not Be, JavaJesus, Mr.Pinklesworth, Meekywiki, Artist7337, Drmies, Firth m, Mild Bill Hiccup, Boing! said Zebedee, CounterVandalismBot, Kitty9992, VandalCruncher, Harland1, Otolemur crassicaudatus, Jackey0105, Andwor9, AikBkj, Another Matt, RogerEllman, Puchiko, Shocky95, Gakusha, DragonBot, Joshisamazing, Chris earl 89, Alexbot, Mccann tom, Icreaser, Robbie098, Sct5333, Shnilokobwe, Fsunka, Itel94, Ola Hansen, Ice Cold Beer, Aurora2698, Jotterbot, PhySusie, Glacialvortex, Razorflame, Handcannonbeast, Applejacks47, Chaosdruid, Thingg, Venera 7, Wnt, Myagooshki666, Deproduction, MasterOfHisOwnDomain, Jaykay0424, Ioannes93, TimothyRias, Misterbeal, InternetMeme, Mhamhamha, XLinkBot, Royboturso, Gwandoya, Crustillicus, BodhisattvaBot, Rror, Ance.cdas, Baconlover13, Ost316, Srossman07, Noctibus, JinJian, Truthnlove, Ttimespan, Airplaneman, Infonation101, EEng, Freestyle-69, Kbdankbot, Xerbaycom, Addbot, Nyw195, Willking1979, NateDres23, Bobocheese, Betterusername, Non-dropframe, Olli Niemitalo, Cre84u, TutterMouse, Presentabsent, Eedlee, Veraptor, WFPM, Ashanda, MrOllie, Lost on Belmont, FerrousTigrus, Ld100, Delaszk, Glass Sword, Maddox1, Jasper Deng, Harvardstudent, Kicka, Tide rolls, EugeneKantarovich, Cesiumfrog, Krano, WikiDreamer Bot, Hartz, Narutolovehinata5, Legobot, Drpickem, Luckas-bot, Dov Henis, Senator Palpatine, II MusLiM HyBRiD II, JustWong, Becky Sayles, THEN WHO WAS PHONE?, AnakngAraw, Solo Zone, Azcolvin429, Atqueamemus, MacTire02, AnomieBOT, Cantanchorus, Jdiyef, Ahmediq152, Jt16733, Gatoradeparade, Kanwaraj, Piano non troppo, Chaosmaker39, Gsd65, LlywelynII, Rhlowe, Unicornlad, VCleemput, Kanat Abildinov, Typesships, Are you ready for IPv6?, Dje 8, Citation bot, Oddball.bfi, Nyrox395, Maxis ftw, Persistent76, Chase4813, ArthurBot, SnorlaxMonster, Gravityforce, ChococatR, Xqbot, TinucherianBot II, Wikidushyant, TechBot, Millahnna, Gap9551, Markell West, GrouchoBot, Celebration1981, Gsard, Amaury, Charvest, Doulos Christos, A. di M., Acannas, CES1596, Ninjainventor, LucienBOT, Paine Ellsworth, Tobby72, TedlyW, Lookang, Lipsquid, ⬚⬚⬚, Sławomir Biały, ArkianNWM, Allen Jesus, Parvons, Cannolis, Orion 8, Citation bot 1, Pinethicket, Tom.Reding, RedBot, IceBlade710, SpaceFlight89, Rohitphy, Jauhienij, RockSolidCosmo, FoxBot, TobeBot, Yunshui, Gafferuk, Comet Tuttle, Le Docteur, Schwede66, MitchLay, Oms22, Earthandmoon, Tbhotch, RobertMfromLI, Brambleclawx, Mean as custard, RjwilmsiBot, Androstachys, DASHBot, EmausBot, John of Reading, Mnkyman, Atwarwiththem, Qurq, Gfoley4, Kueller1, Dewritech, Baguettes, Syncategoremata, Jmencisom, Dcirovic, Sheeana, The Blade of the Northern Lights, Solomonfromfinland, JSquish, Ryan.vilbig, ZéroBot, Crua9,

GoldRenet, Dffgd, Everard Proudfoot, Quondum, Wikfr, Confession0791, JosJuice, Ocaasi, Wtsbeynon, BrokenAnchorBot, Brandmeister, L Kensington, Zayzya, Ally1604, Checkmark56, Fluctuating metric, RockMagnetist, Teapeat, Laned130, Rememberway, ClueBot NG, Nebulosus, CocuBot, PoqVaUSA, Jj1236, Lilptrsn, Megalobingosaurus, MerlIwBot, Helpful Pixie Bot, BG19bot, Negativecharge, Furkhaocean, GKFX, Bryanpiczon, Ninney, Cadiomals, Mr.viktor.stepanov, RGloucester, BattyBot, Tutelary, David.moreno72, GravityForce, Padenton, Khazar2, JYBot, Davidlwinkler, LightandDark2000, Mogism, Rudrene, Reatlas, CsDix, Everymorning, Vanderoops, Yuan Jullian Morales, Gacman67, DavidLeighEllis, Prokaryotes, Jwratner1, My name is not dave, Mfb, Konveyor Belt, Mproncace, Mahusha, Monkbot, Opencooper, Filedelinkerbot, Gronk Oz, 3primetime3, WaryLouka, Oiyarbepsy, Stefania.deluca, Loraof, Danmcz, Ps20231131, Gladamas, Pishcal, Freedom2003, Tetra quark, James 123234, Inorout, Supdiop, KasparBot, MusikBot, The oracle 2015, Sweepy, Jeffryan123, ImperatorRomanorvm, Addycrisp, Sir Cumference, The golden colten, CAPTAIN RAJU, J.A.Witt (Tony), Hiimemilylol, Dan6233, Sunshine2night, H.dryad, Dr Peter Donald Rodgers, L3erdnik, Worldandhistory, Rekzy FFA, New Speech Killer, BIGMANFI17, Crazyguys123, Ahmadhammo2, WikipediaTranslator, Bear-rings, Dat nuke tho, ApokryItaroes, John David Best, FabulousFerd, Arbor Fici, Ozi67864, Todd Troll, Namaneinstein, Josh Everitt, Kayallwestfall, Coolkid63, Sonicsword and Anonymous: 1288

- **Newton's law of universal gravitation** *Source:* https://en.wikipedia.org/wiki/Newton'{}s_law_of_universal_gravitation?oldid=735749216 *Contributors:* XJaM, William Avery, Caltrop, Patrick, Michael Hardy, Dcljr, SebastianHelm, Pizza Puzzle, Charles Matthews, Geraki, Robbot, Piels, Ancheta Wis, Giftlite, Sj, Wolfkeeper, Everyking, Antandrus, Beland, MisfitToys, EricJamesStone, Shotwell, Mike Rosoft, Rfl, JimJast, Discospinster, Rich Farmbrough, FT2, Vsmith, Dbachmann, WegianWarrior, Bender235, Lycurgus, Spoon!, Bobo192, Stesmo, Smalljim, I9Q79oL78KiL0QTFHgyc, Jojit fb, Kjkolb, Jeodesic, Sam Korn, Nsaa, Danski14, Alansohn, Arthena, MarkGallagher, Snowolf, Wtmitchell, Wtshymanski, Allen McC.~enwiki, Gene Nygaard, Vadim Makarov, Brookie, WilliamKF, Linas, Camw, StradivariusTV, Drostie, Mpatel, JRHorse, Zzyzx11, Paxsimius, Ashmoo, Thierry Dugnolle~enwiki, Kbdank71, Sjö, Coemgenus, Sdornan, Salix alba, Ligulem, Ems57fcva, Matt Deres, Anskas, Latka, GnuDoyng, Rbonvall, Gurch, Chris D'Amato, Wgfcrafty, Jared Preston, DVdm, Random user 39849958, Sanpaz, Gwernol, YurikBot, TexasAndroid, Sceptre, RussBot, Bhny, Madkayaker, Howcheng, Gadget850, Dna-webmaster, Wknight94, Enormousdude, Ketsuekigata, Josh3580, GraemeL, CWenger, MagneticFlux, RG2, AssistantX, FyzixFighter, Sbyrnes321, SmackBot, MattieTK, Yamaguchi⁇, Gilliam, Iskander32, Thumperward, Complexica, Tianxiaozhang~enwiki, Baa, Sbharris, Firetrap9254, NYKevin, Krich, Fuhghettaboutit, Kingdon, T-borg, Pwjb, Greg.collver, Xiutwel, Filpaul, Sigma 7, Stefano85, ElizabethFong, Sadi Carnot, Yevgeny Kats, SashatoBot, Lambiam, Nishkid64, Gobonobo, WhiteHatLurker, Mr Stephen, Onionmon, Mthsmith, Paul venter, Newone, Pathosbot, Chetvorno, Chris55, Mellery, McVities, Karenjc, Cydebot, Rracecarr, BishopOcelot, SteveMcCluskey, Thijs!bot, Epbr123, Daa89563, TonyTheTiger, Andyjsmith, Headbomb, Marek69, Codee240, John254, Kathovo, Invitatious, D.H, Stannered, Luna Santin, Deeplogic, Mogzig, Storkk, Myanw, Shambolic Entity, JAnDbot, Dan D. Ric, Seddon, Acroterion, Magioladitis, VoABot II, JNW, Appraiser, Twsx, Aa35te, Cardamon, CodeCat, Causesobad, Duckysmokton, Ptrpro, FisherQueen, MartinBot, Nthitz, Jay Litman, J.delanoy, Trusilver, Ali, TheSeven, P.wormer, NewEnglandYankee, Wesino, Sarregouset, Baseball-bob, Timboyk12, Jarry1250, Yecril, Quiet Silent Bob, T prev, VolkovBot, Larryisgood, ABF, Jeff G., Philip Trueman, TXiKiBoT, Oshwah, Asuffield, Red Act, Hqb, Gian-2, Canaima, Jackfork, PDFbot, Windrixx, Maxw41, Gunnar Berlin, Falcon8765, Newsaholic, Brianga, SieBot, Paradoctor, Gerakibot, JerrySteal, Darkwarlord95, Zbvhs, Quest for Truth, Masgatotkaca, Captain Yankee, James.Denholm, Paolo.dL, Prestonmag, Pac72, WWStone, Kenkenko, OKBot, Duae Quartunciae, Anchor Link Bot, Rotovia, Denisarona, Escape Orbit, 678right, ClueBot, ICAPTCHA, IceUnshattered, Spoladore, Donteras, Drmies, CounterVandalismBot, Make91, Maymay, GrapeSmuckers, Abdullah Köroğlu~enwiki, Ktr101, Excirial, Jusdafax, Erebus Morgaine, SpikeToronto, Iohannes Animosus, Muro Bot, Thehelpfulone, CarlosPatiño, Thingg, Redrocketboy, SoxBot III, DumZiBoT, MadameBouvier, AgnosticPreachersKid, Terry0051, Little Mountain 5, Jakezing, Kbdankbot, Addbot, Proofreader77, Willking1979, D c weber, Guoguo12, Kung foo masta, Ronhjones, TutterMouse, MrOllie, Glass Sword, Tron78, Favonian, Jasper Deng, Barak Sh, Tide rolls, Lightbot, Krano, مانی, Alfie66, Angrysockhop, Johncolton, Luckas-bot, Yobot, Stamcose, ArchonMagnus, KamikazeBot, DrTrigon, عالم, محبوب, Tempodivalse, AnomieBOT, JackieBot, Rudolf.hellmuth, Csigabi, Materialscientist, E2eamon, Alex Yuwen, ArthurBot, SnorlaxMonster, Xqbot, ამჯერად, Tranxodox, The sock that should not be, Capricorn42, Grim23, Ched, AbigailAbernathy, NOrbeck, GrouchoBot, AVBOT, RibotBOT, SassoBot, Mathonius, Amaury, Rickproser, 78.26, Kongkokhaw, A. di M., A.amitkumar, Ashik, GliderMaven, Paine Ellsworth, Lookang, HJ Mitchell, Steve Quinn, XXx xD LeGeNd MoJo xXx, HamburgerRadio, Citation bot 1, Pinethicket, I dream of horses, PrincessofLlyr, Elpiades, Tom.Reding, MastiBot, SpaceFlight89, Jauhienij, ActivExpression, TobeBot, Info4sina, EdFalzer, Vrenator, J0z777, Bluefist, Reaper Eternal, JLincoln, Diannaa, I change stuff ha, Sjsharksfan, Seanlacroosed, Onel5969, Mean as custard, TjBot, Hajatvrc, Jbwhitmore, Deagle AP, EmausBot, Gfoley4, Razor2988, Vjga, Infringement153, Winner 42, Mmeijeri, Wikipelli, Dcirovic, K6ka, Zegod, Thecheesykid, Mz7, JSquish, John Cline, PRABHAT PINGREJA, Josve05a, Urgent01, Sakapraia, Druzhnik, Quondum, Christina Silverman, Dennis Kwaria, Jay-Sebastos, L Kensington, Donner60, Danlevy100, Zueignung, Sailsbystars, Carmichael, RockMagnetist, LGSlayer1127, DASHBotAV, Nathanlongan, Daniel55423, ClueBot NG, Iiii I I I, CocuBot, IjrmoneyI, Heimdallen, Millermk, Rokkyo13, EnglishTea4me, Braincricket, Thomask0, Widr, Moooo1234, Helpful Pixie Bot, BG19bot, Jwchong, Northamerica1000, Tiger42653, Mifter Public, Dan653, AwamerT, Piguy101, Jobin RV, Mark Arsten, Joydeep, Glevum, Snow Blizzard, YVSREDDY, Glacialfox, BattyBot, David.moreno72, Toumajk, Zhaofeng Li, BrightStarSky, Webclient101, Elec junto, SteenthIWbot, New York Resident, Sozopol, Sokotrof10, Hillbillyholiday, Reatlas, Cfarrar27, Vanamonde93, Melonkelon, PhantomTech, Tentinator, Kogmaw, Gmantheladiesman, Alexbassist, DavidLeighEllis, Glaisher, Prokaryotes, Jwratner1, Ginsuloft, Physikerwelt, Macks2008, Technomite, Internetconservationist, KhaiYuen.TAN, Monkbot, Deshmukhswapnil97, Kinetic37, Jr513, G6w000, Eurodyne, Stefania.deluca, SA 13 Bro, Tylerrigss01, Ducketduckduck, Poopnoop123, Sir IssacNewton2yy73y843yywy3, Arghyadeep Acharya, TheUniversalist, Superdupersmartdude, Inorout, KasparBot, Akhil1033, Abgohil, Epgggggggg and Anonymous: 729

- **Weight** *Source:* https://en.wikipedia.org/wiki/Weight?oldid=728635575 *Contributors:* AxelBoldt, Dreamyshade, CYD, Vicki Rosenzweig, Bryan Derksen, MarXidad, The Anome, Tarquin, Ed Poor, Alex.tan, Josh Grosse, SimonP, Peterlin~enwiki, Heron, Patrick, Michael Hardy, Bcrowell, Ahoerstemeier, William M. Connolley, Angela, Evercat, Rob Hooft, BRG, Nikola Smolenski, PS4FA, Charles Matthews, Crissov, Phys, Indefatigable, Chris K~enwiki, JorgeGG, Donarreiskoffer, Robbot, Schutz, Altenmann, Modulatum, Pingveno, StefanPernar, Overlord359, Lacerta~enwiki, Anthony, SaltyPig, Alanyst, GreatWhiteNortherner, Peter L, Enochlau, Giftlite, Wolfkeeper, Nunh-huh, Ævar Arnfjörð Bjarmason, Tom harrison, Mark Richards, Monedula, Wwoods, Bensaccount, Micru, Mckaysalisbury, Edcolins, Kandar, Yath, Antandrus, Alteripse, OverlordQ, Vanished user 1234567890, Jossi, Vina, Icairns, Gscshoyru, Arosa, Marcasireland, Mike Rosoft, Monkeyman, Jørgen Friis Bak, Discospinster, Rich Farmbrough, Vsmith, AlanBarrett, Paul August, Bender235, BruceRD, Spoon!, Femto, Bobo192, Vdm, Army1987, Chapium, Nicop (Usurp), CoolGuy, Shereth, Luckyluke, Espoo, Jumbuck, Alansohn, Riana, Ynhockey, Gene Nygaard, Nuno Tavares, Mindmatrix, Madmardigan53, Jok2000, Ch'marr, Dysepsion, Graham87, BD2412, Chun-hian, Lasunncty, FreplySpang, Virtualphtn, JHMM13, SeanMack, The wub, FlaBot, RexNL, Fresheneesz, Jmorgan, Chobot, DVdm, Mhking, UkPaolo, YurikBot, Wavelength, Zaidpjd~enwiki, RobotE,

- **Weak interaction** *Source:* https://en.wikipedia.org/wiki/Weak_interaction?oldid=725260635 *Contributors:* AxelBoldt, Chenyu, Sodium, Bryan Derksen, Tarquin, AstroNomer, Andre Engels, XJaM, Heron, JohnOwens, Gdarin, Delirium, Andrewa, Andres, Emperorbma, Timwi, Fibonacci, Phys, Phil Boswell, Lowellian, Mayooranathan, Tea2min, Giftlite, Sj, Herbee, Xerxes314, Jcobb, Mckaysalisbury, Munkee, Toby Woodwark, Beland, Bbbl67, Icairns, AmarChandra, Lumidek, Jørgen Friis Bak, Discospinster, ArnoldReinhold, Roybb95~enwiki, Gianluigi, Joanjoc~enwiki, Shanes, AJP, AtomicDragon, Danski14, Alansohn, Arthena, Axl, SidneySM, Hwefhasvs, DV8 2XL, Nightstallion, Kazvorpal, Linas, StradivariusTV, Benbest, Bbatsell, Palica, Tevatron~enwiki, Graham87, BD2412, Ketiltrout, Rjwilmsi, Strait, Erkcan, The wub, FlaBot, Naraht, Itinerant1, Srleffler, Chobot, Krishnavedala, DVdm, YurikBot, Borgx, Bambaiah, Hairy Dude, Jimp, Sillybilly, Conscious, Epolk, JabberWok, Gaius Cornelius, Shaddack, SCZenz, Irishguy, Shimei, Willtron, RG2, Phr en, That Guy, From That Show!, Luk, SmackBot, David Kernow, Tom Lougheed, WookieInHeat, Dauto, Chris the speller, Philosopher, Moshe Constantine Hassan Al-Silverburg, Complexica, DHN-bot~enwiki, Zirconscot, BIL, Wen D House, "alyosha", Maxwahrhaftig, Akriasas, Vina-iwbot~enwiki, Bdushaw, TTE, SashatoBot, Fontenello, Herr apa, Condem, Tony Fox, MottyGlix, JRSpriggs, Heartofgoldfish, Calmargulis, Green caterpillar, Joelholdsworth, Cydebot, Michael C Price, Mtpaley, Thijs!bot, ChKa, Kichwa Tembo, Headbomb, Hcobb, Icep, Escarbot, AntiVandalBot, Jimeree, Steelpillow, JAnDbot, Magioladitis, Swpb, مساع, Wormcast, DAGwyn, Giggy, Khalid Mahmood, Gah4, Tarotcards, Mstuomel, 2help, Lighted Match, DorganBot, Halmstad, Idioma-bot, VolkovBot, Jcuadros, Hilarious Bookbinder, TXiKiBoT, Brocq 18, Rei-bot, CaptinJohn, Awl, Shenanegins, BotKung, Wingedsubmariner, Antixt, Xxxlilbritxxx, Ptrslv72, Monty845, AlleborgoBot, SieBot, Paolo.dL, Skyentist, Ptr123, ClueBot, Bondchic007, SuperHamster, Erudecorp, Rotational, Jackey0105, Alexbot, Cenarium, Zomno, Zahnrad, He6kd, TimothyRias, InternetMeme, Timo Metzemakers, Stephen Poppitt, Addbot, Some jerk on the Internet, Markdman, ChenzwBot, Ehrenkater, Tide rolls, Luckas-bot, Yobot, Les boys, Kilom691, THEN WHO WAS PHONE?, Rifter0x0000, Duping Man, Dickdock, Magog the Ogre, AnomieBOT, Materialscientist, Citation bot, Quebec99, Kreigiron, Xqbot, Drilnoth, BurntSynapse, GrouchoBot, Omnipaedista, RibotBOT, Workanode, Jaz1305, Mnmngb, Dave3457, FrescoBot, Charles.walker, LucienBOT, Ionutzmovie, Grandiose, Pinethicket, Boulaur, Rameshngbot, RedBot, 23790AD, Tea with toast, Jauhienij, FoxBot, Earthandmoon, RjwilmsiBot, Itamarhason, Newty23125, EmausBot, WikitanvirBot, GA bot, GoingBatty, Dcirovic, Splibubay, StringTheory11, Braswiki, Git2010, Wayne Slam, Jsayre64, Maschen, ChuispastonBot, ClueBot NG, VinculumMan, Physics is all gnomes, Fjpyanez, Widr, Mouse20080706, Helpful Pixie Bot, Geo7777, Bibcode Bot, Junaid2754, Bolatbek, B wik, Phbarnacle, Neutral current, Glevum, Zedshort, Idenshi, Marioedesouza, Cup o' Java, Dexbot, Spray787, Reatlas, Bfaster, CsDix, Jamesmcmahon0, Ihatedirac2k13, Kharkiv07, Jwratner1, YimmyYohnson, Monkbot, BalderdashVonDrivel, ASCarretero, Malerisch, Cpt Wise, Lachlan Newland, Tetra quark, KasparBot, Megaraptor12345, Fmadd and Anonymous: 164

- **Mechanics of planar particle motion** *Source:* https://en.wikipedia.org/wiki/Mechanics_of_planar_particle_motion?oldid=723519020 *Contributors:* The Anome, Bender235, Dominic, GiovanniS, Rjwilmsi, Sanpaz, RussBot, Gilliam, Colonies Chris, Mets501, Coffee2theorems, Magioladitis, JerroldPease-Atlanta, Dolphin51, Brews ohare, SchreiberBike, Download, Halberdo, 11kravitzn, Dcirovic, Maschen, Zfeinst, Snotbot, Helpful Pixie Bot, BG19bot, BattyBot, Mogism, Alexander.j.mead, JJMC89 and Anonymous: 3

- **Centrifugal force** *Source:* https://en.wikipedia.org/wiki/Centrifugal_force?oldid=733474345 *Contributors:* The Anome, Dcljr, Bearcat, Gandalf61, Wolfkeeper, Icairns, Vsmith, Bender235, Wtmitchell, Velella, LFaraone, Woodstone, Sjö, DVdm, Sanpaz, RussBot, Marcus Cyron, Dbfirs, BOT-Superzerocool, Reyk, FyzixFighter, SmackBot, Gilliam, Thumperward, Sbharris, Colonies Chris, BullRangifer, JorisvS, Isaacl, Dicklyon, Dr.K., DangerousPanda, MarsRover, Gregbard, Rracecarr, Martin Hogbin, Headbomb, Widefox, Ianare, Global Cerebral Ischemia, Transcendence, Cardamon, David J Wilson, AlexiusHoratius, J.delanoy, Hans Dunkelberg, Uncle Dick, Izno, Wilhelm meis, JohnBlackburne, Jehan60188, Philip Trueman, Oshwah, Jackfork, Andy Dingley, Brianga, Ronald S. Davis, Flyer22 Reborn, IdreamofJeanie, Dolphin51, ClueBot, GorillaWarfare, The Thing That Should Not Be, Piriczki, Bbanerje, Djr32, PixelBot, Brews ohare, TimothyRias, Jovianeye, Addbot, Yoenit, Leszek Jańczuk, FDT, MrOllie, Download, CarsracBot, Glane23, Tide rolls, Luckas-bot, Yobot, Fugal, AnomieBOT, Götz, Mintrick, Materialscientist, Citation bot, LilHelpa, MauritsBot, Xqbot, Acebulf, Magicxcian, Gap9551, AbigailAbernathy, Austlang, Abce2, Omnipaedista, Chongkian, Mushushu, Nagualdesign, Citation bot 1, WQUlrich, AMSask, Overthinkingly, Serols, Codelyoko14, FoxBot, DixonDBot, Yen3000, Brian the Editor, Unbitwise, RjwilmsiBot, DRAGON BOOSTER, Beyond My Ken, Duncanb1234, Jeraxbeckwith, Vinnyzz, EmausBot, John of Reading, WikitanvirBot, Ajraddatz, Dcirovic, Solomonfromfinland, JSquish, Mrfredmister, Urgent01, Sonez1113, Shoobloo, Starcubs4, BrokenAnchorBot, L Kensington, Deed89, Donner60, Aayush.shaurya, ChuispastonBot, Decodicil, Rememberway, ClueBot NG, Anagogist, Widr, Anupmehra, Helpful Pixie Bot, BravoNovemberGolf, Hat guy7769, Lowercase sigmabot, BG19bot, Pablete85, Cncmaster, Snow Blizzard, PlasmaTime, Cyberbot II, Webclient101, User276, Pinapples2000, Wywin, Ruby Murray, Jakec, EvergreenFir, TheDoctor2012, Davidtheapple, MaximilianChocolatemeister98, Montyv, My name is not dave, Ginsuloft, Sam Sailor, Jeffwash, Submariner80, אבנ״ר, MRD2014, Rahulsharma96, Samu carkey, CV9933, KasparBot, JJMC89, Zaza PhD, Nirmal Diaz, Gamerthatpro, Boxama, Claysmithpa, Entranced98, Hamlery and Anonymous: 183

- **Centripetal force** *Source:* https://en.wikipedia.org/wiki/Centripetal_force?oldid=734923813 *Contributors:* Trelvis, Sodium, Bryan Derksen, The Anome, Tarquin, AstroNomer, Ap, Rjstott, Peterlin~enwiki, Patrick, Michael Hardy, Chris-martin, Ixfd64, Minesweeper, Looxix~enwiki, AugPi, BenKovitz, Andrewman327, Gutza, The Anomebot, Dtgm, Phil Boswell, Donarreiskoffer, Robbot, Rfc1394, Caknuck, Xanzzibar, Snobot, Giftlite, Wolfkeeper, Xerxes314, Guanaco, Opera hat, Karol Langner, Sam Hocevar, Mike Rosoft, Discospinster, Rich Farmbrough, Guanabot, Smyth, Bender235, RoyBoy, Kine, Bobo192, Smalljim, Ency, Drw25, Larryv, Grilo-TC, Alansohn, Gary, User6854, Yummifruitbat, Snowolf, Wtmitchell, Bucephalus, Cburnett, Danhash, ZakuSage, Oleg Alexandrov, Woohookitty, Cleonis, Bennetto, Sjö, Rjwilmsi, Vuen, Yamamoto Ichiro, Ian Pitchford, Musical Linguist, Margosbot~enwiki, Crazycomputers, Ewlyahoocom, Gurch, Principia~enwiki, Weeclarky, Tedder, King of Hearts, Chobot, DVdm, Sanpaz, Meadsteve, Hede2000, JabberWok, NawlinWiki, DeadEyeArrow, JoanneB, Snaxe920, FyzixFighter, Attilios, SmackBot, RDBury, InverseHypercube, Harald88, Gilliam, DiabloDan, ERcheck, Kurykh, MrDrBob, Silly rabbit, Dustimagic, VirEximius, Hatchetfish, Lifeeth, Rainmonger, Fuhghettaboutit, Bowlhover, Hoof Hearted, Mini-Geek, Henning Makholm, Kendrick7, Mion, Stefano85, Yevgeny Kats, Aïki, JorisvS, Minglex, Michel M Verstraete, Plvekamp, 16@r, Stwalkerster, Dicklyon, Dr.K., Iridescent, Stephen-Buxton, Igoldste, Aceawh979, Chris55, MarsRover, Myasuda, WillowW, Rracecarr, LeFrog, Nein~enwiki, Chrislk02, Gpd209, AndrewDressel, N5iln, Oerjan, Headbomb, John254, A3RO, Dgies, CTZMSC3, AntiVandalBot, Lklundin, JAnDbot, Utkarsh sinha, Roleplayer, VoABot II, Swpb, UnaLaguna, Inklein, Catgut, MartinBot, Supernatent, Lilac Soul, LedgendGamer, J.delanoy, AAA!, Choihei, Pisanidavid, McSly, Luke13f, KylieTastic, The enemies of god, VolkovBot, Larryisgood, JohnBlackburne, Suhel1992, Philip Trueman, TXiKiBoT, Ballz4kidz, Yeokaiwei, Irtrav, LeaveSleaves, Natural Philosopher, Fuzzywallaby, Maxno, VanBuren, Insanity Incarnate, Euicho, VinHantran, SieBot, Krawi, Danielleb32, Flyer22 Reborn, Radon210, Belinrahs, Paolo.dL, Aly89, Oxymoron83, Svick, Kerrio, Atif.t2, ClueBot, The Thing That Should Not Be, Uxorion, Cambrasa, Ndenison, Harland1, DragonBot, Djr32, Diagramma Della Verita, Addasc, Sun Creator, Brews ohare, PhySusie, SchreiberBike, 7, Versus22, Johnuniq, Tobywharne, HappyJake, Jht4060, Addbot, Narayansg, Ronhjones, CanadianLinuxUser, FDT, Laaknor-

Bot, Favonian, Kyle1278, Peti610botH, Kisbesbot, مانى, Fryed-peach, Legobot, Yobot, AnomieBOT, Exp HP, Ggsgas, 1exec1, Letmebefell, Jim1138, JackieBot, Materialscientist, Jpc4031, Citation bot, Williamsburgland, Zad68, Capricorn42, Tad Lincoln, Dutch chatty, GrouchoBot, Seeleschneider, Jhylands, Thehelpfulbot, Savig, Dger, Erikbrice, Haen45, Citation bot 1, Pezzells, SendTripItAll, Pinethicket, I dream of horses, Jschnur, Jauhienij, FoxBot, Sumone10154, Vrenator, DARTH SIDIOUS 2, TjBot, Inluminetuovidebimuslumen, EmausBot, WikitanvirBot, Mrthebeast, Laurifer, Hotsaucekd, Slightsmile, Dcirovic, Ironpole, Solomonfromfinland, Hhhippo, AvicBot, JSquish, ZéroBot, Gclink, Ocaasi, L Kensington, Maschen, Donner60, Puffin, Just granpa, MaxLupton, ClueBot NG, Unpilot15, Movses-bot, Andrew Kurish, Helpful Pixie Bot, Calabe1992, Vinle2, Niigabod, Biggdaddy69, Glevum, Cookedphysics, Achowat, News Historian, Shaun, TheCalculus, BattyBot, Ema--or, David.moreno72, Astronaut shivam, BrightStarSky, Nnnghv143, Vanamonde93, Xentricity, W. P. Uzer, JustComeHonorFace, Kaylaw61, Ammarzaidi82, Physics kids, JJMC89, Chriscalandro, GSS-1987 and Anonymous: 372

- **Coriolis force** *Source:* https://en.wikipedia.org/wiki/Coriolis_force?oldid=735506548 *Contributors:* AxelBoldt, Trelvis, Mav, Bryan Derksen, Szopen, The Anome, 0, Gareth Owen, Rjstott, Ed Poor, XJaM, Rodrigob, LapoLuchini, Heron, Karl Palmen, Steverapaport, Frecklefoot, Michael Hardy, TakuyaMurata, Looxix~enwiki, Ellywa, Ahoerstemeier, DavidWBrooks, Jimfbleak, William M. Connolley, Theresa knott, Suisui, Pbn~enwiki, Julesd, Glenn, AugPi, Rossami, Nikai, Cherkash, RodC, Dankelley, Denni, Kbk, The Anomebot, IceKarma, Saltine, Echidna, Morn, Tonderai, Shantavira, Bearcat, Robbot, Jxg, DHN, Hadal, Robinh, JerryFriedman, Splatt, Xanzzibar, Buster2058, Ancheta Wis, Psb777, Giftlite, Marnanel, Mikez, Sj, Wolfkeeper, Wwoods, Eequor, Brockert, SWAdair, Bobblewik, Mooquackwooftweetmeow, Chowbok, Utcursch, Andycjp, J. 'mach' wust, Antandrus, Bob.v.R, Karol Langner, Elektron, PeR, Mschlindwein, Vitaleyes, Jh51681, Eisnel, Esperant, Discospinster, Zaheen, Rich Farmbrough, Hydrox, Vsmith, Colin Angus Mackay, Berkut, Tinus, Pavel Vozenilek, DcoetzeeBot~enwiki, Bender235, Kaszeta, LordHarris, Livajo, Pt, El C, Ruyn, Laurascudder, Tom, Spoon!, Bobo192, Army1987, Smalljim, ParticleMan, La goutte de pluie, Kjkolb, Obradovic Goran, Haham hanuka, Pearle, Perceval, Jumbuck, Zachlipton, Alansohn, Riana, Mrtomh, Wtmitchell, SidP, Amorymeltzer, Spoonless, Cmapm, Woodstone, Gene Nygaard, Ghirlandajo, Capecodeph, Netkinetic, Yurivict, Dan100, Kenyon, Oleg Alexandrov, Firsfron, Linas, LostAccount, Swamp Ig, Aaron McDaid, Kzollman, Ilario, Cleonis, Benbest, Cbdorsett, Sf222, Eteq, Spot Color Process, Saperaud~enwiki, Rjwilmsi, Nightscream, BlueMoonlet, Jmcc150, Boccobrock, Brighterorange, Tomtheman5, AySz88, Titoxd, Gelo71, Old Moonraker, Mathbot, Gark, Rune.welsh, RexNL, Nimur, OrbitOne, Carrionluggage, Chobot, Elpaw, GangofOne, DVdm, Sanpaz, Bgwhite, Gwernol, Yurik-Bot, Wavelength, Hairy Dude, Deeptrivia, Jimp, Arado, Red Slash, Conscious, Chuck Carroll, Shell Kinney, Gaius Cornelius, Tavilis, Marcus Cyron, EWS23, DavidH, Introgressive, Howcheng, Krea, Irishguy, Johndarrington, CPColin, Semperf, Epipelagic, Ligand, Mysid, Bantosh, Saulpwanson, 2over0, Tcsetattr, StuRat, Fang Aili, KGasso, Chrishmt0423, Magic.dominic, AlexD, MagneticFlux, FyzixFighter, Eog1916, Luk, Dfloren1, Itub, Attilios, RupertMillard, SmackBot, Frogital, Lestrade, CarbonCopy, KiwiKittyBoy, Melchoir, NineEighteen, Bigbluefish, WilyD, Midway, Jrockley, Eaglizard, Ajm81, Richard B, Dna26, Canthusus, HalfShadow, Gilliam, Ohnoitsjamie, Skizzik, Eug, Valley2city, Chris the speller, Kurykh, @modi, Jprg1966, SchfiftyThree, Bazonka, Cornflake pirate, Thick as a Planck, Sbharris, Hugh24, BW95, Gyrobo, Javalenok, Can't sleep, clown will eat me, Makewa, Bsodmike, Gbuffett, Tamfang, Zelda Simpson, Xiner, Rrburke, Addshore, Tomytalker, Makemi, Mr Minchin, Blake-, Dreadstar, Amosslee, ILike2BeAnonymous, Sigma 7, Blahm, Skankboy, Vildricianus, Atkinson 291, Kuru, General Ization, Philosophus, Loodog, Fev, Shadowlynk, JorisvS, Bjankuloski06, IronGargoyle, Pflatau, Masiano, Thegreatdr, PseudoSudo, Plvekamp, Roflcopter123abc, BillFlis, Dr Smith, Dicklyon, Afogarty, That CS Guy, Jbuford39, KevinDM84, Kvng, Dan Gluck, Iridescent, Paul venter, Joseph Solis in Australia, Myrtone86, Nfutvol, Gavintlgold~enwiki, Woodshed, Tawkerbot2, Chetvorno, JForget, DangerousPanda, CmdrObot, Avanu, Lighthead, Vyznev Xnebara, Runningonbrains, CWY2190, Nilfanion, Jowan2005, Mhs5392, Logicus, Vectro, Cydebot, Monzonda, Nick.hardman, Asknine, Svance, MC10, Xaariz, Gogo Dodo, Xxhopingtearsxx, Rracecarr, Dancter, Viscious81, Christian75, Quadrius, Thrapper, Zalgo, Lo2u, FrancoGG, Thijs!bot, Epbr123, Ante Aikio, Martin Hogbin, Headbomb, John254, Bowfee, Davidhorman, Philippe, Lajsikonik, Dawnseeker2000, Escarbot, Oreo Priest, DewiMorgan, AntiVandalBot, Gioto, Seaphoto, Wolf grey, Golgofrinchian, Raylopez99, JAnDbot, Inks.LWC, Vandymorgan, EKindig, 100110100, Mdoc7, Magioladitis, Diablod666, Bongwarrior, VoABot II, A4, Swpb, Ling.Nut, PIrish, Gabriel Kielland, Dirac66, User A1, DerHexer, Strider01, Rakesh Dhanireddy, Stephenchou0722, MartinBot, Getztashida, David J Wilson, Jonathan Hall, CommonsDelinker, J.delanoy, Nvaccaro, Rmotz, Heycobber15, Drphysics, Jeepday, Plasticup, Farbror Erik~enwiki, White 720, Jake roman, Juliancolton, Ross Fraser, Oz1sej, CardinalDan, Funandtrvl, Spellcast, Lights, VolkovBot, Cjolly92, Pleasantville, Seattle Skier, AlnoktaBOT, Jdchamp31, SergeyKurdakov, Pparazorback, Philip Trueman, Maximillion Pegasus, Davehi1, Eddiehimself, Ann Stouter, Ridernyc, Ask123, L fle, Glennd83, Pishogue, Ian Strachan, Benua, Ball of pain, Falcon8765, Enviroboy, Turgan, MCTales, Elg26, Maskedskulker, Logan, Noillirt, Petergans, FlyingLeopard2014, HowardMorland, Charles Benham, James599, Malcolmxl5, Menschenfresser, Andrewjlockley, Flyer22 Reborn, Arbor to SJ, Mimihitam, Oxymoron83, Harry-, Jarvi006, Techman224, Udirock, Timbercat, Svick, Correogsk, Cloudjunkie, Dolphin51, Kanonkas, Faithlessthewonderboy, Toxicity^2, Sfan00 IMG, ClueBot, Procellarum, GorillaWarfare, Turboguppy, CurlyGirl93, The Thing That Should Not Be, Maniac18, Mild Bill Hiccup, Herbert Dingle, Stuart.clayton.22, Awickert, Excirial, Jptate, Nimbus1947, Lartoven, Brews ohare, Tyler, The Albino Alligator, Aitias, ForestDim, Warren oO, SoxBot III, JKeck, Gnowor, Nathan Johnson, BodhisattvaBot, Eejey, Rror, Facts707, Itwilltakeoff, Padfoot79, WikiDao, Chanerdar, MarcM1098, D.M. from Ukraine, Alabastair, Addbot, Johngorno, Willking1979, Dgroseth, Laserpointergenius, TheNeutroniumAlchemist, Ronhjones, The2dayslate, Mrniceguy85020, Scientus, Vishnava, Bobelehman, CanadianLinuxUser, WFPM, NjardarBot, FDT, Rehman, Tide rolls, Lightbot, OlEnglish, Kein Einstein, Legobot, Yobot, Jahon whahite, Fenderbenderstrat, Anypodetos, KamikazeBot, Samtar, Azcolvin429, VectorField, AnomieBOT, Climatedragon, Archon 2488, 1exec1, Jim1138, Piano non troppo, Aviast, AdjustShift, LlywelynII, Kingpin13, Jrobinjapan, Materialscientist, Citation bot, Er Cicero, WitchDrSmith, Ashhley!, V35b, Xqbot, Tripodian, Wperdue, Procyon11, Nasnema, NFD9001, RadiX, Oliballz, Tescobar, Maddogbrgs1, Tom is short, Shadowjams, A. di M., Deundre, A.amitkumar, Thehelpfulbot, FrescoBot, Blackguard SF, Dger, NCS2004, Polatrite, Citation bot 1, Pinethicket, Jonesey95, Hoo man, A Thousand Clowns, Σ, Vectornaut, Zazou25, Richard, Nashpur, Urineography, Corinne68, Meier99, FoxBot, Trappist the monk, Ilikemen123456789, Zvn, Flowirin, Windatheels, Diannaa, Starpine, Sideways713, RjwilmsiBot, NerdyScienceDude, Perspeculum, Citrab121, EmausBot, John of Reading, Orphan Wiki, Immunize, Gfoley4, Ajraddatz, Racerx11, RA0808, Dcirovic, Lou1986, Daonguyen95, Hpfeil, Bersibot, Midas02, Reality Bent, Wikfr, Nanju.murthy, Hropod, Donner60, Scientific29, ChuispastonBot, RockMagnetist, Targaryen, Teapeat, 28bot, Rememberway, ClueBot NG, IllicitDolmar, CocuBot, Satellizer, Victorindia87, Tideflat, Widr, Helpful Pixie Bot, Plswinford, Gob Lofa, Bibcode Bot, BG19bot, Gustavesarkozy, Ehines1, Wiki9-2-11, Cfullmer, Enrymather, ArrakisFrance, Cyberbot II, Kb*babe128, ChrisGualtieri, JYBot, EagerToddler39, Dexbot, FoCuSandLeArN, Mr. Guye, Lugia2453, Watjen, Jamesx12345, MagnusOxlund, Watchwolf49z, Rudrene, Nakashchit, The Anonymouse, JustAMuggle, Reatlas, Leon00, Spencer62, Tentinator, Dustin V. S., Ilustros, Spyglasses, Zenibus, AddWittyNameHere, Gingeroscar, JaconaFrere, Abhishekkumar9418, Elenceq, Opencooper, Gorgeandsmokey, Rapaulsen11, Trackteur, Myapello, Whistlemethis, Crystallizedcarbon, TuxedoMonkey, Eteethan, Ashwin bla bla, Gravityassist1, Axial Compressor, KasparBot, Cruithne9, Notsodumbasuthink, Aravindpvk, Andrewg4oep, Inoahguy24, Sunny X5, EatEn, GreenC bot and Anonymous: 817

- **Friction** *Source:* https://en.wikipedia.org/wiki/Friction?oldid=733263830 *Contributors:* The Anome, Peterlin~enwiki, Heron, Isis~enwiki, Bernfarr, Ram-Man, Stevertigo, D, Michael Hardy, Shellreef, Kku, Dgrant, Ellywa, Ahoerstemeier, Jschwa1, Evercat, Andrevan, Aravindet, Dysprosia, Malcohol, Hao2lian, DJ Clayworth, Saltine, Raul654, Finlay McWalter, Jni, Phil Boswell, Robbot, Chris 73, Boffy b, Yelyos, Ashley Y, Sverdrup, Ojigiri~enwiki, Diderot, Hadal, Diberri, Jimqode~enwiki, Dina, Enochlau, Psb777, Giftlite, Djinn112, Wolfkeeper, BenFrantzDale, Herbee, Leflyman, SheikYerBooty, Everyking, Zaphod Beeblebrox, Quamaretto, PlatinumX, Andycjp, Antandrus, Jossi, Karol Langner, Rdsmith4, DragonflySixtyseven, FrozenUmbrella, Icairns, Richard Stephens, Jcw69, Ukexpat, Grm wnr, Adashiel, Trevor MacInnis, Freakofnurture, Discospinster, Eb.hoop, Rich Farmbrough, Vsmith, Wk muriithi, Paul August, Bender235, ESkog, Kaisershatner, JoeSmack, Brian0918, Rick MILLER~enwiki, Rgdboer, Shanes, Shamilton, RoyBoy, Agamennon, Deanos, Sajt, Spoon!, Thu, Bobo192, Smalljim, Shenme, Wisdom89, Elipongo, Nk, Physicistjedi, Brainy J, Zetawoof, Minghong, Haham hanuka, Tatari-kun, Alansohn, Gary, Mo0, Babajobu, Ricky81682, Blues-harp, Riana, XB-70, Evil Monkey, VivaEmilyDavies, Sketchee, RainbowOfLight, Randy Johnston, Shoefly, Mikeo, Krubo, Ceyockey, Feezo, StuTheSheep, Nuno Tavares, Mel Etitis, Justinlebar, LOL, Swamp Ig, Borb, Pixeltoo, MONGO, Tylerni7, Rtdrury, Sir Lewk, Keta, Bbatsell, TotoBaggins, Isnow, Zzyzx11, Wayward, PhilippWeissenbacher, Dysepsion, Tslocum, Graham87, Rjwilmsi, Vary, JHMM13, Tawker, Nneonneo, Sampson~enwiki, Yamamoto Ichiro, Baddox, Titoxd, Nihiltres, Nivix, RexNL, Ewlyahoocom, Gurch, Fresheneesz, Alphachimp, Bmicomp, King of Hearts, Chobot, DaGizza, Krishnavedala, Sanpaz, Cactus.man, The Rambling Man, YurikBot, Wavelength, Anuran, Kafziel, Brandmeister (old), Wolfmankurd, Pip2andahalf, Anonymous editor, JabberWok, DanMS, Stephenb, Wimt, NawlinWiki, Astral, Zephyr9, Kdkeller, Retired username, Dhollm, Syrthiss, Dbfirs, DeadEyeArrow, FF2010, Enormousdude, Zzuuzz, StuRat, Closedmouth, Josh3580, Dspradau, GraemeL, JoanneB, Smurrayinchester, JLaTondre, Wbrameld, Caco de vidro, Allens, Kungfuadam, DVD R W, Eog1916, SmackBot, Lcarsdata, KnowledgeOfSelf, Blue520, KocjoBot~enwiki, Jagged 85, Alksub, Eskimbot, Chych, Motorneuron, TheDoctor10, Swerdnaneb, Macintosh User, Gilliam, The Gnome, Dauto, ERcheck, Des1974, Andy M. Wang, Hraefen, Squiddy, Guess Who, Chris the speller, Persian Poet Gal, JDCMAN, Sirex98, Silly rabbit, Dlohcierekim's sock, DHN-bot~enwiki, Darth Panda, Wazzzup7up, KieferSkunk, VDZ, Can't sleep, clown will eat me, DHeyward, SieG KilleR, TheKMan, EvelinaB, Mechj, Addshore, RedHillian, SundarBot, Nakon, T-borg, J.Wolfe@unsw.edu.au, Brownan, Jbergquist, Hammer1980, Acdx, Where, BrotherFlounder, TenPoundHammer, Rory096, Srikeit, Kuru, Rigadoun, Scientizzle, Danlina, Loodog, Michael L. Hall, Dumelow, Herr apa, Peterlewis, Tasc, George The Dragon, Fasdewx, Waggers, Spiel496, Avant Guard, LaMenta3, Sgstarling, LostTemplar, Ginkgo100, JYi, BranStark, Wizard191, Iridescent, Dreftymac, Joseph Solis in Australia, Shoeofdeath, GoZags, J Di, Igoldste, Benplowman, Octane, Happy-melon, Alan Joe Skarda, Quodfui, Tawkerbot2, Chetvorno, JForget, Ale jrb, Dycedarg, Megaboz, Anakata, Basawala, Dgw, Xakuzzah, McVities, Lazulilasher, Smeschia, Wingman358, Equendil, Cydebot, Abeg92, Rifleman 82, Michaelas10, Zgystardst, JFreeman, ST47, Rracecarr, Strongbad1982, Synergy, Tawkerbot4, Christian75, Chrislk02, JoshHolloway, Optimist on the run, Zanhsieh, Raoul NK, JamesAM, Thijs!bot, Epbr123, Mercury~enwiki, AndrewDressel, Blah3, Mojo Hand, Headbomb, A3RO, Kathovo, RickinBaltimore, Yettie0711, DaveJ7, CharlotteWebb, Uruiamme, Dzubint, AntiVandalBot, Luna Santin, Quintote, Prolog, Jj137, Farosdaughter, Alphachimpbot, Legare, JAnDbot, Husond, NapoliRoma, MER-C, Skomorokh, Arch dude, BCube, Db099221, EdsGodma, Hut 8.5, Lawilkin, Steveprutz, Acroterion, Penubag, Magioladitis, Jmbrock, Bongwarrior, VoABot II, LOLDSFAN, Swpb, Raggiante~enwiki, Animum, 28421u2232nfenfcenc, Allstarecho, User A1, Strikehold, DerHexer, JaGa, Deathwing23, Waninge, Seika7, Patstuart, Kurtwz, 1234567890123, JunKen, Qe2eqe, Philcohen, Neonblak, MartinBot, Zachplaysguitar5, R'n'B, Pbroks13, Igit~enwiki, Player 03, Tgeairn, Rabidanimals, J.delanoy, Petergress, Trusilver, Pursey, Fleiger, Dellarb, Katalaveno, Jeepday, Samtheboy, Pyrospirit, Goingstuckey, TomasBat, Bobianite, Gregfitzy, Juliancolton, KudzuVine, Fltchr, JavierMC, Barak181, Dorfrottel, Annax3, Martial75, CardinalDan, Idioma-bot, Aznskill101, 28bytes, VolkovBot, ABHISHEKARORA, Thedjatclubrock, ABF, Orphic, Jeff G., Soliloquial, Barneca, Philip Trueman, Fjhs, TXiKiBoT, Davehi1, Andrew153, DWLink, Anonymous Dissident, Escalona, Melsaran, Raymondwinn, Psyche825, BotKung, Ilyaroz, Sdman923, Andy Dingley, Meters, HopsonRoad, Falcon8765, Pedicabo ego vos et irrumabo, WatermelonPotion, Stephen J. Brooks, Insanity Incarnate, Chickyfuzz14, AlleborgoBot, Nagy, Kalivd, FlyingLeopard2014, SieBot, Nolafolk, Whiskey in the Jar, Ivan Štambuk, Dusti, Ttony21, AquaDTRS, Tiddly Tom, ToePeu.bot, Winstonio, Viskonsas, RJaguar3, 8056alt, Smsarmad, Yintan, Stonejag, Keilana, Flyer22 Reborn, Tiptoety, Radon210, Jtwhetten, Oda Mari, Jojalozzo, Paolo.dL, Allmightyduck, Oxymoron83, Harry-, Beast of traal, Romain.guises, Tombomp, Redmercury82, Hobartimus, Nskillen, Hyuu~enwiki, OKBot, Thatkiwiboy, Ward20, Verethor, Capitalismojo, Geoff Plourde, Joshschr, Pinkadelica, Dolphin51, Denisarona, Danesaw, Muhends, Rat at WikiFur, ClueBot, Traveler100, Avenged Eightfold, Snigbrook, The Thing That Should Not Be, Wolfch, Kafka Liz, Ndenison, Mild Bill Hiccup, J8079s, CounterVandalismBot, Firestonetireguy, Jonathanrcoxhead, Master188, Friction child, Shakejunt420, Bleachgangsta, PixelBot, Sentriclecub, Ludwigs2, Abrech, Gtstricky, Arjayay, JamieS93, ChrisHodgesUK, Thingg, Some fat dude, Little cistheman, Porridgebowl, Apparition11, Crazy Boris with a red beard, Will-B, Rror, Nepenthes, Mwaxman11, Subversive.sound, RyanCross, C Rockefeller, CalumH93, Jtknowles, Cunard, Addbot, Eric Drexler, Some jerk on the Internet, Melab-1, Fyrael, Syedrazahabbasjilani, Seanjp8329, Fieldday-sunday, SpillingBot, MrOllie, Download, EconoPhysicist, LAAFan, Skittleskittle, Rehman faiz11, Wtfily, CUSENZA Mario, SamatBot, LinkFA-Bot, Jasper Deng, BrucetonCATS, Tassedethe, Tide rolls, QuadrivialMind, Onixman, Legobot, Luckas-bot, Yobot, SuperColbertFan, THEN WHO WAS PHONE?, IW.HG, محبوب عالم, م, AnomieBOT, 1exec1, Bsimmons666, Jim1138, IRP, JackieBot, Materialscientist, Paranormal Skeptic, Citation bot, Another Stickler, GB fan, Rockoprem, ArthurBot, Johnbobkins, The Firewall, Zad68, Sir Stig, Nasnema, Haraldwallin, Cacaisgood, Tyrol5, Omniapaedista, Iων, WaysToEscape, Mike Dill~enwiki, Imveracious, FengRail, Paisiello2, HJ Mitchell, Ammatsun, Citation bot 1, Rotje66, Pinethicket, Abductive, Tom.Reding, Jschnur, Trappist the monk, Pvkwiki, Gregcohen79, Greg.Hartley, Vrenator, Reaper Eternal, Tbhotch, Minimac, DARTH SIDIOUS 2, Mean as custard, Rjwilmsi-Bot, TjBot, Fearstreetsaga, DASHBot, EmausBot, Fuujuhi, Gfoley4, Fly by Night, Wikipelli, Dcirovic, Hhhippo, PBS-AWB, BAICAN XXX, Knight1993, Nolanjshettle, Gregzore, AvicAWB, Cobaltcigs, H3llBot, Suslindisambiguator, EWikist, Wayne Slam, Hunocsi, Alrino, Jesanj, Zayzya, Baygenz, GrayFullbuster, TYelliot, DASHBotAV, Ljhappy1201, Adawson88, Cgt, ClueBot NG, Youngrod123, UniEmpPime, Cwmhiraeth, Meredithm33, Stararthagirl, SSMG-ITALY, Bped1985, Ulrich67, Tideflat, Cntras, Braincricket, Seadogburger, Edwinv1970, Oddbodz, Helpful Pixie Bot, Iste Praetor, Bibcode Bot, Doorknob747, Lowercase sigmabot, BG19bot, Meldraft, Nollid0331, Lexikon-Duff, Lynskyder, M.Sokolow, Exploding Peanuts, GSMOL, Zedshort, Glacialfox, Gantvik, PsiEpsilon, Joshopoke, CarnivorousGnomeCatuse, Cimorcus, Toasted Onion, Sloppyts, ChrisGualtieri, Ducknish, Stoney2stone, G.Kiruthikan, CuriousMind01, Lugia2453, Jamesx12345, Sgtdawn, Reatlas, Coffeecartoon, Shubh96365530, Tc456, Tentinator, Duchifat, Lollilollilollipuppies, Smart brain3232, Caronwilliams, 17aturguie, Croinop, Monkbot, Shabab Rahman, Filedelinkerbot, Grand'mere Eugene, KasparBot, RATLAM, Jayesh07, Jhwiebenga, Mindotaur and Anonymous: 1163

- **Shear force** *Source:* https://en.wikipedia.org/wiki/Shear_force?oldid=729487967 *Contributors:* Dmmaus, Rsrikanth05, Matt Heard, Gilliam, Yogesh Khandke, Peter Horn, Ibadibam, HitroMilanese, Wiae, Dthomsen8, Se'taan, Willondon, Materialscientist, Mmm333k, Donner60, ClueBot NG, Widr, Felixandhisthoothbrush, MusikAnimal, ChrisGualtieri, KatieBoundary, JaconaFrere, Owais Khursheed, Ezra best and Anonymous: 21

- **Tidal force** *Source:* https://en.wikipedia.org/wiki/Tidal_force?oldid=720726241 *Contributors:* Paul Drye, Mav, Bryan Derksen, Tarquin, XJaM, JeLuF, Fxmastermind, Patrick, Michael Hardy, Bcrowell, Alfio, William M. Connolley, BRand11, Theresa knott, Susurrus, Itai, Robbot, Sverdrup, Dbenbenn, Wolfkeeper, Mark.murphy, Joe Kress, Espetkov, Mmm~enwiki, Fpahl, Icairns, Urhixidur, Qutezuce, Vsmith, Bender235, Kop, Paulroyaux, Viriditas, Sasquatch, Daniel Arteaga~enwiki, Burn, Jheald, Gene Nygaard, Isfisk, Woohookitty, Jannex, StradivariusTV, RuM, FlaBot, Mathbot, Gurch, Chobot, Krishnavedala, DVdm, ErkDemon, Nick, Zwobot, Epipelagic, Bota47, Light current, Serendipodous, Deuar, SmackBot, NickyMcLean, Fireworks, Ashill, Jrockley, IstvanWolf, Chris the speller, Kostmo, Hongooi, Hgrosser, Occultations, Yevgeny Kats, J 1982, Ajay5150, Breno, Mgiganteus1, Goodnightmush, Arctic-Editor, Chandu15, Thijs!bot, Headbomb, Widefox, JAnDbot, SiobhanHansa, WolfmanSF, Dividor, A3nm, Leyo, Polomarco, Chiswick Chap, Idioma-bot, Deor, Amikake3, Piperh, Hapsi, Michael H 34, Zain Ebrahim111, Philmac, Codymcarlson, GrNephrite, Mseliw, Paolo.dL, Svick, Canglesea, Ferred, Renacat, Roibeird, Rockfang, Excirial, Brews ohare, Njardarlogar, Chadoh, XLinkBot, Terry0051, Kbdankbot, Addbot, Fgnievinski, Download, Tide rolls, מלמד כ"ץ, Angrysockhop, Luckas-bot, Yobot, Vini 17bot5, AnomieBOT, Materialscientist, Citation bot, Xqbot, TinucherianBot II, JimVC3, Srich32977, NOrbeck, GrouchoBot, Legato33, Tom.Reding, Lithium cyanide, RedBot, Serols, December21st2012Freak, IVAN3MAN, RjwilmsiBot, Hajatvrc, EmausBot, Tomukas, Rami radwan, Jmencisom, Solomonfromfinland, Just an astrophysicist, Moorsmur, Byeong Soo Park, Teapeat, ClueBot NG, Gareth Griffith-Jones, Widr, Helpful Pixie Bot, CitationCleanerBot, Rfassbind, Melonkelon, Crwaterhouse, Striker19065, TCMemoire, Monkbot, Alexnacache, CAPTAIN RAJU, The Quixotic Potato, Wojje and Anonymous: 108

27.7.2 Images

- **File:3199_-_Athens_-_Stoà_of_Attalus_Museum_-_Bronze_weights_-_Photo_by_Giovanni_Dall'Orto,_Nov_9_2009.jpg** *Source:* https://upload.wikimedia.org/wikipedia/commons/4/47/3199_-_Athens_-_Sto%C3%A0_of_Attalus_Museum_-_Bronze_weights_-_Photo_by_Giovanni_Dall%27Orto%2C_Nov_9_2009.jpg *License:* Attribution *Contributors:* Own work *Original artist:* G.dallorto

- **File:A_Swarm_of_Ancient_Stars_-_GPN-2000-000930.jpg** *Source:* https://upload.wikimedia.org/wikipedia/commons/6/6a/A_Swarm_of_Ancient_Stars_-_GPN-2000-000930.jpg *License:* Public domain *Contributors:* Great Images in NASA Description *Original artist:* NASA, The Hubble Heritage Team, STScI, AURA

- **File:Acceleration_vector_plane_polar_coords.svg** *Source:* https://upload.wikimedia.org/wikipedia/commons/5/58/Acceleration_vector_plane_polar_coords.svg *License:* CC0 *Contributors:* Own work *Original artist:* Maschen

- **File:Addition_of_forces.JPG** *Source:* https://upload.wikimedia.org/wikipedia/commons/4/40/Addition_of_forces.JPG *License:* CC BY-SA 3.0 *Contributors:* Own work *Original artist:* Ilevanat

- **File:Andre-marie-ampere2.jpg** *Source:* https://upload.wikimedia.org/wikipedia/commons/7/74/Andre-marie-ampere2.jpg *License:* Public domain *Contributors:* Practical Physics, Millikan and Gale, 1920, scanned by B. Crowell *Original artist:* Benjamin Crowell --Bcrowell 16:51, 26 May 2005 (UTC)

- **File:Antarctic_bottom_water.svg** *Source:* https://upload.wikimedia.org/wikipedia/commons/e/e7/Antarctic_bottom_water.svg *License:* CC BY-SA 4.0 *Contributors:* <a href='//commons.wikimedia.org/wiki/File:Antarctic_bottom_water_hg.png' class='image'><img alt='Antarctic bottom water hg.png' src='https://upload.wikimedia.org/wikipedia/commons/thumb/2/2c/Antarctic_bottom_water_hg.png/100px-Antarctic_bottom_water_hg.png' width='100' height='77' srcset='https://upload.wikimedia.org/wikipedia/commons/thumb/2/2c/Antarctic_bottom_water_hg.png/150px-Antarctic_bottom_water_hg.png 1.5x, https://upload.wikimedia.org/wikipedia/commons/thumb/2/2c/Antarctic_bottom_water_hg.png/200px-Antarctic_bottom_water_hg.png 2x' data-file-width='3300' data-file-height='2550' /></a> *Original artist:* Fred the Oyster

- **File:Apollo_15_feather_and_hammer_drop.ogg** *Source:* https://upload.wikimedia.org/wikipedia/commons/3/3c/Apollo_15_feather_and_hammer_drop.ogg *License:* Public domain *Contributors:* Taken from Spacecraftfilms.com DVD "Apollo 15: The Great Explorations Begin" *Original artist:* NASA

- **File:ArealVelocity.svg** *Source:* https://upload.wikimedia.org/wikipedia/commons/9/9b/ArealVelocity.svg *License:* CC-BY-SA-3.0 *Contributors:* Transferred from en.wikipedia to Commons. *Original artist:* The original uploader was Xyzzy n at English Wikipedia

- **File:Aristoteles_Louvre2.jpg** *Source:* https://upload.wikimedia.org/wikipedia/commons/7/75/Aristoteles_Louvre2.jpg *License:* CC BY-SA 2.5 *Contributors:* Derivative work of File:Aristoteles_Louvre.jpg; originally from en.wikipedia; description page is/was here. *Original artist:* Original image: Eric Gaba (User:Sting) - derivative work: Interstate295revisited at en.wikipedia

- **File:Banked_turn.svg** *Source:* https://upload.wikimedia.org/wikipedia/commons/5/5b/Banked_turn.svg *License:* GFDL *Contributors:* <a href='//commons.wikimedia.org/wiki/File:Banked_turn.PNG' class='image'><img alt='Banked turn.PNG' src='https://upload.wikimedia.org/wikipedia/commons/thumb/2/20/Banked_turn.PNG/100px-Banked_turn.PNG' width='100' height='122' srcset='https://upload.wikimedia.org/wikipedia/commons/thumb/2/20/Banked_turn.PNG/150px-Banked_turn.PNG 1.5x, https://upload.wikimedia.org/wikipedia/commons/thumb/2/20/Banked_turn.PNG/200px-Banked_turn.PNG 2x' data-file-width='660' data-file-height='805' /></a> *Original artist:* Fred the Oyster

- **File:Bar_magnet.jpg** *Source:* https://upload.wikimedia.org/wikipedia/commons/d/d8/Bar_magnet.jpg *License:* CC-BY-SA-3.0 *Contributors:* ? *Original artist:* ?

- **File:Bascula_9.jpg** *Source:* https://upload.wikimedia.org/wikipedia/commons/2/27/Bascula_9.jpg *License:* CC-BY-SA-3.0 *Contributors:* **self made**, http://commons.wikimedia.org/wiki/Image:Bascula_9.jpg *Original artist:* L.Miguel Bugallo Sánchez (http://commons.wikimedia.org/wiki/User:Lmbuga)

- **File:Bcoulomb.png** *Source:* https://upload.wikimedia.org/wikipedia/commons/0/04/Bcoulomb.png *License:* Public domain *Contributors:* http://en.wikipedia.org/wiki/Image:Bcoulomb.png *Original artist:* ?

- **File:Beta-minus_Decay.svg** *Source:* https://upload.wikimedia.org/wikipedia/commons/a/aa/Beta-minus_Decay.svg *License:* Public domain *Contributors:* This vector image was created with Inkscape. *Original artist:* Inductiveload

27.7.3 Content license

www.ingramcontent.com/pod-product-compliance
Lightning Source LLC
Chambersburg PA
CBHW081113170526
45165CB00008B/2439